Recent Trends in
INVERTEBRATES

Vol. 1

H.S.Bhamrah
Kavita Juneja

ANMOL PUBLICATIONS

ANMOL PUBLICATIONS
4378/4B, Ansari Road,
Darya Ganj, New Delhi - 110002.

Recent Trends in Invertebrates

ISBN 81 7041 473 3 (Set)
ISBN 81 7041 474 1 (Vol.I)
ISBN 81 7041 475 X (Vol.II)
ISBN 81 7041 476 8 (Vol. III)
ISBN 81 7041 477 6 (Vol. IV)
ISBN 81 7041 478 4 (Vol. V)
ISBN 81 7041 479 2 (Vol. VI)
ISBN 81 7041 480 6 (Vol. VII)
ISBN 81 7041 481 4 (V0l. VIII)

PRINTED IN INDIA

Published by J.L. Kumar for Anmol Publications and Laser Typeset by D.V. Computers, Delhi, Printed at Efficient Offset Printers, New Delhi.

PREFACE

The present title has been designed to meet the requirements of the undergraduate students. In the Recent Trends in Invertebrate all phyla have been discussed. The representative types of each phylum have been described in detail. This is followed by the classification of the phylum with a description of its common animals. Recent findings and work have been incorporated in the text. Special care has been taken to make the language as simple and lucid as possible. Simple and accurate diagrams accompany the text. In the interest of the students efforts have been made to cover the syllabi of nearly all Indian Universities.

The book features both a text and a laboratory guide. It deals with the structures and physiological phenomenon of each system. General topics are also dealt in details. It is hoped that this book will not only meet the requirement of undergraduate students but also be useful as a guideline to the teachers in their teaching. It is also hoped that the general public will find that the book is not riddled with too much of scientific and technical jargon.

It is difficult to acknowledge adequately the assistance of all who have contributed to the preparation of this book. First of all the authors express their thanks to their friends and colleagues whose continuous inspiration have initiated them to bring out the present volume.

The authors are grateful to Dr. J. Bahadur, Professor and Head, School of Studies in Zoology, Jiwaji University Gwalior, Dr. A.K. Pandey, Professor and Head, School of Studied in Zoology, Vikaram University, Ujjain for their valuable suggestions.

Constructure criticism from students and teachers will be highly appreciated by the authors.

H.S. Bhamrah

Kavita Juneja

PREFACE

[illegible]

[illegible]

[illegible]

[illegible]

[illegible]

[illegible]

CONTENTS

Character and classification 1- 67

Chilomonas, peranema, chlamydomonas, volvox, Noctiluca, Ceratum, Proterospongia, Mastigamoeba, Bodo, Trichomonas, Giardia, Trichonympha, Pelomyxa, Arcella, Dihlugia, Euglypha, Globigerina, Allogromia, Thalasicola, Collozoum, Lithocircus, Actinosphaerium, Actinophrys, Calthrulina, Babesia, Sarcocystis, Nosema, Porodon, Coleps, hirtus, Didinum, Colpoda, Balantidium, Nyctotherus, Frontonia, carachesium, Stentor, Spirostomum, Stylonchia, Euplotes, Ephelota, Podophrya, Acineta,

2. **Englena** 68-82

Habit and habitat, culture, Morphology, Locomotion Nutrition, Respiration, Excretion, Osmoregulation, Behavior, Reproduction, Position of *Englena,*

3. **Trypanosoma** 83-93

Morphology, Polymorphism in *Trypanosoma,* Locomotion, Nutrition, Respiration, Excretion, Osmoregulation, Reproduction, Life-cycle, Disease, Treatment, Control measure.

4. **Leishmania** 94-97

Morphology, Reproduction, Life-cycle, Pathogenicity,

5. **Amoeba Proteus** 98-115

Habits and habitat, culture, Morphology, Physiology, Locomotion, Nutrition, Respiration, Excretion, Osmoregulation, Behavior, Reproduction, Regeneration

6. **Entamoeha histolytica.** 116-124
Distribution, Habits and habitat, Morphology, Physiology (Locomotion, Nutrition, Respiration, Excretion) , Reproduction, Pathogenic effects, Treatment, species of Entamoeba.

7. **Elphidium** 125-132
Habits and habitat, Morphology, physiology (Locomotion, Nutrition, Respiration and Excretion, Osmoregulation), Reproduction and life-cycle, Alternation of generation.

8. **Radiolaria** 133-136

9. **Monocystis** 137-146
Habits and Habitat, Morphology, Physiology (Locomotion, Nutrition, Respiration, Excretion, Reproduction and life cycle, Mode of Inspection, Effect of parasitism, conclusions from the life- history of *Monocystis*.

10. **Plasmodium** 147-159
Habits and habitat, Morphology, Life-cycle, symptoms, Control of Malaria.

11. **Eimeria** 160-165
Habits and habitat, Morphology, physiology, life-cycle. Alternation of generation, Effect on host.

12. **Gregarina** 166-169
Habits and habitat, Morphology and life-cycle.

13. **Opalina** 170-173
Habits and habitat, Morphology, Physiology, Life-cycle, systematic position of *Opalina*.

14. **Paramecium.** 174-205
Habits and habitat, Morphology (External feature, Internal feature), Physiology (Locomotion, Nutrition, Respiration, Excretion, Osmoregulation, Behaviour), Reproduction,

15. **Vorticella** 206- 217
Habits and habitat, Morphology, physiology (Locomotion, Nutrition, Egestion, Respiration, Excretion, Osmoregulation), Reproduction

16. **Suctoria** 218-222
Morphology, Nutrition, Reproduction, Affinities

17. **Skeleton is protozoa** 223 - 229
18. **Locomotion in Protozoa** 230 - 242
19. **Osmoregulation is Protozoa** 243 - 245
20. **Nucleus in Protozoa** 246 - 251
21. **Nutrition in Protozoa** 252 - 259
22. **Reproduction in Protozoa** 260 - 267
23. **Economic Importance of Protozoa** 268 - 275

1

CHARACTERS & CLASSIFICATION

The protozoa are a heterogenous assemblage of some 50,000 *acellular* or single-call organisms possessing typical (eukaryote) membrane bound cellular organelles. Being the simplest in structure, protozoa are regarded most primitive or first animals (Gr. *protos* - first; 30 zoon = animal) of Nature. The acellular level of organization is the only characteristic by which the protozoa as a whole can be described; in all other respects they display extreme display. They exhibit all types of symmetry, a great range of complexity, and adaptations for all types of environmental conditions Protozoa were first seen by *Leeuwenhoek* (1677). The other pioneers, in the field of Protozoology, are *Joblott* (1718), *Trembley* (1744), *Linnaeus* (1758), *Muller* (1773), *Dujardin* (1845), *Haeckel* (1862), *Pasteur* (1870), *Butschli* (1881), *Golgi* (1889), *Ronald Ross* (1898), etc. Term Protozoa was coined by *Goldfuss* (1817). *Grassi* (1952), *Hall* (1953), *Kudo* (1954), *Pitelka* (1953-59), *Allen* (1962) etc. have made their valuable contribution to the protozoology in present century.

GENERAL CHARACTERS

(1) Protozoa occur wherever moisture is present-in the sea (marine) in all types of fresh water, and in the soil. There are commensal, mutualistic, and many parasitic species. In fact the sporozoans are entirely parasitic.

(2) All though most protozoa occur as solitary individuals, there are numerous colonial forms. Some colonial forms, such as *Volvex*, attain such a degree of cellular interdependence that they approach

a true multicellular level of structure. Both solitary and colonial species may be either free-moving or sessile.

(3) The great majority of protozoa are microscopic. *Anaplasma,* a blood parasite, is so smalls that it occupies only 1\6 to 1\10th of a red blood corpuscles. At the other hand, the freshwater ciliate, *Spirostomum,* may reach a length of 3.0 mm. and be seen with the naked eye. A fossil forminiferan, *Nummulites,* was 19 cm. across, probably the largest size that has been attained by any protozoan.

(4) Body symmetry non, radial, spherical or bilateral.

(5) The body is either naked or covered by *pellicle.* Sometimes an exoskeleton is also present.

(6) The shape of body is usually constant but in some cases it is unstable and in others it may change with the environment or age.

(7) Their locomotory organelles are finger-like *pseudopodia, flagella, cilia* or absent as in sporozoans.

(8) Nutritian may be *holozoic* (animal like), *holophytic* (plant-like), *saprozoic* or *parasitic.* The organelles of ingestion and egestion may be present or absent. Digestion occurs intracellularly inside the food vacuoles in most cases.

(9) No specific respiratory and excretory organs are present. Both are carried through general body surface by *diffusion* Excretion sometimes carried with the help of contractiloe vacuele.

(10) In fresh water protozoa excess water is usually eleminated by a contractile vacuole. Thus these are mainly osmoregulatory in function.

(11) Asexual reproduction by *binary fission, multiple fission* or *budding* occurs at some time in the life history of almost all protozoa. Meiosis, gamete formation, and fertilization have been observed in many species, but the nature of these events and their occurrence in the life cycle of the organism is highly variable. Encystment is common.

(12) Life-history is often complicated in some cases with alternation of asexual and sexual phases.

(13) Their body is not distinctly differentiated into *somatoplasm* and *germplasm,* hence natural death does not take place in them.

Special characters

(1) Their acellular body regulates all the life activities thus their is no physiological division of labour. The division of labour is restricted to the organelles which are intracellular or subcellular structures.

(2) The animals give their respond to various stimuli. This gives an indication of the beginning of the nervous system of metazoans.

(3) The microconjugants and macroconjugants in the conjugation of *Vorticella* indicate the beginning of *sexual dimorphism*.

Classification of Phylum-Protozoa

There is disagreement among. Protozoologits on the ways of classifying protozoa. Modern workers like *Kazloff* (1972) and *Robert D. Barnes* (1980) tend to believe that this group includes more than one unicellular phyla. *Barnes* (1980) raised the four major groups viz. *Mastigo phora, Sarcodina, Sporozoa* and *Ciliophora* to the rank of independent phylum. Here we are following the classification given by "Committee on Taxonomy and Taxononic Problems of the Protozoologists" (*Honiberg et al.*, 1964).

According to *Honiberg et al.*, protozoa have been classified into four subphyla.

Subphylum	I	SARCOMASTIGOPHORA
Subphylum	II	SPOROZOA
Subphylum	III	CNIDOSPORA
Subphylum	IV	CILIOPHORA

SUBPHYLUM .I SARCOMASTIGOPHORA

(1) Organelles of locomotion are preudopodia or flagella.

(2) Nucleus monomorphic (single type).

(3) Nutrition hologoic or holophytic.

(4) Asexual reproduction by binary or multiple fission

(5) Sexual reproduction by syngamy; no spore formation.

This subphylum is divided into three superclasses:

SUPER CLASS A - MASTIGOPHORA

1. Simple and primitive. Body with firm pellicle.
2. One to many flagella for locomotion.
3. Asexual reproduction by binary, more or less symmetrogenic, fission.
4. Autotrophic or heterotrophic or both.

Class-I. Phytomastigophorea

1. Mostly free-living.
2. Plant-like flagellates with or without chromoplasts.
3. Usually one or two flagella, the nucleus vesicular.
4. Reserve food as starch or paramylon.

Order. 1. Chrysomonadia

1. Small flegellate with yellow or brown chromoplasts.
2. One to three flagella are present.
3. Gullet absent; stigma present.
4. Nutrition holophytic. Starch absent but leucosin and fat droplets are present.
5. Silicious cysts. Marine or fresh water.

   Examples. *Chromulina, Synura, Ochromonas, Uroglena* etc.

Order. 2. Coccolithophorida

1. Tiny marine flagellates covered by calcareous platelets-coccoliths.
2. Two flagella and yellow to brown chromoplasts.
3. No endogenous silicious cysts.

Examples: *Coccolithus, Rhabdosphaera.*

Order 3- Heterochlorida

1. Two unequal flagella and yellow-green chromoplasts.

2. Silicious cysts.

   Examples: *Heterochloris, Myxochloris.*

Order-4-Cryptomonadida

1. Comprassed with a rigid pellicle.
2. Gullet reaches upto the middle of the body.
3. They have two unequal flegella. Green, yellow, brown or colourless chromatophores which form starch.
4. Stigma often present.
5. Marine or fresh water

   Examples: *Chilomonas, Cryptomonas.*

Order-5- Dinoflagellata

1. Small and planktonic. Body naked or with a thick pellicle or cellulose theca.
2. Two flagella, one lying transversely and the other pointing backwards.
3. Chromatophores are green, yellow, or brown.
4. Reserve food is starch or oil.
5. There are complex vacuoles which are not contractile.
6. Some are bioluminescent.
7. Mostly marine, some are parasitic.

   Examples: *Noctiluca, Ceratium; Oodinium, Gonyaulax.*

Order-6 Chloromonadida

1. Small, with a delicate and dorso-ventrally flatenned pellicle.
2. Two flagella, one trailing.
3. Chromatophores green and numerous.
4. Reserve food is oil.

5. Stigma absent; complex Contractile vacuole present
6. Largely freshwater.

Examples: *Gonyostomum, Vacularia.*

Order-7-Euglenida

1. Large with a thick and firm pellicle.
2. Anterior end with gullet leading into reservoir.
3. One or two flagella arising from an anterior recess.
4. Stigma present in coloured form.
5. Chromatophores green and numerous sometimes colourless.
6. Reserve food is paramylon and oil.
7. Mostly fresh water.

Examples: *Euglena, Phacus, Peranema, Rhabdomonas.*

Order-8. Volvocida

1. Small with rigid cellulose covering theca.
2. Gullet absent.
3. Flagella usually two sometimes more.
4. Body with green, usually single cup-shaped chromoplast.
5. Reserve food is starch and oil.
6. Stigma present.
7. Mostly fresh water forms. Many colonial species.

Examples: *Volvox, Chlamydomonas, Polytomella, Gonium, Pandori*

Class-2 Zoomastigophorea (Zoomastigina)

1. Flagellates with neither chromoplasts nor leucoplasts.
2. One to many flagella, in most cases with basal granule complex.
3. Reserve food glycogen.
4. Sexual stage unknown.

5. Many Commensals, symbionts and parasites.

Order-1. Choano flagellida.

1. Fresh water flagellates, with a single flagellum surrounded by a collar.
2. Sessile, sometimes stalked, sometimes with lorica.
3. Solitary or colonial.

Examples: *Codosiga, Salpingoeca, Proterospongia.*

Order-2. Bicosoecida

1. Largely fresh water flagellate encased within a lorica.
2. Two flagella, one free, the other attaching posterior end of body or shall.

Examples: *Bicosoeca, Poteriodendron.*

Order-3. Rhizomastigida

1. Small and amoeboid forms.
2. Flagella one or more upto four.
3. Locomotion by pseudopodia and flagella.
4. Chiefly freshwater.

Examples: *Màstigamoeba, Dimorpha*

Order - 4. Kinetoplastida

1. Small and more or less amoeboid in forms; no gullet.
2. Flagella basically one, but up to four.
3. Mostly parasitic forms, some are freshwater. Freshwater forms sessile, sometimes stalked.

Examples: *Bodo, Trypanosoma, Leishmania.*

Order. 5. Retortamonadida

1. Gut parasites of insects or vertebrates,

2. One to four floagella. One glagellum associated with ventrally located cytostome.

Example: *Chilomastix.*

Order-6- Diplomonadida

1. Small with a pellicle and often with a cytostome.
2. Bilaterally symmetrical flagellates, with two nuclei, each nucleus associated with four flagella.
3. Mostly parasitic.

Examples: *Giardia, Hexamita.*

Order -7. Oxymonadida

1. Symbiotic fagellates having one to many nuclei. Each nucleus associated with four flagella, some of which are turned posteriorly and adhere to body surface.

Examples: *Oxymonas, Pyrsonympha.*

Order-8. Trichomonadida

1. Four to six flagella, one of which is trailing.
2. Parasites in genital passage.

Example: *Trichomonas.*

Order - 9. Hypermastigida

1. Highly specialized forms with numerous flagella.
2. Kinetosomes arranged in a circle, plate or longitudinal or spiral rows.
3. Uninucleate or multinucleate but never binucleate.
4. Food ingestion by pseudopodia.
5. Symbionts in the guts of termites, cockroaches and wood roaches.

Examples: *Lophomonas, Trichonympha, Barbulanympha.*

SUPER CLASS - B - OPALINATA

1. Body covered by longitudinal, oblique rows of cilia arising from anterior subterminal rows.
2. Cytostome is absent.
3. Two or many monomorphic nuclei.
4. Binary fission generally symmetrogenic.
5. Sexual reproduction involves syngamy with flagellated gametes (anisogamous).
6. Parasitic found in gut of frogs and toads; less commonly in fishes, salamanders and reptiles.

Examples: *Opalina, Zelleriella.*

SUPER CLASS - C - SARCODINA

1. Protozoa with pseudopodia as feeding and locomotor organelles.
2. Flagella, when present, only in development.
3. Asexual reproduction by binary fission.
4. Mostly solitary and free-living, some parasitic and colonial.
5. Skeletons of various forms and composition characteristic of some groups.

Class - 1- Rhizopoda

1. The locomotory organelles are *lobopodia, filopodia* or *reticulopodia* but never axopodia.
2. Generally creeping forms.

Sub.-Class-(i) Lobosia

1. Pseudopodia usually lobopodia.

Order-1. Amoebida

1. Body is naked without any exoskeleton.
2. Typically uninucleate, ectoplasm and endoplasm are clearly differentiated.

3. Nucleus with honeycomb lattice.
4. Largely fresh water, some marine; many parasites.

Examples: *Amoeba, Entamoeba, Pelomyxa.*

Order-2. Arcellinida

1. Body enclosed in a shall or test with an aperture through which the pseudopodia protrude.
2. They are free living mostly in fresh water.

Examples: *Arcella, Difflugia, Euglypha.*

Sub (ii) Granuloreticulosia

1. They have fine granular reticulopodia

Order-Foramaniferida

1. Chiefly marine species with mostly multichambered shells.
2. Shells may be composed of tectin, which may incorporate foreign particles, but most commonly the shells are calcareous.
3. Shalls with one or more openings through which reticulopodia emerge. The reticulopodia are fine and branched forming network.
4. Chiefly marine.

Examples: *Elphidium, Globigerina, Orbulina, Spirillina.*

Sub Class (iii) Filosa.

1. Pseudopodia as filopodia, tapering and branching.
2. Body naked with a shall with single aperture.
3. Marine and freshwater.

Examples: Allogromia, Gromia

Sub Class (iv) Mycetozoia

1. Body large, amoeboid, multinucleate forming plasmodia.
2. Life-cycle complex and has sexual reproduction, usually sporangia are formed which liberate spores.

3. Pseudopodia numerous and blunt.

Examples: *Arcyria, Plasmodiophora.*

Class - 2. Actinopodea

1. The locomotony organelles are delicate *axopodia* with axial fila ments
2. They are primarily sessile or floating forms.
3. Test may be present or absent.
4. Reproduction both asexual and sexual.

Subclass- (i) Radiolaria

1. Body comparatively larger; central capsule is perforated by one or many pores.
2. Siliceous skeleton.
3. Axopodia or sometimes reticulopodia or filopodia.
4. Exclusively marine.

The radiolarians can be divided into two groups or classes: The Polycystina has a central capsule with evenly distributed pores or a pore field. The Phaeodaria have three pore fields and the organic nature of the capsule wall is different from that of Polycystina.

Examples: *Collozoum, Acanthometra, Aulocantha.*

Subclass (ii) Acantharia

1. Imperforate, nonchitinoid central capsule without pores.
2. Anisotropic skeleton of strontium sulphate.

3. Axopodia present. Marine.

Example: *Acanthometra.*

Subclass (iii) Heliozoida

1. Spherical animal without central capsule commonly called *Sun-animalcules*

2. Axopodia radiating.
3. Body cytoplasm is differentiated into outer vacuolated ectoplasm and inner dense endoplasm.
4. Skeleton absent (naked); if present, it is made up of siliceous scales and spines.
5. Primarily in fresh water.

Examples: *Actinophrys, Actinosphaerium, Camptonema.*

Subclass (iv) Proteomyxida

1. Largely marine and freshwater parasites of algae and higher plants.
2. Pseudopodia are filopodia and reticulopodia.
3. Reproduction by binary or multiple fission; multiple fission in cyst.

Examples: *Vampyrella, Pseudospora.*

Class - 3 - Piroplasmea

1. Small, round, rod-shaked or amoeboid.
2. Parasite in red blood corpuscles of vertebrates.
3. Do not produce spores.

Example: *Babesia*

SUBPHYLUM - II SPOROZOA

1. Exclusive endoparasites.
2. The locomotory organelles are absent in adults.
3. Cilia or flagella may be present in gametes.
4. Nutrition saprozoic.
5. Asexual reproduction by multiple fission and sexual by syngamy, followed by spore formation.
6. Sporozoites are infective stage; nucleus is of single type.

Class - 1, Telosporea

1. Pseudopodia are absent and locomotion is by gliding or body flexion.

2. Spores without capsules or filaments, naked or encysted.
3. Adult trophozoites with one nucleus.
4. Reproduction is both asexual and sexual.

Subclass - (i) Gregarinia

1. Mature trophozoites are large and extracellular
2. Reproduction is sexual with sporogony; spores contains eight sporozoites.
3. These are parasite of gut and body cavity of invertebrates.

Examples: *Gregarina, Monocystis*

Subclass (ii) coccidia

1. Mature trophozoites are small and intracellular.
2. Gametocytes are dimorphic.
3. Sporzoites multiply by schizogony in tissue cells.

Order 1. Eucoccida.

1. Asexual and sexual both the phases occur in life-cycle.
2. They are parasite in epithelial and blood cells of invertebrates and vertebrates.

Suborder (a) Eimeriina

1. Macrogametes and microgametes develop independently.
2. Syzygy is absent: Zygote is non motile.
3. Sporozoites are encased in sporocyst.

Example: *Eimaria*

Suborder (b) Haemosporina

1. Macrogametes and microgametes develop independently.
2. Syzygy is absent. Zygote (ookinate) motile.
3. Sporozoites are naked.

4. Schizogony in vertebrate host and sporogony in invertebrate hosts.

Example: *Plasmodium.*

Class - 2 - Toxoplasmea

1. Spore formation is absent.
2. Reproduction asexual by binary fission.
3. Cysts are formed which have many naked sporozoites. No flagella or pseudopodia.

Examples: *Toxoplasma, Sarcocystis.*

Class - 3- Haplosporea

1. Spores are present.
2. Pseudopodia may be present but flagella absent.
3. Reproduction only by asexual means.
4. Parasites of fish and invertebrates.

Examples: *Ichthyosporidium, Haplosporidium.*

SUBPHYLUM III. CNIDOSPORA

1. Adult trophozoite has many nuclei.
2. Spore formation occurs throughout life.
3. Spores have several cells having one or more polar filaments which are coiled threads and can be shot out.
4. Zygote gives rise to one or more trophozoites without sporogony.

Class-1 Myxosporidea

1. Spores develop from several nuclei and enclosed in two or three valves. Polar capsule present.

Order. 1. Myxosporida

1. Spores are large with a bivalved membrane.
2. Trophozoite amoeboid and not intracellular.
3. Polar capsule 1, 2 or 4 with filaments.

Example: *Myxidium*

Order - 2 - Actinomyxida

1. Spores large with a trivalved membrane.
2. Polar capsule three, each with a filament.

Example: *Triactinomyxon.*

Class-2. Microsporidea

1. Spores small, with a univalved membrane.
2. Polar capsule absent or present with 1 or 2 filaments.

Example: *Nosema.*

SUBPHYCUM -IV CILIOPHORA

1. The ciliates comprise the largest subphylum of Protozoa. Some 8000 species have been described and many groups are still not well known. They are also the most animal like and exhibit a very high level of organelle development.
2. The shape of body is definite. The body wall is a complex living pellicle, containing alveoli, trichocysts and other organelles, in addition to infraciliature.
3. The body surface is covered with uniform cilia which function in locomotion.
4. The cilia around the cytostome region have become specialized as compound ciliary organelles called membranelles and undulating membranes in many ciliates that employ fitter feeding.
5. Most of them have two types of nuclei-larger macronucleus and smaller micronucleus.
6. One or more contractile vacuoles are present in fresh water forms.
7. Reproduction asexual by transverse binary fission.
8. Conjugation also takes place with fusion of nuclei.
9. The anal aperture or cytopyge is a permanent aperture.

10. No alternation of generation in the life-cycle.

Class-1 Ciliata

1. Cilia are present during the whole or a part of their life cycle.
2. Mouth or cytostome and Cytophanynx are often present.
3. Nutrition is holozoic.
4. Reproduction takes place asexually by binary fission or budding and sexually by conjungation.
5. Mostly free living in fresh water, marine but parasites; colonial and sedentary forms also occur.

Class ciliata is divided into four subclasses:

Sub class-(i) Holotricha.

1. Cilia usually short and uniform, arranged in longitudinal rows all over the body.
2. No adoral spiral of long cilia or memberanelles.
3. Buccal ciliature either absent or, if present, usually inconspicuous.

The sub class Holotricha is divided into following orders:

Order 1, Gymnostomatida

1. Chiefly large ciliates with no oral ciliature.
2. Cytostome opens directly to out side.

Examples: *Didinium, Nassula, Coleps.*

Order 2.- Trichstomatida

1. With vestibutar but no buccal ciliature.
2. Free living and parasitic forms.

Examples: *Colpoda, Balantidium.*

Order 3 - Chonotrichida

1. Vase-shaped ciliates lacking body cilia.
2. A funnel at the free end of the body bears vestibular cilia.

3. Chiefly marine and ectocommensel on crustaceans.

Examples: *Spirochona, Chilodochona.*

Order - 4- Apostomatida

1. Body with spirally arranged ciliation.
2. Cytostome mid-ventral.
3. Marine parasite. Life. cycle complex usually involving two hosts, one of which is crustacean.

Example: *Hyalophysa.*

Order - 5 Astomatida

1. Body ciliation uniform.
2. Cytostome absent.
3. Commensals or endoparasites living in the gut and body cavity of oligochaete worms.

Examples: *Hoplitophyra, Anoplophyra.*

Order - 6. Hymenostomatida.

1. Mouth usually lies on the side of the body at the end of peristome, remains permanently open and leads into cytopharynx.
2. Buccal cavity present with membrances.
3. Small body with uniform cilia.

Example: *Paramecium*

Order 7 - Thigmotrichida

1. A small group of fresh and marine water ciliates found in association with bivalve molluscs.
2. Anterior end of body bears a tuft of thigmotaitic cilia.

Examples: *Boveria, Thigmophrya.*

Subclass (ii) Peritrichia

1. Adults without body cilia.

2. The apical end of the body typically bears a conspicuous buccal ciliature.

3. Mostly sessile bearing a stalk.

   It bears a single order-Peritrichida.

Example: *Vorticella, Carchesium.*

Subclass (iii) suctoria

1. Sessile and stalked body.
2. Young with cilia. Adult with few to many tentacles.
3. Adult completely devoid of any ciliature.

   This subclass bears only one order-Suctorida.

Examples: *Ephelota, Podophrya, Acineta.*

Sub-Class (iv) Spirotrichia

1. Reduced body cilia.
2. Well developed conspicuous buccel ciliature.

Order. 1- Heterotrichida.

1. With uniform body cilia or body encased in a lorica and body cilia absent.

Examples: *Stentor, Halteria.*

Order. 2. Oligotrichida

1. Body cilia reduced or absent.
2. Conspicuous buccal membranelles, commonly extending around apical end of body.

Examples: *Halteria, Strombidium.*

Order 3. Tintinnida

1. Loricete, mostly free swimming ciliates with inconspicuous oral membranelles when extended.
2. Largely marine.

Examples: *Favella, Codonella.*

Order- 4- Entodinimorphida

1. Ectocommensel in the digestive tract of herbivorous mammals.
2. Body cilia reduced or absent.
3. Prominent buccal ciliature often in separate anterior clumps
4. Posterior end may be drawn out into spines.

Examples: *Entodinium, Cycloposthium.*

Order 5- Odontostomatida

1. Body laterally compressed and wedge-sheped, with carapace.
2. Reduced body and buccal cilia.

Example: *Saprodinium.*

Order - 6 - Hypotrichida

1. Body dorso-ventrally flatenned.
2. Body cilia at ventral side, forming cirri.

Examples: *Euplotes, Urostyla, Oxytricha.*

CHILOMONAS

Phylum	-	Protozoa
Subphylum	-	Sarcomastigophora
Super class	-	Mastigophora
Class	-	Phytomastigophorea
Order	-	Cryptomonadida
Genus	-	*Chilomonas*

1. It occurs in abundance in stagnant water and is common in laboratry culture.

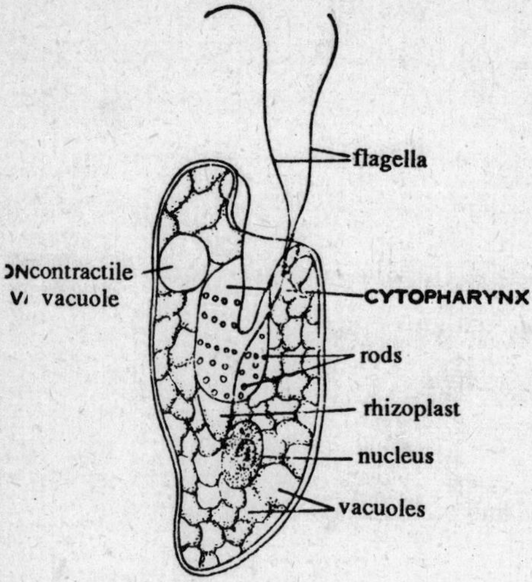

Fig. 1.1 Chilomonas.

2. Body is oval, colourless, about 35 μ long
3. Body with a deep anterior pit, the cytopharynx.
4. Two flagella projecting forwards from the cytopharynx.
5. Endoplasm is distinctly alveolar. Contractile vacuole lies on one side near the anterior end.
6. Nucleus spherical with endosome found at posterior end.
7. Mode of nutrition saprozoic.

PERANEMA

Phylum	-	Protozoa
Subphylum	-	Sarcomastigophora.
Super class	-	Mastigophora
Class	-	Phytomastigophorea
Order	-	Euglenoidida
Genus	-	*Peranema*

1. It is widely distributed in stagnant freshwater.
2. Body is covered over by an elastic and spirally striated pellicle.
3. Body is broad and truncated postteriorly and gradually tapers anteriorly.
4. Stigma, para flagellar body and chromatophores are absent.

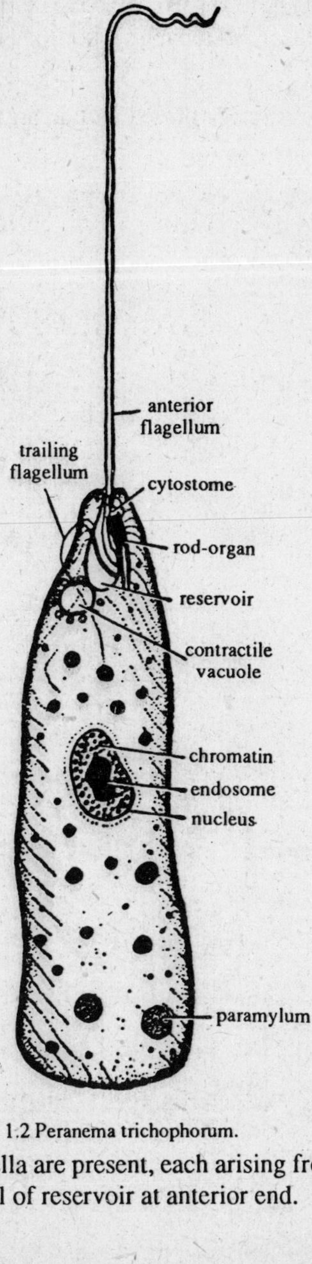

Fig. 1.2 Peranema trichophorum.

5. Paired, unequal flagella are present, each arising from its own blepharoplast in the wall of reservoir at anterior end.

6. A pair of hyaline rod-like structures forming the rod organ, runing alongside the reservior and supports the gullet which can be opened or closed.

7. Nucleus is centrally placed with a large; sømetimes irregular endosome.

8. It feeds phagotrophically. The prey is siezed by the rod organ and pushed into the cytostome.

CHLAMYDOMONAS

Phylum	-	Protozoa
Subphylum	-	Sarcomastigophora
Super class	-	Mastigophora
Class	-	Phytomastigophorea
Order	-	Volvocida
Genus	-	*Chlamydomonas*

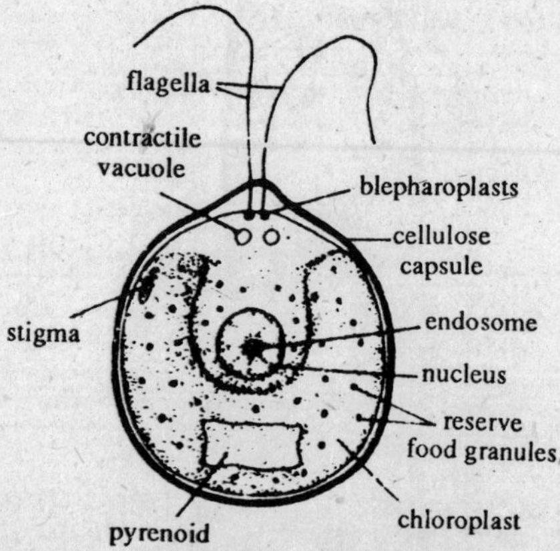

Fig. 1.3 Chlamydomonas.

1. It is a solitary animal commonly found in stagnant freshwater. Cosmopolitan.

2. Body is ovoid or flatenned surrounded by a thick cell-wall made of cellulose.

3. It has two flagella, originaling from the blepharoplasts situated in endoplasm.
4. Body is green due to presence of a cup-shaped chloroplast. A pyrenoid is embedded in the based part of chloroplast.
5. Nucleus is vesicular.
6. Red stigma and a pair of contractile vacuoles are found near the base of flagella.
7. Nutrition holophytic i.e. plant-like.
8. Reserve food as starch grain.
9. Asexual reproduction by binary fission, often in the palmella stage.
10. The sexual reproduction by the fusion of flagellate isogamy.

VOLVOX

The systematic position is same as that of *Chlamydomonas*.

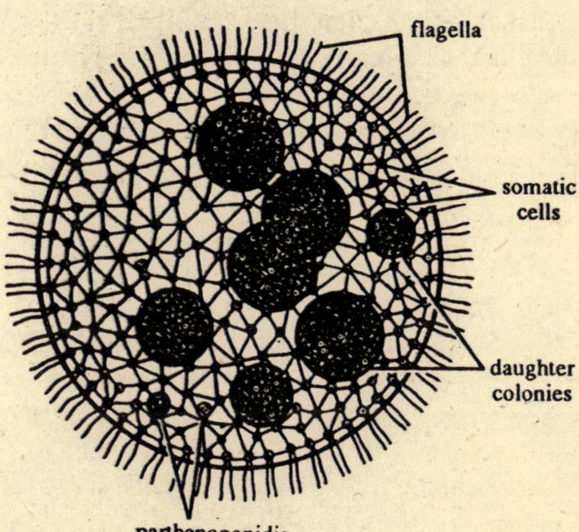

Fig. 1.4 Volvox globator

1. It is a pelagic freshwater, colonial organism found in ponds and lakes along with plankton.

2. It has cosmopolitan disfribution.
3. The coloney looks like a small green, hollow sphere filled with a watery jelly.
4. Coloney contains thousands of zooids (somatic cells) arranged in a single peripheral layer. The cells are connected by protoplasmic threads and not differing in size.
5. Each zooid has two flagella, two or more contractile vacuoles, cup-like chloroplast, a single nucleus, a red stigma but no gullet.
6. Nutrition is holophytic.
7. The colonees are polarised since they swim always in one region hence called anterior pole. Locomotion by combined action of flagella.
8. The coloney increases in size by binary fission. The zooids may be sometic and reproductive. The reproductive zooids found posteriorly take part in reproduction.
9. Some of these zooids are called *parthenogonidia* repeatedly divide to form daughter colonies which are liberated by the disintigration of parent coloney.
10. During sexual reproduction some specilized cells, called as *antheridia* give rise to several biflagellate microgametes. Other cells, called *archegonia,* give-rise to non-flagellate macrogametes. Both fuse in water to form zygote that forms new coloney after dormancy.

NOCTILUCA

Phylum	-	Protozoa
Subphylum	-	Sarcomastigophora
Super class	-	Mastigophora
Class	-	Phytomastigophorea
Order	-	Dinoflagellida
Genus	-	*Noctiluca*

1. *Noctiluca* is a marine and pelagic organism, occuring near the shores often in large number.
2. It is abundantly found in Atlantic and Pacific seas.

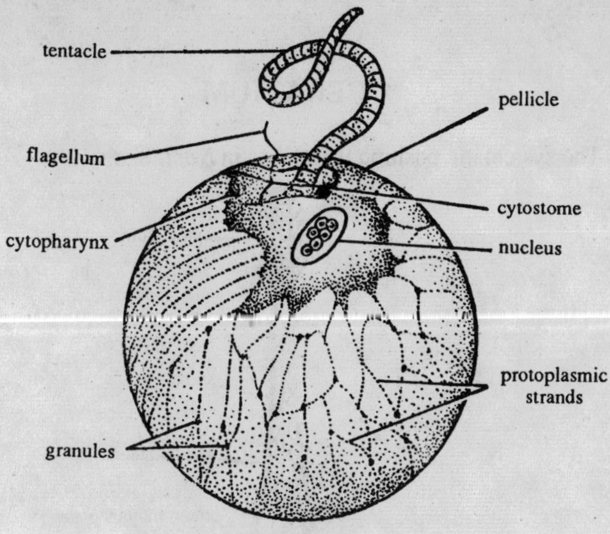

Fig. 1.5 Noctiluca

3. It has a large, melon-like, bilaterally symmetrical body measuring 1-15. mm. in diameter and enclosed in a firm cuticular pellicle.

4. The cytoplasm is highly vacuolated and differentiated into ectoplasm and endoplasm.

5. In a depression at the ventral pole is a longitudinal oral groove representing the sulcus. This groove has an oval mouth leading into the gullet and two flagella; a long, coarse transversely striated flagellum called tentacle, other in delicate flagellum.

6. From the central sulcus and clump of protoplasm containing the nucleus, various delicate strands run through the watery interior to the periphery.

7. Nutrition is holozoic i.e. animal like.

8. Reproduction takes place asexuelly by binary fission, as well as by syngamy. The syngamy involves fusion of two organisms and is followed by formation of zoospores.

9. When agitated, it glows with bluish or greenish light. The *luminescence* is caused by photogenic granules *luciferin,* which under the influence of enzyme, *luciferase,* emits light.

CERATIUM

The systematic position is same as in *Noctiluca*.

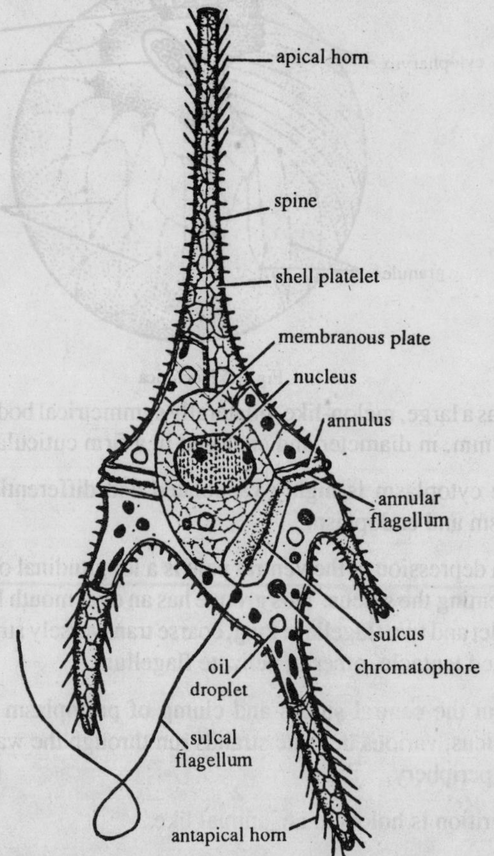

Fig. 1.6 Ceratium hirrundinella, semidiagrammatic, with portion of shell removed.

1. *Ceratium* includes marine as well as fresh water species.
2. It is common in U.S.A.
3. The body is usually covered with a cellulose plate grooved by a transverse and a longitudinal furrow each containing a flagellum, which projects out through a pore in theca.

4. The cellulose plates, which are produced into an anterior or apical and two posterio-lateral processes giving it a triangular shape.
5. The transverse groove is called as annulus running, around spiral girdle. Longitudinal groove is called as sulcus.
6. Cytoplasm is differentiated into ectoplasm and endoplasm.
7. Cytoplasm contains starch granules, chromoplast, nucleus and a complex system of contractile vacuoles.
8. It is mainly holophytic but some species are holozoic.
9. Reproduction is by oblique binary fission.
10. Cyst formation sometimes occurs.
11. Several species are known to be luminescent.

PROTEROSPONGIA

Phylum	-	Protozoa
Subphylum	-	Sarcomastigophora
Super class	-	Mastigophora
Class	-	Zoomastigophorea
Order	-	Choanoflagellata
Genus	-	*Proterospongia*

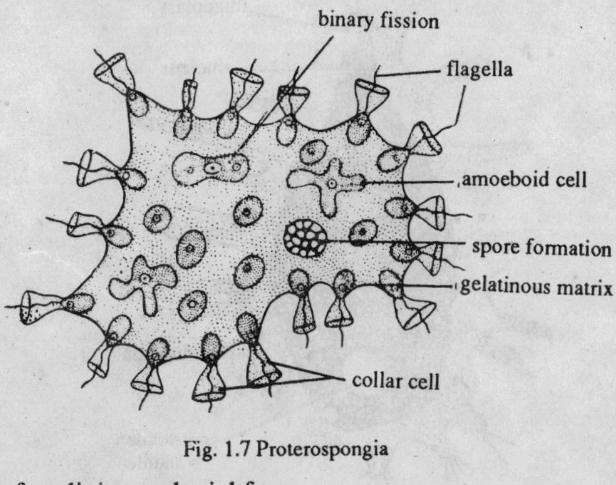

Fig. 1.7 Proterospongia

1. It is a free-living, colonial form.
2. It consists of a gelatinous mass in which upto 60 collared cells lying at periphery.
3. Each bearing a single flagellum surrounded basically by a thin, transparent and funnel-like protoplasmic collar.

4. Collar helps in food capturing or ingestion.
5. Inside the matrix are also some amoeboid zooids.
6. Any collared individuals can migrate inward and change into an amoeboid individual or vice-versa.
7. *Proterospongia* is often regarded as a link between protozoan and sponges.

MASTIGAMOEBA

Phylum	-	Protozoa
Subphylum	-	Sarcomastigophora.
Superclass	-	Mastigophora.
Class	-	Zoomastigophorea
Order	-	Rhizomastigida.
Genus	-	*Mastigamoeba*

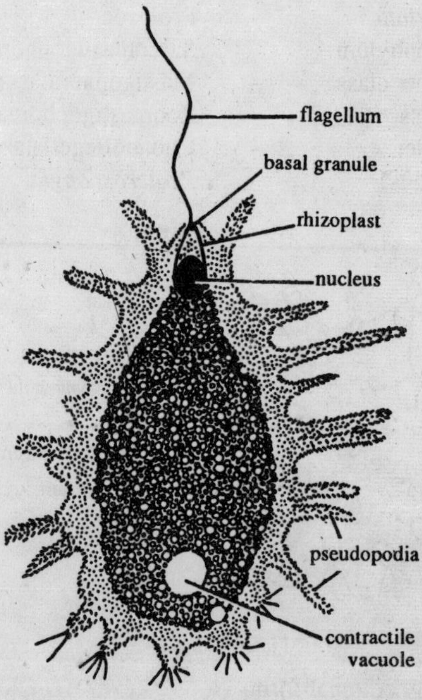

Fig. 1.8 Mastigamoeba.

1. It is a free-living flagellate found in fresh water ponds and lakes.

2. Body is irregular or amoeboid in form pushing out several finger like pseudopodia, which usually disappear when the organism swims. One flagelloum is present.
3. Protoplasm contains single nucleus and contractile vacuole
4. Nutrition is holozoic. Pseudopodia help in ingestion of food.
5. It reproduces by a longitudinal binary fission.
6. It is considered to be a connecting link between Mastigophora and Rhizopoda.

BODO

Phylum	-	Protozoa
Subphylum	-	Sarcomastigophora
Super class	-	Mastigophora
Class	-	Zoomastigophorea
order	-	Kinetoplastida
Genus	-	*Bodo*

Fig. 1.9 Bodo saltans.

1. It is a free-living animal found in the faecal matter.
2. Body is small, naked, more or less elongated, and with one surface more convex than the other.
3. A pair of unequal flagella arise from a depression at the anterior end, one flagellum trailing behind and used for anchoring.
4. Ingestion takes place at a spot near the base of flagella. A permanent cytopyge in not known.
5. Small rostral vacuole and near it is a conspicuous contractile vacuole.
6. Single rounded nucleur with centrally placed endosome lies near anterior end of the body.

TRICHOMONAS

Phylum	-	Protozoa
Subphylum	-	Sarcomastigophora
Super Class	-	Mastigophora
Class	-	Zoomastigohorea
Order	-	Trichomonadida
Genus	-	*Trichomonas.*

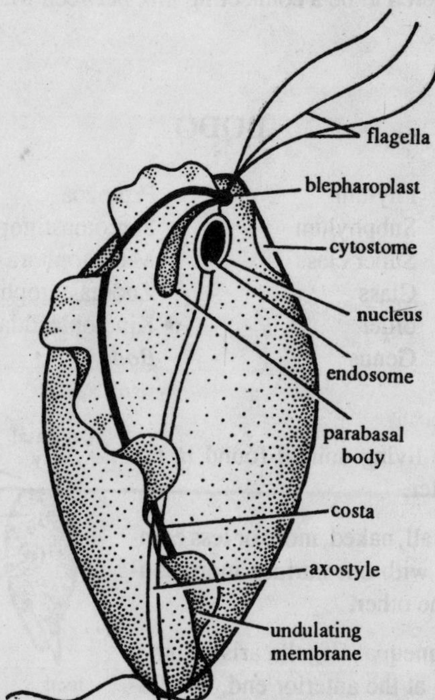

Fig. 1.10 Trichomonas

1. It is one of the most common intestinal multiflagellats, living in the intestine of all the vertebrates and also in leeches and termites. They exist only in the trophozoite phase and have no cystic phase.

2. It has cosmopolitan distribution.

3. The body is bear shaped taporing posteriorly, provided with four flagella of which one is directed backwards united to the body by an undulating membrane. Body measures 5-20 μ in length.

4. Cytostone is present anteriorly, used for ingestion of food.

5. Blepharoplast is minute situated at the anterior end. Extending posteriorly from the blepharoplast is club-shaped parabasal body.

6. A contractile fibre, the *costa* runs along the line of attachment of the undulating membrane.

7. The body is supported by a cylinder of microtubules, the *axostyle,* which extend backwards and projects from posterior end as *spike.* The spike helps to anchor the animal while feeding.

8. The endoplasm contains a large nucleus situated anteriorly. The endosome is prominant.

9. Feeding is saprozoic but most species ingest bacteria, yeast and other solid food.

10. Reproduction takes place exclusively by longitudinal binary fission.

11. No cyst formation as transmission occurs directly in the active, free trophozoite stage.

Pathogenecity In man *Trichomonas buccalis* is found in mouth; *T.hominis* in colon and *T. vaginalis* in vagina of female (*T. vaginalis* is found in large number in the leucorrhoeic discharge of females. It causes inflamation of the vaginal mucosa.) Of other animals *T. foetus* of cattles, *T. gallinae* of birds are important. The disease caused by them is called *trichomoniasis.*

GIARDIA

Phylum	-	Protozoa
Subphylum	-	Sarcomastigophora
Super class	-	Mastigophora
Class	-	Zoomastigophorea
Order	-	Diplomonadida
Genus	-	*Giardia*

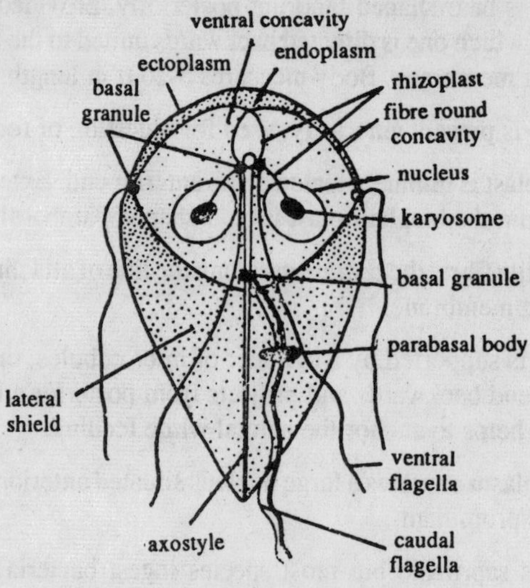

Fig. 1. 11 Giardia

1. *Giardia* is a parasite in the small intestine and colon of man, rats, mice, dogs, cats and frogs. *G. lamblia* is found in the small intestine of man, especially children.
2. Cosmopolitan distribution.
3. Body is oval in shape or half bear-shaped with broad forward anterior end and narrow posterior end.
4. The body is bilateral-symmetrical.
5. The dorsal side is convex, but the ventral surface contains a slight concavity, forming a sucking disc to the wall of intestine.
6. There are two vesicular nuclei and four pairs of long flagella. The basal granules of foremost flagella are inter-connected and also connect with a pair of nuclei.
7. Axostyle forms the mediam longitudinal axis of the body.
8. Just behind the sucking disc is a deep staining parabasal body.
9. Cytopharynx is absent and the rhizoplast make complicated loops through the cytostome.
10. Reproduction takes place by longitudinal binary fission.
11. It forms oral thick-walled cysts. Division occurs in the cyst so that the cyst has four nuclei.

12. It prevents absorption of fats by the host, the unabsorbed fat causes diarrhoea.

    Pathogenecity: It interfers with the absorption of food and causes chronic enteritis, fever, anemia, allergy and acute enterocolitis.

TRICHONYMPHA

Phylum	-	Protozoa
Subphylum	-	Sarcomastigophora
Super class	-	Mastigophora
Class	-	Zoomastigophorea
Order	-	Hypermastigida
Genus	-	*Trichonympha*

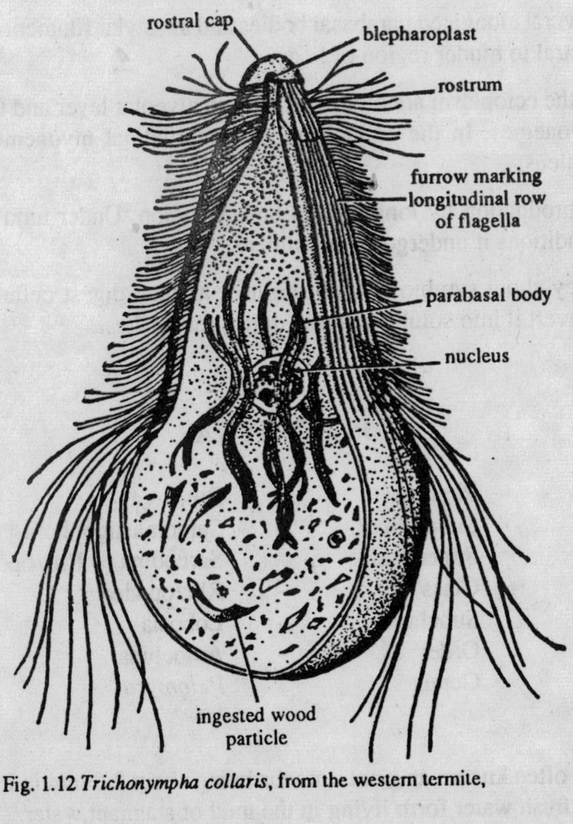

Fig. 1.12 *Trichonympha collaris*, from the western termite,

1. It is found in the alimentary canal of termites and cockroaches.
2. It is cosmopolitan in distribution.
3. The body is large, more or less oval in shape measures 50-300 μ in length.
4. Body is divisible into three regions: a tapering anterior flagellated rostrum, bearing apical cap; an intermediate bell-like portion, bearing numerous flagella; and a posterior rounded non flagellated region.
5. All the flagella are connected through fibrils with a hemispherical blepharoplast, lying at the base of the cap. Some flagella are arranged in longitudinal rows.
6. Several elongated parabasal bodies and axostylar filament run from rostral to hinder region of body.
7. In the ectoplasm are oblique fibres, an alveolar layer and tranverse myonemes. In the endoplasm are longitudinal myonemes and a nucleus.
8. Reproduction by longitudinal binary fission. Under unfavourable conditions it undergoes encystment.
9. They show symbiosis with the host as they digest cellulose and convert it into soluble carbohydrates for the host.

PELOMYXA

Phylum	-	Protozoa
Subphylum	-	Sarcomastigophora
Super class	-	Sarcodina (= Rhizopoda)
Class	-	Rhizopodea
Subclass	-	Lobosia
Order	-	Amoebida
Genus	-	*Pelomyxa*

1. It is often known as giant-amoeba being about 2.5 mm in length. It is a fresh water form living in the mud of stagnant water.

2. The body is asymmetrical and the body shape changes constantly.

3. It has a single large, hyaline and blunt pseudopodium.

4. The cytoplasm is highly vacuolated and contains 100-1000 nuclei, all of which are alike.

5. It contains a number of peculiar, some-what rounded refringent bodies, which are of albuminous nature and which constantly harbour bacteria of *Cladothrix* sp. (symbiotic).

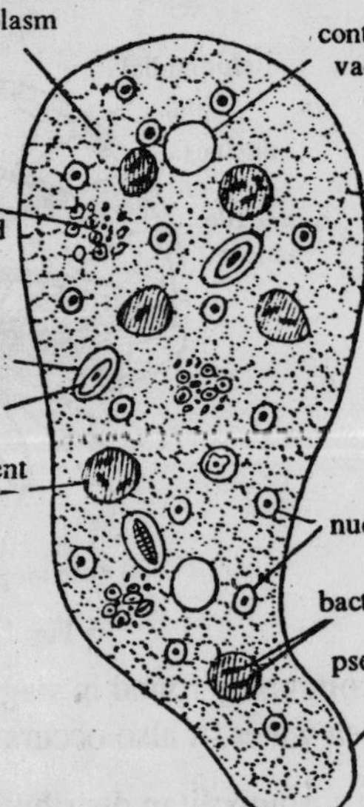

Fig. 1.13 Pelomyxa.

6. Nutrition is holozoic. Diatoms form the chief food. Numberous food vacuoles, and many fluid filled vacuoles but not contractile vacuoles. Reserve food material in the form of glyeogen granules.

7. Asexual reproduction by plasmotomy, in which the cytoplasm divides 2-6 times and each containing few nuclei.

8. Sexual reproduction by the formation of gametes which conjugate in pair. Encystment in some species.

ARCELLA

Phylum	-	Protozoa
Subphylum	-	Sarcomastigophora
Super class	-	Sarcodina
Class	-	Rhizopodea
Sub class	-	Lobosia
Order	-	Arcellinida
Genus	-	*Arcella*

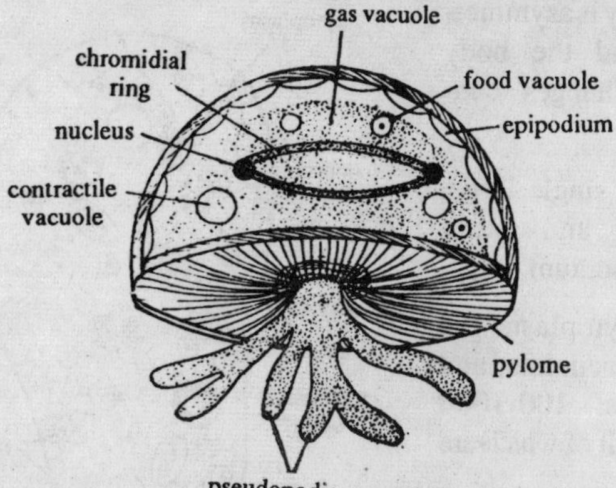

Fig. 1.14 Arcella.

1. It is commonly found in stagnant water attached to weeds growing in fresh water. It also occurs in moist forest soil, moss etc.
2. It has cosmopolitan distribution.
3. The body is covered by a yellow to brown, thick, hard, transparent and disc-shaped shall or test secreted by animal itself.
4. The shall is made of pseudochitin containing silica and iron.
5. The ventral surface has an inverted funnel-like depression, leading to central mouth or *pylome*.
6. 1-6 small, hyaline, finger-like, simple or branched lobopodia are extended through the pylome, which function as locomotory organelles and also helps in feeding.
7. The cytoplasm is differentiated into nonvacuolated ectoplasm and endoplasm. The body is attached to the inner test by filamentous ecto--plasmic strands known as epipodia.
8. Endoplasm contains food vacuoles, contractile vacuoles, reserve food material and two prominant nuclei in *A. vulgaris* and upto 15 in *A. podophrya.*
9. Nutrition is hologoic. Food includes mainly green and colourless flagellates.
10. A characteristic feature is the presence of a ring of granules called *chromidia.*

11. Asexual reproduction by binary fission. Two nuclei divide forming four nuclei, two of which extruded from pylome. This mass secretes a new shall. Multiple fission also occurs.

DIFFLUGIA

Systematic poisition is same as in *Arcella*.

1. It is a common, free-living animal found on muddy floor of small leaf chacked fresh water puddies, ponds, ditches or even moist soil.
2. It has world wide distribution.
3. It possesses a beautifully symmetrical test usually made up of intricately placed sand grains of a definits size range. In certain localities, diatom shalls or spicules of fresh water sponges may be used to suppliment the sand grains.

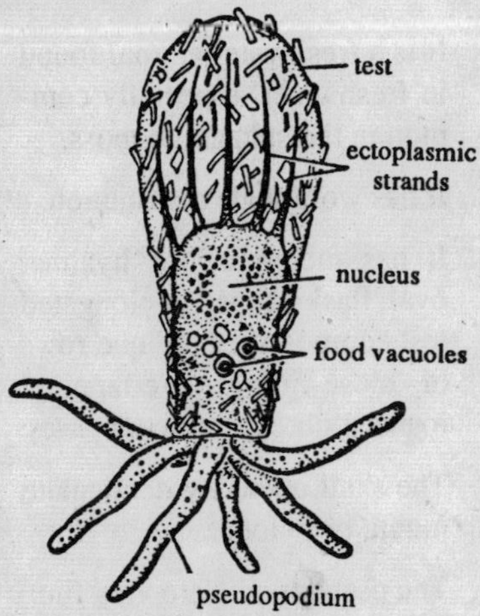

Fig. 1.15 Difflugia oblonga.

4. The shape of the test (varies with species) may be vase-shaped, or flask-shaped and always has a distinct neck with a terminal mouth or *pylome*.
5. About six slender, hyaline lobose pseudopodia may protrude out from pylome.
6. Cytoplasm is differentiated into ectoploasm and endoplasm. A nucleus surrounded by chromidia; numerous contractile vacuoles and food vacuoles are scattered around the periphery of the endoplasm.
7. Often the animals are a vivid green because of zoochlorellae in their cytoplasm as symbionts.
8. Mode of nutrition holozoic, it feeds upon micro-organisms and filamentous algae such as *Spirogyra*.

9. It reproduces by binary fission, but multiple fission has also been reported.
10. *Difflugia oblonga* is very active in the spring and most of the summer, but becomes encysted within its test in the late fall.

EUGLYPHA

Systematic position is same as in *Arcella.*

1. It is a freshwater animal found in freshwater, especially common in the sphagnum moss.
2. It has word wide distribution.
3. It has membraneous hyaline, oval, flask-shaped or elongated test composed of oblique rows of close-fitting, overlapping, rounded siliceous scales or plates.
4. The shall opens by denticulate mouth or pylome.
5. The pseudopodia are very fine, filiform, filopodia, which are often branched.

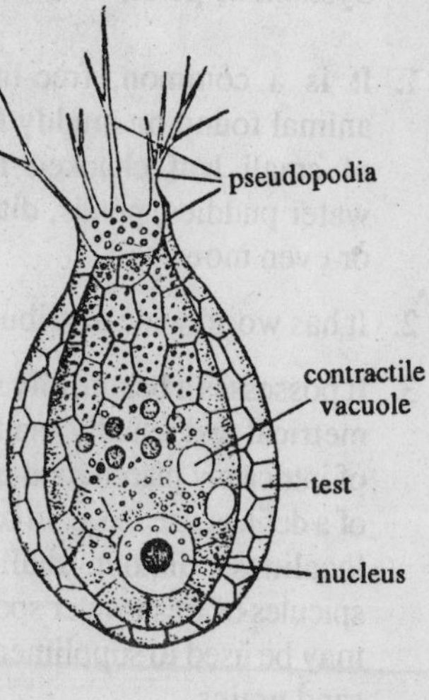

Fig. 1.2 Euglypha,

6. Ectoplasm is indistinct. Endoplasm contains shall-plates, vacuoles, one or two large centrally placed nucleus surrounded by perinuclear chromidia.
7. It reproduce by encystment. Multiple fission is not evident.

GLOBIGERINA

Phylum	-	Protozoa
Subphylum	-	Sarcomastigophora
Super class	-	Sarcodina
Class	-	Rhizopodea
Sub class	-	Granuloreticulosia
Order	-	Foraminiferida
Genus	-	*Globigerina*

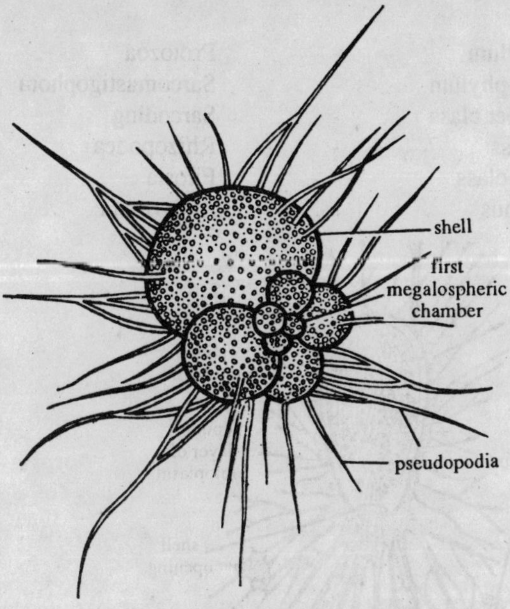

Fig. 1.17 Globigerina

1. It is one of the most common and abundant pleagic forminiferan.
2. It has cosmopolitan distribution.
3. It has a polythalamus perforated and calcareous shall.
4. The chambers are globular, and are arranged in a helicoid spiral.
5. The cytostome occupies all the chambers passing through the opening in the septa.
6. From the shall, long, filamentous, viscous and contractile psenudopodia extend all around. They form pseudopodial network. The pseudopodia are very important. They exhibit streaming movement.
7. Shells of dead individuals sink to the bottom of the sea like rain, forming there a grey mud called the *globigerina ooze.*

ALLOGROMIA

Phylum	-	Protozoa
Subphylum	-	Sarcomastigophora
Super class	-	Sarcoding
Class	-	Rhizopodea
Subclass	-	Filosia
Genus	-	*Allogromia*

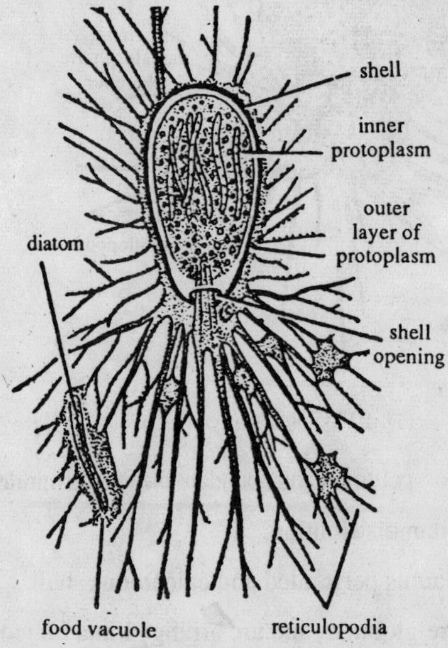

Fig. 1.18 Allogromia oviformfs.

1. It is found in both fresh water and marine water.
2. Body of animal is covered by a shall which is one chambered.
3. The cytoplasm flows out from the terminal opening of shall and flows around the shall, in this way shall becomes internal.
4. Cytoplasm contains a single nucleus, a contractle vacuole and a food vacuole.

5. The pseudopodia are very long and delicate, and join to form a network, hence called *reticulopodia*. They help in capturing prey.

6. Reproduction by multiple fission.

THALASSICOLA

Phylum	-	Protozoa
Subphylum	-	Sarcomastigophora
Super class	-	Sarcordina
Class	-	Actinopodea
Subclass	-	Radiolaria
Genus	-	*Thalassicola*

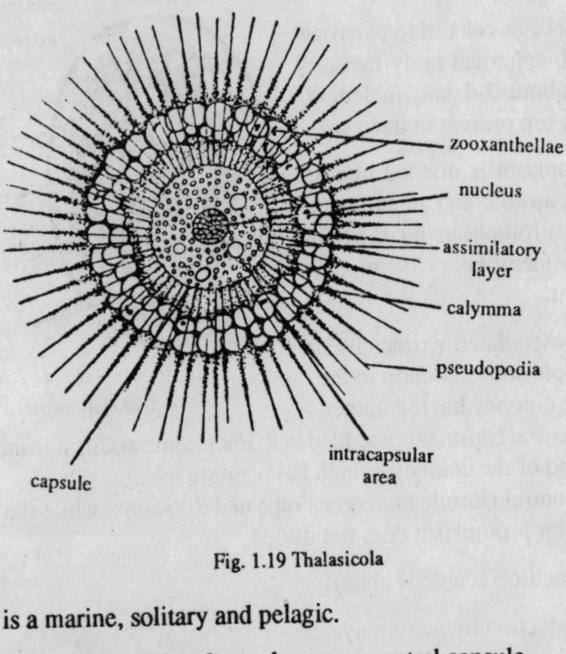

Fig. 1.19 Thalasicola

1. It is a marine, solitary and pelagic.

2. It has a perforated and membraneous central capsule.

3. The intercapsular cytoplasm contains a single large nucleus and a number of small vacuoles. The extracapsular cytoplasm has the

usual three regions: inner thin assimatory zone; middle thick vacuolar zone or calymma with large closely-packed, fluid-filled vacuoles lying in gelatinous material and often having yellow symbioitic dinoflagellates; and outer thin pseudopodial region.

4. Contractile vacuole is absent.
5. Nutrition is holozoic
6. Reproduction by binary fission and swarm-spore formation.

COLLOZOUM

The Systematic position is the same as that of *Thallassicola*

1. It is a large, colonial form having a sub-spherical body measuring about 3-4 cm. in length. They are present in deep sea.
2. Protoplasm is divided into an intracapsular and an extracapsular protoplasm by a perforated membrane. Skeleton is absent.

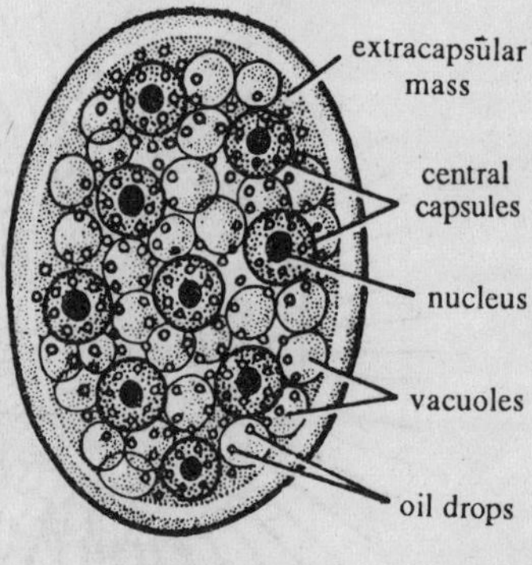

Fig. 1.20 Collozoum

3. The vacuolated extracapsular protoplasm is common to the entire coloney having numer--ous central capsules embedded in it. Each centtral Capsule indicates a zooid of the coloney, which has a single nucleus.
4. The central capsule undergoes repeated divisions, while the extra-capsular protoplasm does not divide.
5. Contractile vacuole is absent.
6. Reproduction by merogony.

LITHOCIRCUS

The systematic position is the same as that of Thallassicola

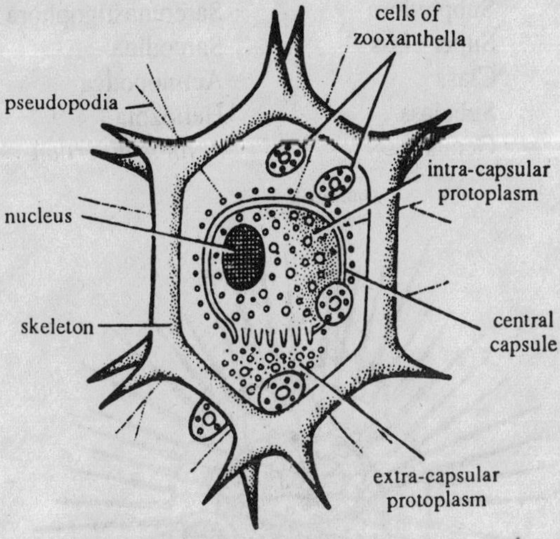

Fig. 1.21 Lithocircus.

1. It is a deep sea, solitary animal with a skeleton.
2. Body in covered by a shall of silicious skeleton.
3. The animal is prorided with a central Capsule embedded in the protoplasm.
4. The Capsule divides the protoplasm into extra capsular and intracapsular protoplasm.
5. Extra capsular protoplasm gives rise to the radiating thread-like pseudopodia. It has symbiotic yellow dinoflagellates.
6. Intra capsular protoplasm contains a large nucleus. Contractile vacuoles is absent.
7. Asexual reproduction by binary fission.

ACTINOSPHAERIUM

Phylum	-	Protozoa
Subphylum	-	Sarcomastigophora
Super class	-	Sarcodina
Class	-	Actinopodea
Subclass	-	Heliozoia
Genus	-	*Actinosphaerium*

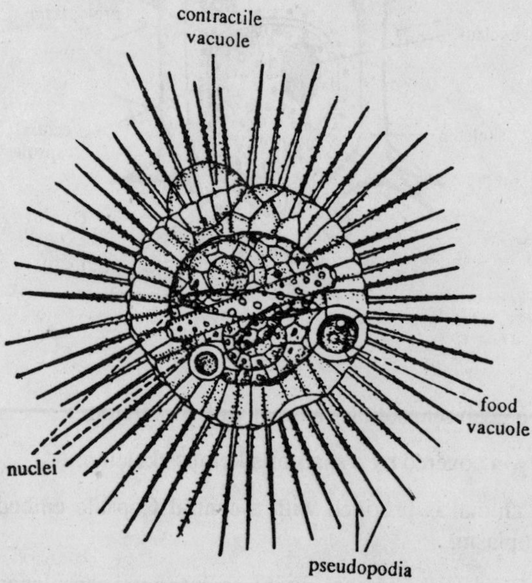

Fig. 1.22 Actinospharium.

1. It is a fresh water form, abundently found among floating vegetation. It is commonly called *Sun animalcule.*

2. Cosmopolitan distribution.

3. Several stiff needle-like axopodia radiate from the central spherical body of frothy protoplasm.

4. Protoplasm is divided into a highly vacuolated or frothy ectoplasm and a deuser granular endoplasm.

5. Several contractile vacuoles in the ectoplasm.
6. Several nuclei are present in endoplasm.
7. The axial filament of pseudopodia do not reach up to nuclei, but terminate freely in the deeper granular layer of ectoplasm.
8. Nutrition is holozoic, voracious in feeding. Axopodia help in feeding rather locomotion.
9. The animal reproduce asexually by plasmotomy. The whole organism divides into two daughter spheres, each contains several nuclei.
10. Autogamy and encystment also observed.

ACTINOPHRYS

The systematic position is same as that of *Actinosphaerium.*

1. It is found in fresh and marine waters.

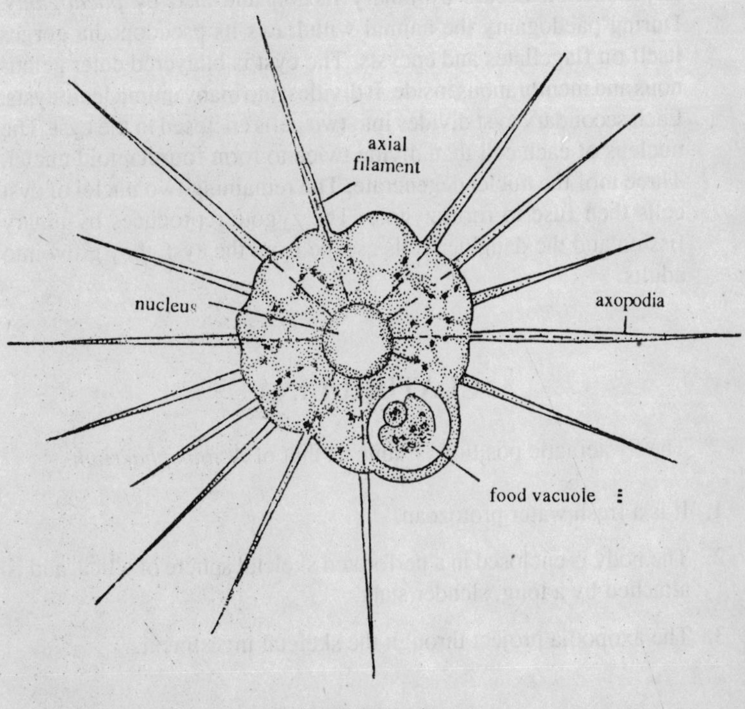

Fig. 1.23 Actinophrys

2. It has cosmopolitan distribution.
3. The body is round from which radiate thin, long pseudopodia, each with a central axial filament covered with an adhesive granular ectoplasm. The are called axopodia.
4. The axial filaments of axopodia are attached to the nuclear membrane.
5. The cytoplasm is not distinctly divided into ectoplasm and endoplasm.
6. The ectoplasm have many vacuoles.
7. Endoplasm is gelationous and contains various vacuoles and food particles besides a single nucleus.
8. Nutrition is holozoic, whenever food or prey touches the axopodia, they attach to it. The axopodia paralyze the prey and at times shorten to convey the prey to central mass of cell.
9. Reproduction occurs by binary fission and also by *paedogamy.* During paedogamy the animal withdraws its pseudopodia gorges itself on flagellates and encysts. The cyst is bilayered enter gelatinous and membranous inside. It divides into many uninucleate cysts. Each secondary cyst divides into two cells enclosed in the cyst. The nucleus of each cell than divide twice to form four haploid nuclei. Three mof the nuclei degenerate. The remaining two nuclei of cyst cells then fuse to form zygote. The zygote reproduces by binary fission and the daughter cells escape from the cyst, they grow into adults.

CLATHRULINA

The Systematic position is same as that of *Actinosphaerium.*

1. It is a fresh water protozoan.
2. The body is enclosed in a perforated skeletal sphere of silica, and is attached by a long, slender stalk.
3. The axopodia project through the skeletal investment.

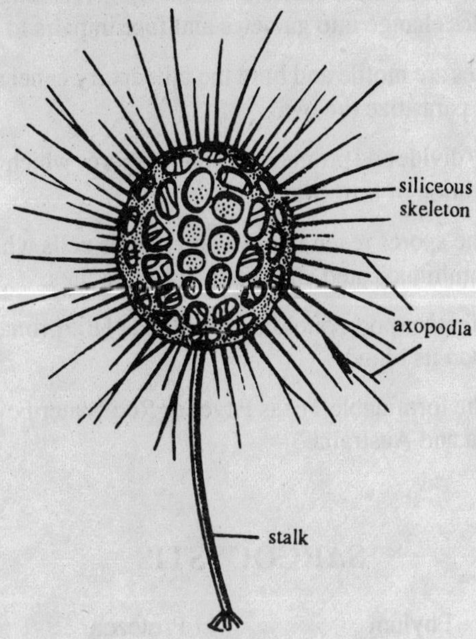

Fig. 1.24 Clathrulina

4. Cytoplasm contains single nucleus.
5. It undergoes multiple fission in the active condition and produce biflagellate merozoites, called *flagellule*. The flagellule latter grow into adults.

BABESIA

Phylum	-	Protozoa
Subphylum	-	Sarcomastigophora
Super class	-	Sarcodina
Class	-	Piroplasmea
Genus	-	*Babesia*

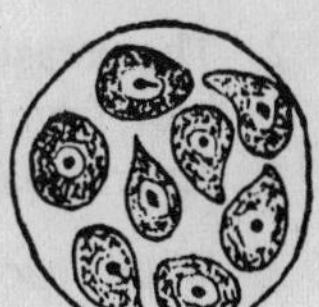

Fig. 1.25 Babesia

1. It is a minute, intracorpuscular blood parasite of cattles, dogs and monkeys
2. The trophozoite lives in the R.B.C. of the host and has a rhizoplast but no flagellum and pigment.

3. Tick is the intermediate host and vector. Upon reaching in its gut, the trophozoites change into gametes and fuse in pairs to form zygote.

4. The zygotes are motile and bore the alimentary canel and reach the ovary and parasitize the ova.

5. In ova they divide and produce amoeboid spores which pass on to the next generation of host (tick).

6. Some of the spores reach the salivary glands cells where the spore becomes multinucleated and produce sporozoites.

7. Tick transfer the sporozoites into the blood of the primary host when it feeds upon its blood.

8. It causes the formidable Texas Fever or Red Water Fever in Cattles in America and Australia.

SARCOCYSTIS

Phylum	-	Protozoa
Subphylum	-	Sporozoa
Class	-	Toxoplasmea
Genus	-	*Sarcocystis*

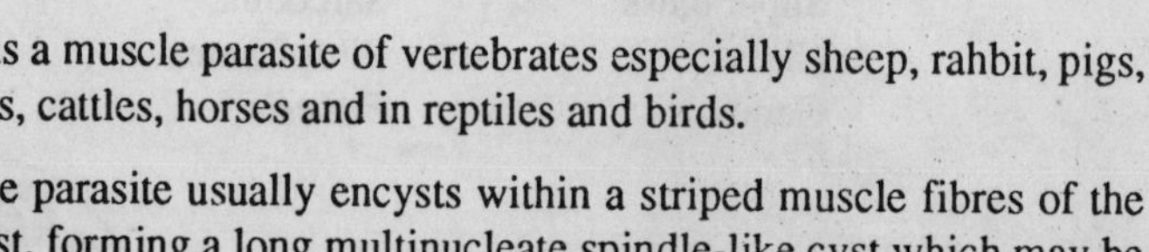

Fig. 1.26 Sarcocystis

1. It is a muscle parasite of vertebrates especially sheep, rahbit, pigs, rats, cattles, horses and in reptiles and birds.

2. The parasite usually encysts within a striped muscle fibres of the host, forming a long multinucleate spindle-like cyst which may be several man in length. It is known as *Miescher's* or *Rainey's tube.*

3. Within each tube a large number of sickle-shaped spores are formed. The spores are called *Rainey's corpuscles.*

4. Very little is known about the mode of transmission and life cycle. When ingested the spores hatches in the intestine of new host liberating a single sporozoite, which penetrates an epithelial cell to grow into a trophozoite through blood circulation to the striped muscle fibres, where they grow into plasmodia.

5. They produce a toxic substance *sarcocystin* and cause a disease known as *sarcosporidiosis*.

NOSEMA

Phylum	-	Protozoa
Subphylum	-	Cnidospora
Class	-	Microsporidea
Genus	-	*Nosema*

1. Nosema bombycis is a parasite in the gut of silk worm larvae where it causes *Pebrine*-a distructive silk-worm disease.

2. Asexual multiplecation of the trophozoites and possibility of some spores developing in the same host cause enormous damege to the hosts's tissue, the result is fatal.

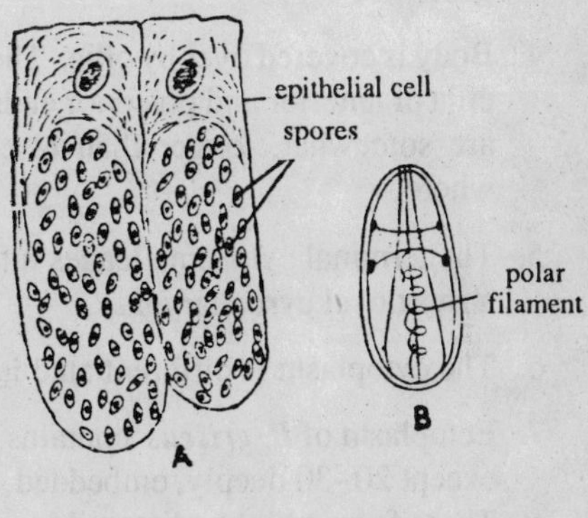

Fig. 1.27 Nosema A-- Spares of Nosema apis in stomach cells. B--Spore of N.bombycis.

3. Its another species *N. apis.* It is parasite in the epithelial tissue and the Malpighian tubules of honey-bees causisng a fatal disease *nosema.*

4. The spores are very small, each containing a coiled thread, the *polar filament,* that represents a polar capsule without a wall.

PRODODON

Phylum	-	Protozoa
Subphylum	-	Ciliophora
Class	-	Ciliatea
Subclass	-	Holorichia
Order	-	Gymnostomatida
Genus	-	*Prorodon*

1. It is a fresh water and marine ciliate.
2. The body is ovoid but sometimes variable in shape, but, when fully extended, it measures 165 - 200μ in length and 40 - 60μ in width.
3. Body is ellipsoidal, with the anterior and slightly wider then the posterior.
4. Body is covered over by cilia. The cilia of anterior and posterior ends are somewhat, longer than elsewhere.
5. The terminal cytosome leades into a short oval cyropharynx.

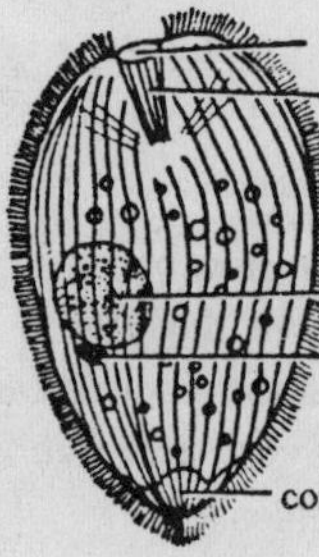

Fig. 1.28 Prorodon griseus.

6. The cytoplasm is differentiated into ectoplasm and endoplasm.
7. Ectoplasm of *P. griseus* contains no trichocysts or other inclusions except 20 -30 deeply, embedded, large, parallel, rod - like trichite. These form a highly distensible oral basket around the subterminal cytostome.
8. Endoplasm is grayish and occassionally may become green because of presence of numerous symbiotic zoochlorellae.
9. Numerous food vacuoles are present. Cytopyge at the posterior end.
10. A large contractile vacuole is present at extreme posterior end.
11. Two nuclei ; the massive macronucleus located centrally, may be spherical or oval containing a prminant nucleolus. A small pear -

shaped micronucleus is found attached externally to the macronucleus.

12. When the pond water dry up, it develops a thick-walled protective cyst around itself. Almost immediately after excysting, the small animals conjugate followed by binary pission. Macronuceus also divides.
13. Macronuclear reorganization takes place only in the permanent protective cyst and requires at least four days.

COLEPS HIRTUS

The systematic position is same as that of *Prorodon*..

1. It is a widely distributed fresh water cliate with an ability to tolrate wide range of temperature, oxygen, salinity and acidity.
2. The body is barrel shaped, brown in colour and is slightly more convex dorsally. It measures about 40-65μ in length and 20-30μ is width.
3. In the alveolar layer of ectoplasm beneath the pellicle, the body is covered with armor plates which are modified into oral teeth at the anterior end, and caudal spines at posterior end.

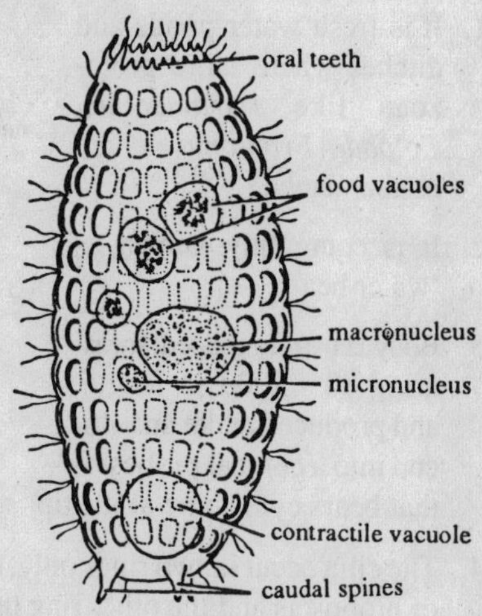

Fig. 1.29 Coleps hirtus.

4. The plates are organic in composition and are arranged in four anterioposteror girdles of 20 plates each which encircle the animal like the staves of a barrle.
5. Each plate bears elevated rectangular areas that give the surface of the cliate the appearence of being divided into more or less symmetrical squares.

6. Shorter cilia located between the longitudinal plates and longercilia occur around the anterior cytostome. Longer clia help in locomotion.

7. The animal is scavanger. It ingests large protozoans and green flagellates.

8. Endoplasm contains many food vacuoles, a posterior contractile vacuole and two nuclei. The macronucleus is centrally placed and is spherical, the micronucleus is small.

9. A subterminal pygosyle is present.

10. Asexual reproduction by binary fission. Conjugation is not reported in this animal.

DIDINIUM NASUTUM

The systematic position is same as that of *Prorodon*.

1. It is fresh water ponds and ditches where large protozoan like *Paramecium, Colpoda, Frontonia* etc. are in abundance.

2. It is commonly known as 'water bear'.

3. Body is barrel - shaped, measuring 80 - 200μ in length, and produced at the anterior end into a conical proboscis that bears cytostome at the tip.

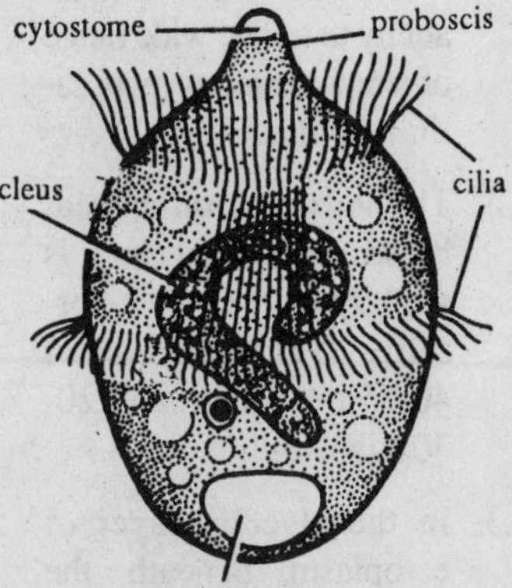

Fig. 1.30 Didinium

4. The cilia occur in two rings only, one ring of cilia is close to the base of proboscis and the other ring on the posterior end.

5. The trichocyst - like boadies are present only in proboseis region.

6. Within the endoplasm, a horse - shoe shaped macro nucleus may be seen; a contractile vacuole and cytopyge are both in the extreme posterior end of the animal.

7. The animal is voracious. Its enormous appetite is satisfied by a diet of large ciliate like *Paramecium* etc.

8. Asexual reproduction by binary fission at a rate of 3-4 divisions per day.

9. Sexual reproduction by conjugation. It encysts under unfavourable conditions.

COLPODA CUCULLUS

Phylum	-	Protozoa
Subphylum	-	Ciliphora
Class	-	Ciliatea
Sub Class	-	Holotrichia
Order	-	Trichostomatida
Genus	-	*Colpoda*
Species	-	*cucullus*

1. It is free - living, freshwater ciliata.

2. The shape of animal superficially resembles a flatenned kidney. It measures 80-120μ in length.

3. A clear defines, diagonal groove is located on the aboral surface extending from the right side of the posterior third of body to the middle where it forms a prominant notch. The groove continues for a short distance on the cytostome lying near the left side.

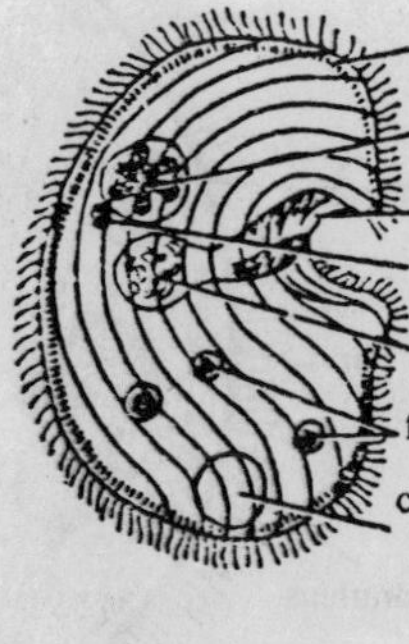

Fig. 1.31 Colpoda cucullus, oral view.

4. Cilia are present on the right side of cytopharynx and rows of cilia on the left side of same organelle give that area a cross - hatched appearence. Delicate cilia through out the body.

5. Ectoplasm posses small clear trichocysts which have a doubtful function in this species.

6. In the endoplasm a spherical macronucleus, a small micronucleus

are present in the middle region. A contractile vacuole present at the posterior end. Many food vacuoles are present.

7. Binary fision takes place in thin walled cyst.

8. Under unfavourable conditions thick walled cyst are formed. The encysted *Colpoda* remains in a dormant state for long periods of time and may become active again even after 38 years.

BALANTDIUM

Phylum	-	Protozoa
Subphylum	-	Ciliophora
Class	-	Ciliatea
Subclass	-	Holtrichia
Order	-	Trichostomatida
Genus	-	*Balantidium.*

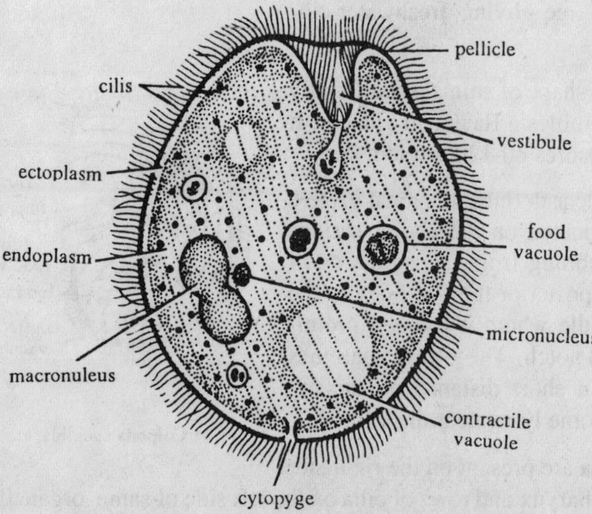

Fig. 1.32 Balantidium coli

1. It is commonly found in the intestine of man, pigs, sheeps, camels, cockroaches etc. It is abundantly found in the rectal contact of the frogs.

2. It has cosmopolitan distribution.

3. It is spherical in shape measuring 60-70μ in length and 40-60μ in breadth.
4. The entire body is covered by longitudinal slightly spiral rows of cilia.
5. These is a large peristomal depression or vestibule leading to cytostome at the anterior end.
6. Broad posterior end bears a cytopyge.
7. Cytoplasm is differntiated into ectoplasm and endoplasm.
8. The endoplasm contains food-vacuoles, two contractile vacuoles and two nuclei. The macronucleus is large rod-like while micronucleus is small and spherical.
9. Nutrition is holozoic. Food consists of erythrocytes, tissue fragments, bacteria and faecel debris.
10. Reproduction takes place by binar fission and conjugation.
11. Cysts are formed, but no multiplection occurs inside the cysts.

Pathogenicity : It causes ulceration of the colon and invdes the mucosa by vitue of *cytolysin,* which it is believed to secrete. It generally cause diarrhoea, but somethimes cause severe or fatal dysentery.

NYCTOTHERUS

The systematic position is same as that of *Balantidium.*

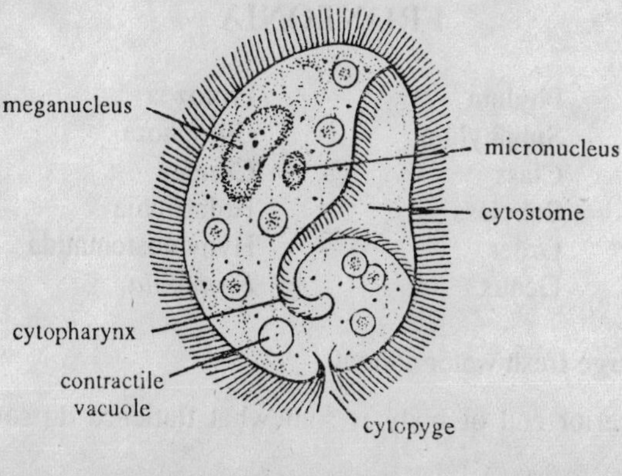

Fig. 1.33 Nyctotherus

1. It is a parasitic ciliate in the rectum of frogs and intenstine of cockroaches.
2. It has worldwide distribution.
3. Body is bean-shaped dorsoventally flatenned and measures 60-120μ in length.
4. The peristome starts from the anterior end and extends down to the middle of left side, where it leads by cytostome into a curved cytopharynx.
5. The entire body of animal is covered with short equal-sized cilia arranged in longitudinal rows. Peristome bears an adoral row of specially large cilia, that extended into the cytopharynx.
6. A permanent cytopyge at the posterior end of the body.
7. Cytoplasm is differentiated into ectoplasm and endoplasm.
8. Endoplasm encloses several food vacuoles, a large contractile vacuole (at the posterior end); a kidney shaped macronucleus and a small micronucleus infront of cytopharynx.
9. Nutrition is holozoic. Food consists of bacteria, undigested food particle etc.
10. Asexual reproduction by binary fission. Sexual by conjugation.
11. Encystment during breeding season of host. The cysts are swallowed by the tadpoles with vegetation.

FRONTONIA

Phylum	-	Protozoa
Subphylum	-	Ciliophora
Class	-	Ciliatea
Subclass	-	Holotrichia
Order	-	Hymenostomatida
Genus	-	*Frontonia*

1. It is a large fresh water animal.
2. The anterior end of body is somewhat flatlened dorsoventrally,

where as the posterior end is more cylindrical. It measures 150-350μ in length and the width 70 - 150μ

3. The body is uniformely covered with cilia except the oral region.
4. Cytoplasm is differentiated into ectoplasm and endoplasm.
5. Embeded in the ectoplasm are numerous fusiform trichocysts, usually right angles to and in contact with pellicle.

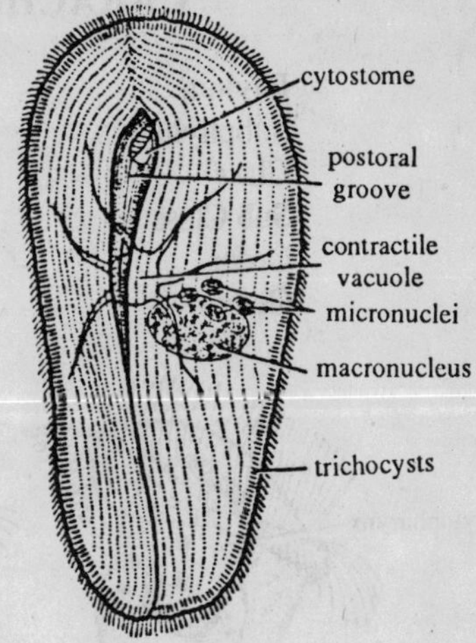

Fig. 1.34 Frontonia leucas,

6. Located ventrally in the anterior fourth or third of the body, the cytostome proper in more or less heart shaped with its point toward the anterior end. From this side a very narrow postoral groove extends almost to the posterior end of the body, then swings toward the opposite side at posterior end.
7. Endoplasm contain a very large contractile vacuole surrounded by 8-10 radial canals.
8. At the same level of the contractile vacuole is a large oval macronucleus. Many (3-5) micronuclei are found near macronuelus.
9. It is Domnivorous feeds on bacteria, alagae, amoebae, arcellae and even rotfers.
10. Symbiotic zoohlorellae sometime occur in the cytoplasm.
11. Reproduction by binary fission and conjugation.

CARACHESIUM

Phylum	-	Protozoa
Subphylum	-	Ciliophora
Class	-	Ciliatea
Sub Class	-	Peritrichae
Order	-	Peritrichida
Genus	-	*Carchesium*

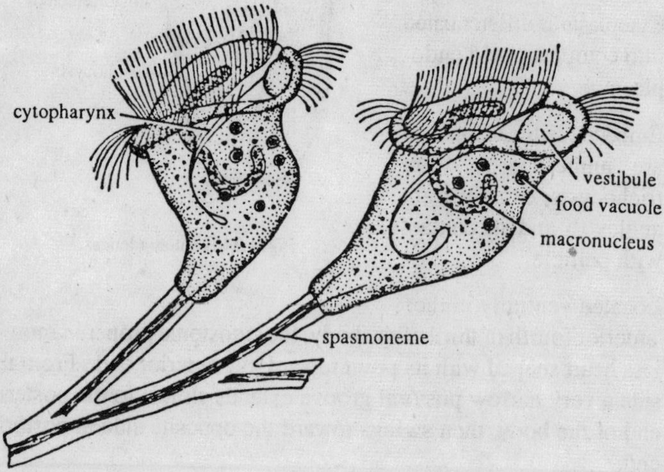

Fig. 1.35 Zooids of Carchesium polypinum.

1. It is fresh water as well as marine colonial ciliate with branching stalks.
2. The zooids are bell-shaped and largely resemble *Vorticella* in structure.
3. The stalk of each zooid has a separate spasmoneme so that it contract independently of the stalks of other zooids in the coloney.
4. The arrangement of cilia reaches it greates complexicity in the adoral or oral ciliature.
5. Endoplasm contains macronucleus and micronucleus, food vacuoles and contractile vacuole.
6. It reproduces by binary fission, conjugation and sometimes encystment occur.

STENTOR

Phylum	-	Protozoa
Subphylum	-	Ciliophora
Class	-	Ciliatea
Subclass	-	Spiroptrichia
Order	-	Heterotrichida
Genus	-	*Stentor*

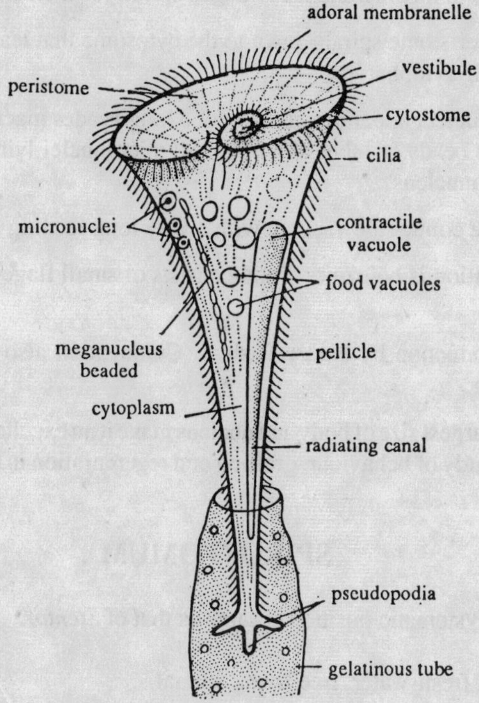

Fig. 1.36 Stentor

1. It is a large, sessile, fresh water ciliate, having a sky - blue pigment called as *stentorin*.
2. It has world-wide distribution.

3. It measures 1-4 mm. in length and is pyriform in shape.
4. The body is covered over by cilia, arranged in longitudinal rows.
5. The animal attaches itself to aquatic plants, animals and other solid objects by its lower narrower end, which gives out pseudopodia to serve as holdfast.
6. In some species the lower end in covered by a loose gelatinous tube.
7. The anterior ends bears a broad peristome encircled by a single row of adoral memberanelles arranged spirally in a clock-wise manner.
8. The peristome spirals down to the cytosome that leads into a short tubaulr cytopharynx.
9. The nuclear apparatus consists of a long beades macronecleus with 15-20 beads or lobes and about 80 micronuclei lying close to the macronucleus.
10. Single contractile vacuole with a very long feeding canal.
11. Nuitrition is holozoic. Food consists of small flagellates and rotifers.
12. Reproduction by binary fission. Conjugation also occurs in late spring.
13. The largest size of body and nucleus make it an excellent material for the study of behaviour, grafting and regerenration in the laboratory.

SPIROSTOMUM

The systematic position is same as that of *Stentor*.

1. It is a fresh-water, free living animal.
2. It has a slim, long, cylindrical and worm-like body, measuring 3-4 mm in length.
3. Body shows anticlock wise spiral striations.
4. Highly developed contractile myonemes run longitudinally in the ectoplasm.

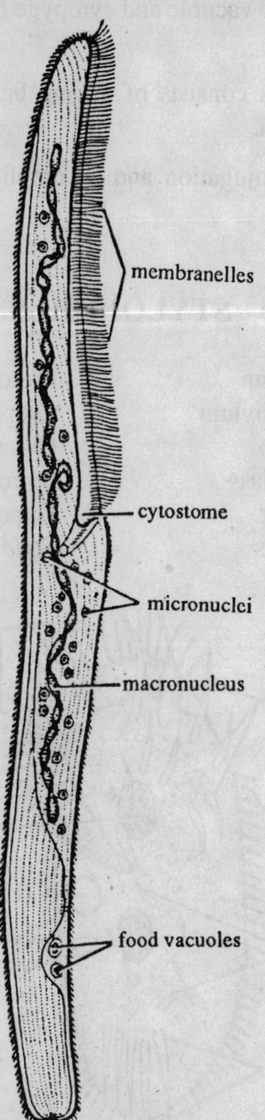

Fig. 1.37 Oral view of *Spirostomum ambigumm.*

5. Peristorm, occupying nearly 1/3 anterior ventral surface of the body.
6. The cilia are short, uniform except for some caudal cilia which are slightly larger secrete mucous.

7. A large contractile vacuole and cytopyge lie at posterior end of the body.
8. Nuclear apparatus consists of a long beaded macronucleus and sevral micronuclei.
9. They undergo conjugation and a complicated process of binary fission.

STYLONCHIA

Phylum	-	Protozoa
Subphylum	-	Ciliphora
Class	-	Ciliatea
Subclass	-	Spiroptrichia
Order	-	Hypotrichida
Genus	-	*Stylonchia.*

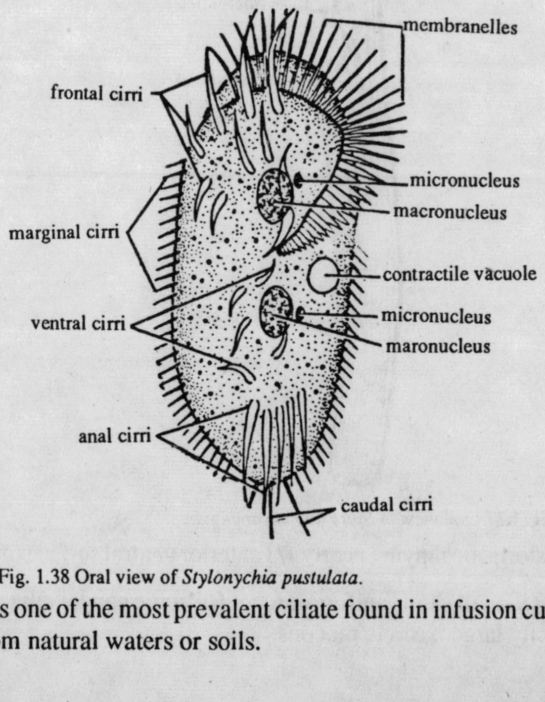

Fig. 1.38 Oral view of *Stylonychia pustulata.*

1. It is one of the most prevalent ciliate found in infusion culture made from natural waters or soils.

2. The body is elongated, oral, rigid and depresed with an arched dorsal surface and flat ventral surface. It measures 150μ in length.
3. The sides of the body are fringed with marginal cirri which are made up of fused cilia in bundles.
4. The ventral surface bears an elaborate system of cirri, including 8 frontals, 5 ventrals, 5 anals, and 3 caudals.
5. The cilia of dorsal surface are represented by few stiff bristles.
6. Peristome is triangular or semicircular. It begins at the anterior end and reaches upto middle of the body.
8. Cytostome remains permanently open.
9. Mode of nutrition holozoic. It takes bacteria, green flagellates. Cannibalism in the scarcity of foods.
10. Two large, oval macronuclei and two small micronuclei. Contractile vacuole is single.
11. Reproduction mainly by transverse binary fission. Conjugation also takes place.

EUPLOTES PATELLA

The systematic position is same as that of *stylonchia*

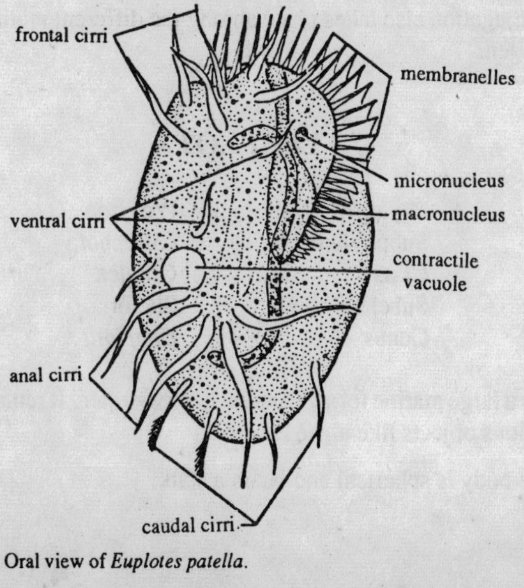

Fig. 1.39 Oral view of *Euplotes patella*.

1. It is found in fresh water and brackish water.
2. Body is subcircular to elliptical with dorsal side convex and the ventral side concave. It measures 90μ in length and 50μ in width.
3. On the dorsal side there are six distinctive ridges with correspinding rows of sensory bristles embedded in rosettes of granules.
4. The ventral surface bears the cirri, of which 6 are frontals, 3 ventral, 4 anals and 4 caudals.
5. A small peristomial field is surrounded by 34-45 memebranelles arranged in a spiral manner. On the right there is somewhat raised peristomial lip.
6. Undulating membrane is absent.
7. Dense endoplasm contain a C-shaped macronucleus. Small micronucleus lies near the macronucleus.
8. A contractile vacuole lies dorsally.
9. An ill-defined cytopyge is present.
10. Mode of nutrition holozoic. It chiefly feeds upon *Chilomonas.*
11. It multiplies by binary fission.
12. Conjugation also takes place among the different mating types of a syngen.

EPHELOTA

Phylum	-	Protozoa
Subphylum	-	Ciliophora
Class	-	Ciliatea
Subclass	-	Suctoria
Genus	-	*Ephelota*

1. It is a large marine form, found in the sea-water. It remains attached various objects like algae etc.
2. The body is spherical and bears a stalk.

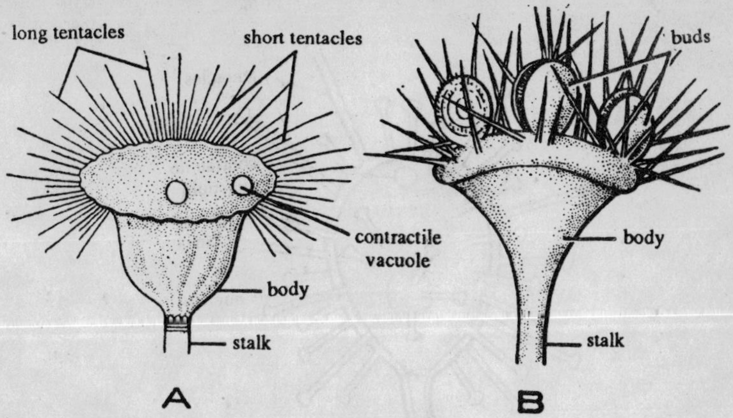

Fig. 1.40 Ephelota

3. Two types of tentacles are present on body; long pointed for piercing the prey and short tentacles are cylindrical for sucking.

4. Cytoplasm contains few contractile vacuoles. The macronucleus is highly branched. Micronucleus is small.

5. The distal end bears a number of buds, each receiving a branch of macronucleus. The neclei behave as in ordinary binary fission. Thus mode of reporduction is by exogeneous buds., The buds detach and acquire cilia on one surface. After a short active free-swimming existence, they lose cilia and grow into tentaculate adults.

6. Cilia only present in buds and not in adults.

PODOPHRYA

The systematic position in same as that of *Ephelota*.

1. It is a fresh water from remains attached to the subsetratum

2. Body is divisible into a globular part and a stalk. the globular part measures 40-50m in diameter.

3. The body bears a number of tentacles that give it a pin-cusshion like appearences. They are evenly distributed. They consist of a core of fluid endodream surrounded by a firm ectoplasm.

4. The cutoplasm contains a large spherical macro - nucleus in the centre and several micronuclei.

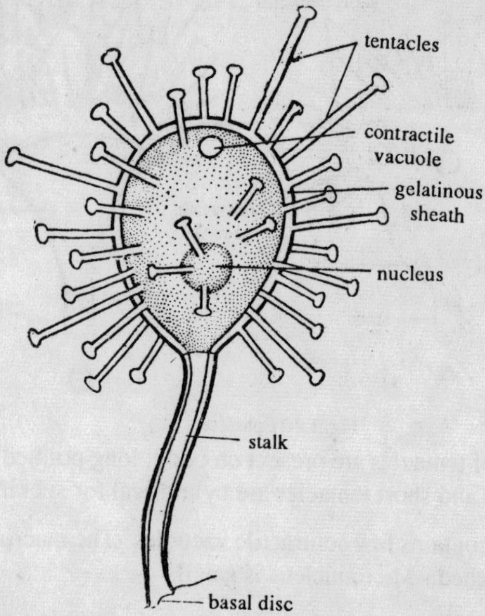

Fig. 1.41 Podophrya

5. A single contractile vacuole near the periphery. The entire body is covered by a gelatinous sheath.
6. Mode of nutrition is holozoic. It feeds other ciliates. The chief food in *Paramecium.* As soon as the tentacles comes in contact with the prey, additional tentacles join in holding and eating it.
7. Reproduction by external budding. A bud is an out growth of the body. It receives parental cutoplasm, a part of parental macronucleus and micronuclei derived from the parental micronuclei by regular division. The bud acquiring cilia and finally separates from parent. After free-swiming the bud comes to rest and attach itself by one end. It grows and given rise to and adult.
8. Conjugation takes place by the fusion of two adjacent stalked adults. Encystment during unfavourable conditions.

ACINETA

The systematic postion is same as that of *Ephelota.*

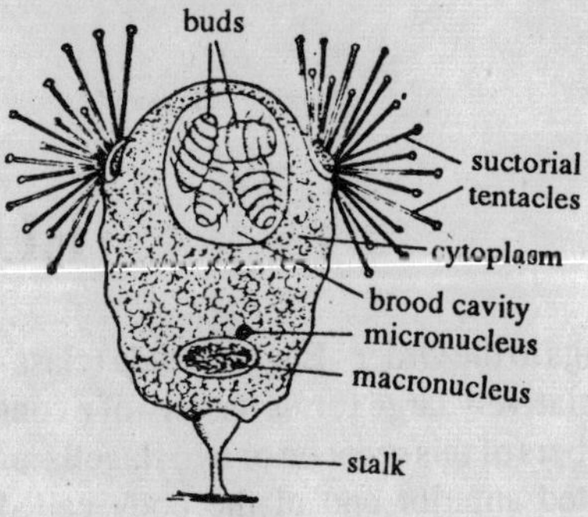

Fig. 1.42 *Acineta tuberosa* with embryos in a brood cavity.

1. It is a fresh water and stalked animal.
2. The body is somewhat Triangular and lies in a shallow cup which is stalked.
3. The knobbed tentacles occur in two or three groups.
4. Reprodcution takes place by endogenous buds. a space called *brood cavity*, develops in the body by invagination, which later closes. Adjacent cytoplasm projects into the cavity and gets detached forming ciliated embryos. These are liberated from brood cavity.
5. After a short free active life, the buds fasten by ventral surface, lose cilia, develp tenacles and grow into adults.

2

EUGLENA

Euglena belongs to the Order - Euglenoidida (class - Mastigophora), which comprises relatively large forms usually of a conspicuously green colour. In the members of this order on or two flagella are present, arising from the invaginated anterior end of the body called *reservoir*. All chlorophyll bearing members posses a light-sensitive organalle called *stigma*. Several non-pigmented form e.,g. *Peranema* etc. are place in this order because of their basic structure.

SYSTEMATIC POSITION

Phylum	-	Protozoa
Sub-phylum	-	Sacromastigophora
Super-class	-	Mastigophora
Class	-	Phytomastigophorea
Order	-	Euglenoidida
Family	-	Euglenoidae
Genus	-	*Euglena*

Habit and habitat

Euglena is a common, solitary and free-living flagellate found in stagnant water containing nitrogenous organic matter. Such water with high organic contents is called *polysaprobic*. Water with medium organic contents is called *mesosaprobic* and with low organic contents as *oligosaprobic*. The pure water is described as *kathrobic*. There are about 100 species of *Euglena* widely distributed all over world in polysaprobic fresh water. It is know to occur in soil, mud and often has been found on the bark of trees. *Pringsheim* (1953) mentioned that some euglenae are marine aǹ some are parasite *Euglenomorpha* lives in the rectum of tadpoles, where they play and unknown symbiotic role. The most common species of *Euglena* is *E.viridis*. The common Indian speçies are *E.agilis, E.orientalis* and *E.fusiformis*. The present description is based on the biology of *E. viridis (Gr. eu* = true; glene = eyeball; and *L., viridis* = green). They are found in such a large number that they produce a green screen on the surface of ponds.

Culture

They can be cultured in laboratory. For preparing culture, some cow or horse dung in boiled in water in a jar and allowed to cool it for two days. Now pour some weeds from the pond containing englenae into the jar and place the jar near the light source. In a few days euglenae develop in the medium. Another method can be utilized for the preparation of culture. Boil the pond or tap water for some times and allow to cool it. Add 20 boiling wheat grains to this water and keep the jar in sunny place. Wheat grains may be added monthly to maintain the culture.

Morphology

Shape and size. Euglena viridis is a minute elongated and spindle - shaped. It is pointed at the posterior, and blunt at the anterior end. the posterior blunt end is called *tailpiece*. It measures 60 µ in length and 15-20µ in breadth at the thickest past of the body.

Pellicle. The shape of the animal is maintained by the pellicle which presents parallel or spiral striations. The striations may or may not be present on the tailpiece. According to *Chadfaud* (1937) the pellicle is two layered, having an outer thin *epicuticle* and deeper and thinner *cuticle*. The pellicle in soft and thin and enabling the animal to change its shape considerably. The pellicle is made of proteins but not of cellulose. Under elctron - microscope the pellicle consists of interlocking longitudinal strips and usually associated with microtubular fibrils, the *myonemes (Leedale, 1964)*.

In some species of *Euglena* just below the pellicle very small elongated or spherical bodies are present. these are believed to secrete mucilage and hence are called *mucous bodies or muciferous bodies*.

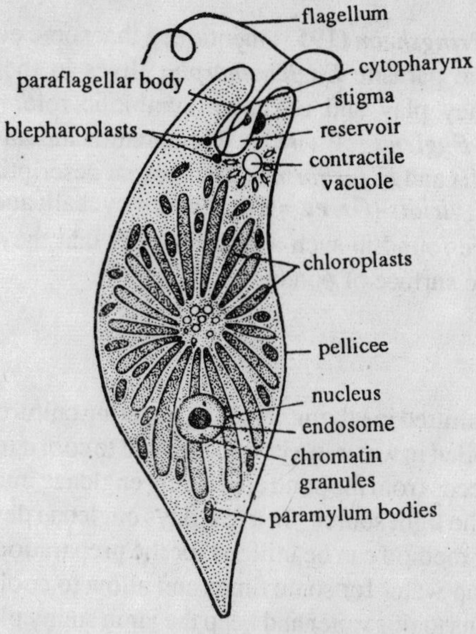

Fig 2.1 Euglena viridis

Cytostome and cytopharynx. Shortly to on side of the middle line at the anterior end is a small opening called *cytostome* or cell mouth. The cytostome leads into a tube called *gullet* or *cytopharynx* which in turn opens at its other and into a more or less spherical chamber called the *reservoir*. The pellicle invaginates into the gullet through the cytostome

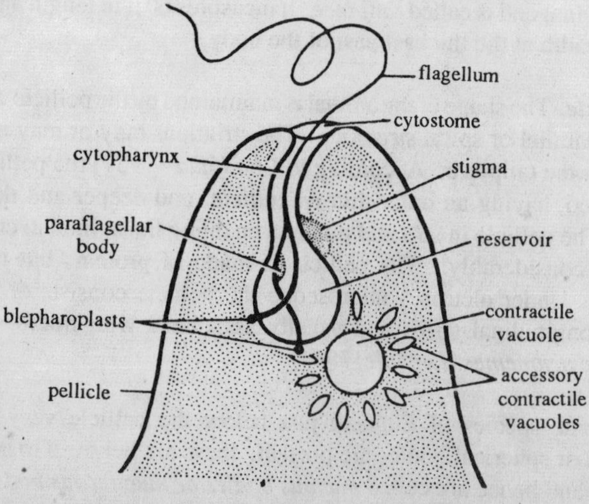

Fig 2.2 Anterior end of *Euglena*.

and lines the reservoir. The pellicle here contains only the epicuticle layer.

The cytostome and gullet are not used for the ingestion of food but as a canal for the escape of fluid from the reservoir.

Contractible vacuole. A large contractile vacuole lies towards one side of the reservoir. It is surrounded by several small accessory vacuoles. The contractile vacuole receives water from the accessory vacuoles and discharge the fluid into the reservoir from where it goes out through the cytostome.

Stigma A conspicuous structure in *Euglena* is the red eye spot or *stigma.* This is placed near the inner end of the gullet close to the reservoir. It consists of protoplasm in which clusters of lipid globules containing orange or red carotinoid pigments. The stigma is cup-shaped with a colourless mass of oily droplets in its concavity which functions as a lens. The stigma and paraflagellar body, together form a photoreceptor apparatus.

Paraflagellar body. A small swelling lies either on one root or at the junction of two roots of the flagellum. It is know as *paraflagellar body. Chadefaud* and *Provasoli* have shown that paraflagellar body and stigma together form the photoreceptor apparatus.

Flagellum. A single, long, whip-like flagellum emerges out of the cytostome through cytopharynx. This whip-like process, by its lasting movements, causes the organism to move in water. The flagellum of *Englena* is also called *tractellum* because it does not propel the body from behind but draws its forward. The flagellum arise from the bottom of the cytopharynx, is forked at the base and is connected by two *rhizoplasts* to, the basal granules or *blepharoplasts.* Each blepharoplasts. is a compact granule that gives rise to a separate filament *(axoneme),* and in most of the species the two axoneme soon unite to form a definite flagellum. In *E.viridis* the flagellum is enlarged at the junction of two axoneme or just behind it to for the paraflagellar body. The paraflagellar body probably contain lactoflavin as sensitiger.

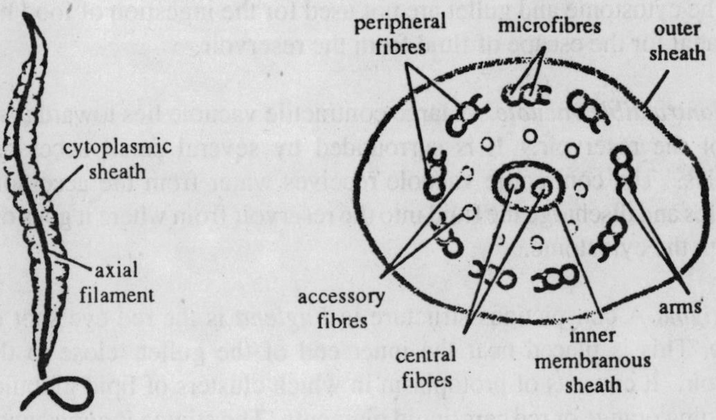

Fig. 2.3 *Euglena*, flagellum & T.S. of flagellum

According to some workers, these are two flagella in *E.spirogyra.*, one long and the other is short. Each arising from the basal granule located at the base of reservoir. The short flagellum does not extend beyond the neck of the reservoir and is aften adharent to the long flagellum giving the appearence of bifurcation.

Each flagellum is cylinderical structure about 0.25 μ in diameter. It is composed of a longitudinal bundle of microtubular fibres (axoneme) enclosed within a unit membrane which is continuous with the cell membrane. The microtubular fibrils of the axoneme are arranged in a precise pattern, best seen in a T.S. Two microtubules occupy the central position. They are called *central fibres* and remain surrounded by inner membraneous sheath. The central fibres are surrounded by nine double microtubular fibrils called *peripherals fibres.* Each peripheral fibre is made of two microfibres. One of the microfibres bears a double row of short *arms* or projections all pointing the same direction. The distal portion of flagellum contains numerous minute fibres, known as *mastigonemes*, which project laterally (*Manton*, 1959). Such a flagellum that bears mastigonemes is called *stichonematic type.* Mastigonemes are believed to project rigidly from the flagellum in two rows arranged in the plane of flagellar undulation.

The structure of blepharoplast is similar to that of the centriole found in animal cells. There are evidences which show that basal body contain DNA and may have some power of self-replication.

Cytoplasm. Two types of cytoplasm can be distinguished in *Euglena,* a dense enter layer, the *ectoplasm* and the central mass, the *endoplasm,* which is more fluid. The endoplasm contains nucleus, chromatophore and paramylum bodies.

Nucleus. Euglena has a single oval nucleus laying in a definite position near the centre of the body. There is a distinct nuclear membrane. The nucleus contains a central body known as *endosome* (Syn. *nucleouls, Karyosome).*

Chromatophores. Nearly all species of *Euglena* have small or large coloured bodies called chromatophores.They are present in the form of little plates or discs, sometimes spindle-shaped, band-like or radiating star-like. The number, shape, size and arrangement of chromatophores vary from one species to another. The colour of chromatophore is green due to the presence of green pigment called chlorophyll. The chlorophyll consists of two pigments, the green chlorophyll *a* and a yellowish β Carotene.

Under electron-microscope the chromatophore is swollen in the centre where it contains two non-pigmented hemispherical bodies called *pyrenophores* Attached to one of each are protein bodies called *py-*

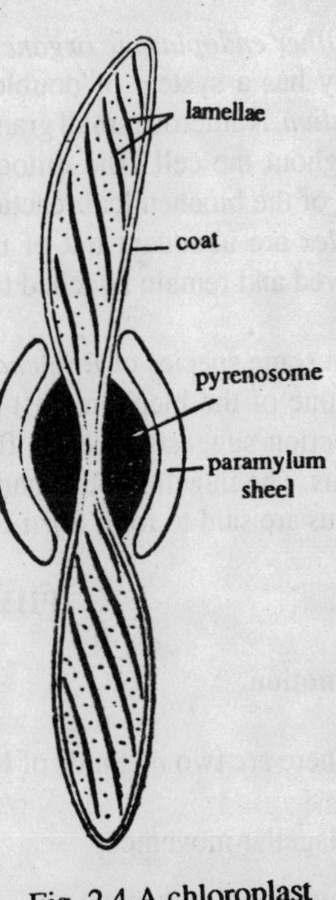

Fig. 2.4 A chloroplast

renosomes. Each pyrenosome is covered with a *paramylum shell.* The two pyrenosomes constitute a *pyrenoid.* Pyrenoids are associated with the formation of a kind of animal carbohydrate called *paramylon.* Paramylon granules thus produced may be found scattered about the cytoplasm.

The chromatophore is surrounded by two membranes. The outer membrane is similar to plasma membrane. The inner membrane is similar to internal membrane system of the chloroplast. The *matrix* or *stroma* contains the chlorophyll-bearing bundles or *thylakoids.* The matrix contains the photosynthetic carbon cycle enzynes, ribosomes and DNA.

Paramylon. Here and there are scattered light bluish-green refractile granules called *paramylon.* It is a polysaccharide similar to starch but it does not give violet colour with iodine solution.

Other endoplasmic organelles. The endoplasm outside the nucleus clearly has a system of double-membranes forming the *endoplasmic reticulum.* Numerous small granules, termed *mitochondria,* are scattered throughout the cell. The mitochondria are important as a location of many of the biochemical reactions of the cell during respiration. *Golgi complex* are abundant but of unknown function. The *ribosomes* also scattered and remain attached to the endoplasmic reticulum.

In some species of *Englena,* a fine fibril called *rhizoplast,* extends from one of the blepharoplast to asmall granule on the nucleus. This connection suggests that the flegellum is under the direct control of nucleus. The flagellum, blepharoplast, rhizoplast and the granule on the nucleus are said to form *infra-ciliary system* or *neuromotor system.*

PHYSIOLOGY

Locomotion.

There are two methods of locomotion is *Euglena:*

1. Flagellar movement
2. Euglenoid movement

1. *Flagellar movement.* The process of the flagellar movement is not clearly understood because, except a few, its movements are not easy to observe in most animals. The whip-like movement of flagellum moves the animal forward and at the same time rotates it around the axis. Thus the animal moves forward in a spiral manner at a fair speed of 3.6 mm per minute. Various theories have been put forward to explain the mechanism of flagellar movements in *Euglena.*

Butschli was first to describe the flagellar movement. He observed that the flagellum undergoes a series of lateral movement and in doing so, a pressure is exerted on the water at right angles to its surface. This pressure creates two forces: one directed parallel and other at right angles, to the main axis of body. The parallel force will drive the animal forward and the other force would rotate the animal on its own axis.

According to *Gray* (1928), the flagellum beats sideways in such a way that its movement proceed from one end to other. These waves create two types of forces, one in the direction of the movement and the

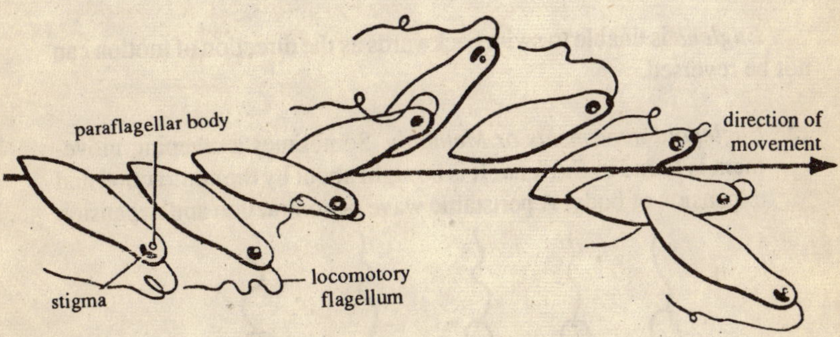

Fig 2.5 *Euglena*. Successive stages in the glagellar movement.

other in the circular direction with the main axis of body. The first force will drive the animal forward and the later would rotate the animal. The posterior end of body describes a smaller circle than the anterior end while the body is propelled forward through the water like a rotating plane which always remains inclined at an angle of 30° to the axis of movement.

Lowends (1941-43) also studied the locomotion in *Euglena.* His views are a little different from those of *Gray.* According to him, the flagellum is always held backwards during swimming and the movement in its occur in a spiral manner from base to its tip. The flagellum beats, on an average, at the rate of 12 beats per second with increasing velocity and amplitude.

Krijisman (1925) observed the movement of flagellum under the dark-field microscope. According to him the normal movement of the flagellum is on the lateral side in a whip-like manner due to which it is called *effective stroke* in which the flagellum remains on outer side and moves quickly downwards. After this in *recovery stroke* the flagellum is brought up in a curved and relaxed state. As a result of these movements the animal moves forwards and rotate around its axis.

While the flagellum is moving it must be dissipating energy (*Brokaw*) as a result of work done against the surrounding water. The energy is supplied by ATP (adenosine triphosphate) present in the mitochondria to the blepharoplasts. The function of mastigonemes present on the flagellum is unknown.

Euglena is unable to swim backwards as the direction of motion can not be reversed.

2. ***Euglenoid movements or Metaboly.*** Sometimes a creeping movement is found in *Euglena.* It is brought about by the contraction and expansion of body. A peristaltic wave of contraction and expansion

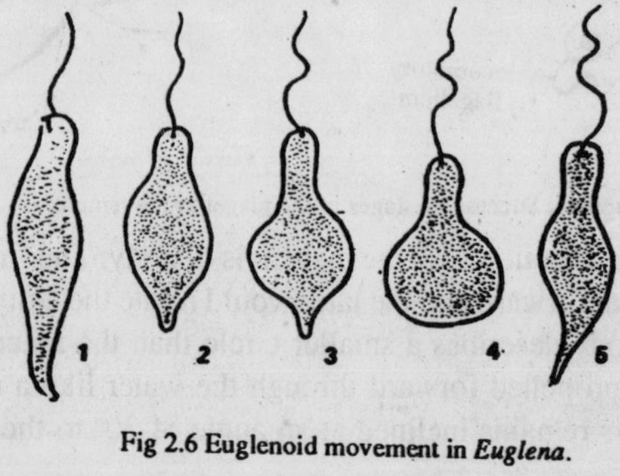

Fig 2.6 Euglenoid movement in *Euglena*.

passes over the entire body from anterior to posterior end and the animal moves forward. The body becomes shorter and wider first at the anterior end, then in the middle and later at the posterior end. The contraction and expansion of the body is due to thin, elastic pellicle and the myonemes present beneath it. This is a very slow movement. *Lowends* found that an *Euglena* of (50-64 μ long) travelled at 168μ

Nutrition

The mode of nutrition in *Euglena* is *mixotrophic* i.e. the nutrition is accomplished either by *holoplytic* or *saprozoic* or by both the modes.

Holophytic or autotrophic nutrition Although *Euglena* has a mouth (cytostome) and gullet, it is doubtful if any food is ingested. The food is manufactured photosynthetically, as in plants, with the aid of light, carbon dioxide and the chlorophyll present in chromatophores. This mode of nutrition is known as *holophyitc* or *autotrophic*. Photosynthesis enables the oxygen of a water molecule to be released, while hydrogen of the water molecule reduces the carbon dioxide present through a series of steps to form carbohydrates and more water. The carbohydrates are stored as a starch-like substance called *paramylum.*

Saprozoic nutrition When the animal is kept in dark for a long time or the chloroplyll is destroyed by treatment with certain drugs, the colourless *Euglena* resulting can continue to live if provided with a medium rich in organic nutrients. This seems to indicate that organic substances in solution are absorbed through the surface of the body, that is, *saprophytic* nutrition.

Myxotrophic nutrition. The euglenae are sometimes holophytic and sometimes saprozoic depending upon the external conditions for maintaining life.

According to recent information, the *Euglena* is *amphitrophic*. It is holophytic in sun light and saprozoic in dark. In the sunlight, it is holophytic but not completely as it needs Vitamin B_{12} or B_1. In the absence of these vitamins it will not multiply. This nutritional versatility is called amphitrophy.

Respiration

Respiration in *Euglena* is *aerobic* i.e. occurs in the presence of oxygen. The exchange of gases takes place through general body surface.

Oxygen diffuses into the cytoplasm from the surrounding water and carbon dioxide diffused out. During active photorynthesis in the organism diffusion of carbon dioxide inwards and the output of oxygen mask the respiratory diffusion.

Excretion

Chemical processes keep on taking place in the body all the times, some of which produce materials that are of no use to the animal and may even prove harmful. During the catabolism of amino acid ammonia is formed. It is highly toxic and dissolved in water. This ammonia must be excreted as soon as it is formed. Some of the ammonia is diffused out through general body surface. Some of the ammonia is excreted out by contractible vacuole. The excretory role of contractible vacuole has gained support in the past few years. (*Chadefaud*). The contractible vacuole is surrounded by specialized granular cytoplasm called *excretory* cytoplasm. Ammonia with large quantities of water first collect in the region of excretory cytoplasm as small droplets as accessory vacuoles which later on fuse to form the contractible vacuole. Periodically, the vacuole reaches its maximum size (*diastole*) and then burst (*systole*) so as to discharge its contents in reservoir, from where waste passes out through the cytostome.

Osmoregulation

The primary function of contractile vacuole is *osmoregulation* or removal of excess of water from the body. The cytoplasm is denser than

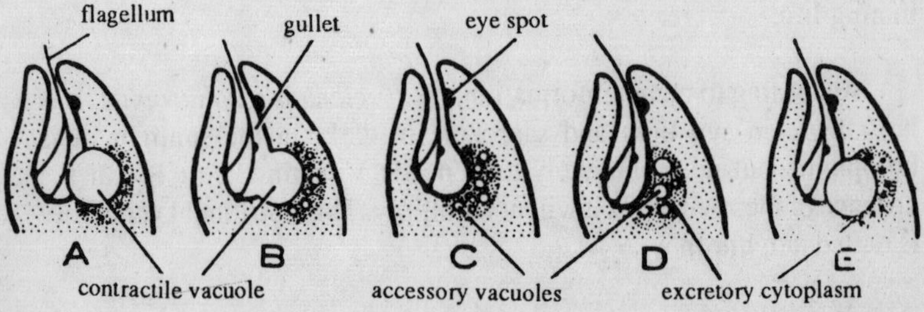

Fig 2.7 Successive stges of contractile vacuole of *Euglena*.

the surrounding water and thus has a relatively higher osmotic pressure. The surrounding water constantly enters the body by endosmosis through the pellicle. Moreover, water is also formed in the body during respiration. The excess of water from all over the body flows towards the excretory cytoplasm (described above) and collects in small accessory vacuoles. These water droplets unite to form the contractile vacuole (diastole) and during systole the water is discharged in the reservoir.

Behaviour

Like other protozoans, *Euglena* possesses a good degree of irritability i.e. the power of responding to external stimuli. This power enables it to avoid unfavourable environmental conditions. While swimming forward rotating and gyrating on its long axis *Euglena* may reach an unfavourable area. When it does it turns violently usually heading into favourable area again. For this the gyration of the anterior end is suddenly widened and a new spiral path is taken. This behaviour gives the appearence of an animal trying a particular direction and correcting the error, if misled. Thus, such a behaviour is called *trial and error* behaviour, but it is better to call it *avoiding reaction.*

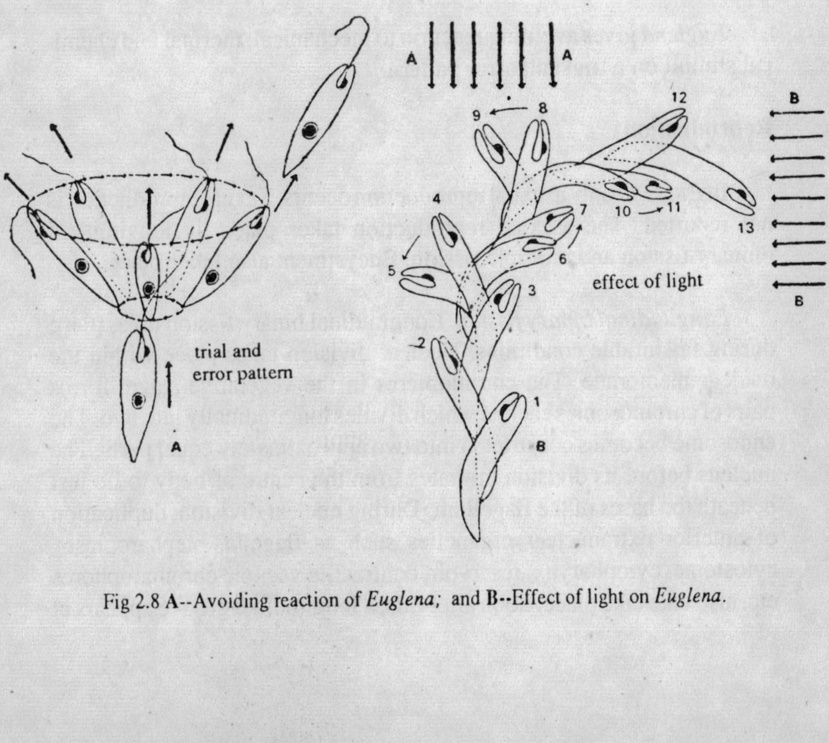

Fig 2.8 A--Avoiding reaction of *Euglena;* and B--Effect of light on *Euglena.*

Reaction to light or phototaxis. Euglena is sensitive to light . It swims towards an ordinary light such at that from a window and avoids strong light. The light is preceived only at the anterior end by the specialized eye-spot. This specialization is important since they depend upon photosynthesis for their nourishment. *Euglena* orientates itself parallel to rays of light whenever the paraflagellar body is shaded by the eye spot or stigma. The animal adjusts its position to the direction of light moving either towards or away from it. It is now known after the experiments of *Tchakotine* (1936) that if the stigma is experimentally destroyed by ultraviolet light, the light-reactions disappear.

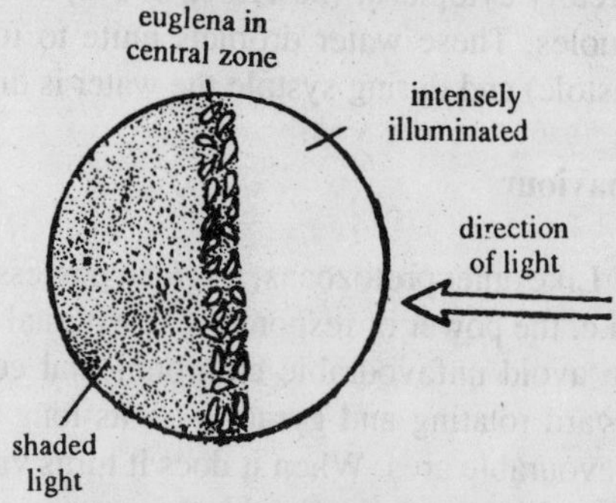

Fig 2.9 Reaction to light in Euglena.

Euglena gives avoiding reaction to mechanical, thermal and chemical stimuli on a trial and error pattern.

Reproduction

In *Euglena* only asexual reproduction occurs. Sexual reproduction is not reported. The asexual reproduction takes place by longitudinal binnary fission and multiple fission. Encystment also takes place.

Longitudinal binary fission. Longitudinal binary fission takes place during favourable conditions. Nuclear division takes place within the nuclear membrane. The chromomeres in the vegetative stage, forms pairs of chromosomes each of which divides longitudinally into two. The endosome becomes constricted into two approximately equal parts. The nucleus before its division migrates from the centre of body to lie just beneath the bases of the flagellum. During nuclear division, duplication of anterior extranuclear organelles such as flagella, blepharoptasts, cytostome, cytopharynx, reservoir, contractile vacuole chromatophores etc. also take take place. Soon after this, a longitudinal groove appears at the anterior end and gradually proceeds backward as a result of which the body of animal is divided longitudinally into two halves. This division is *symmetrogenic* as one daughter is the plane mirror image of the other.

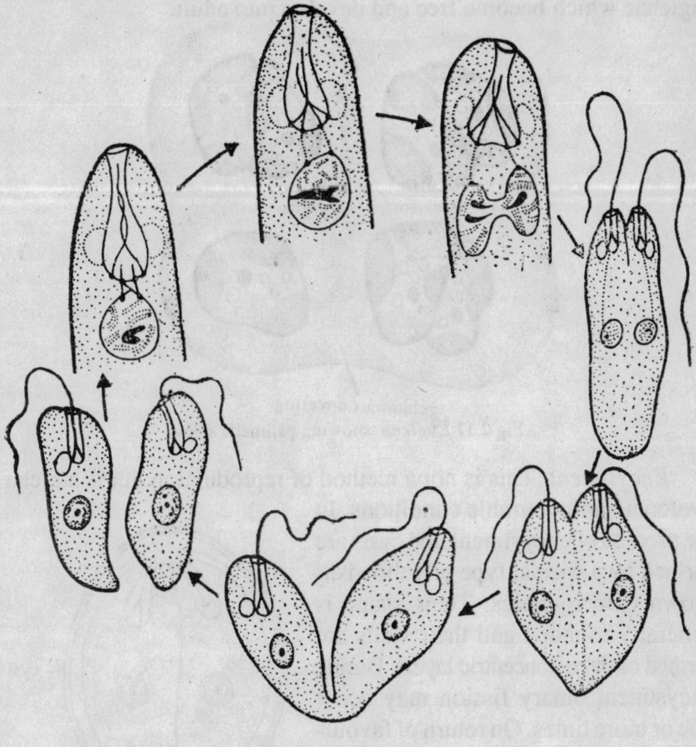

Fig 2.10 *Euglena* showing binary fission.

In some species, it has been observed that the paraflagellar body disappears before the onset of division. In others, the stigma breaks up into small pieces prior to binary fission. According to some workers the flagellum is completely lost and each daughter cell develops its own new flagellum. Often division takes place while the animals are in encysted condition.

Multiple fission and Palmella stage. During unfavourable conditions, multiple fission takes place in *Euglena*. It secretes very rapidly thick or then cyst within which it divides into several (16-32) daughter euglenae. Sometimes, it loses its flagellum and round up into an alga-like cell in which metabolism continues and reproduction occurs by fission, thus forming extensive green scums on the surface of ponds. This condition is called palmella stage. Later flagella grow in these daughter euglenae which become free and develop into adult.

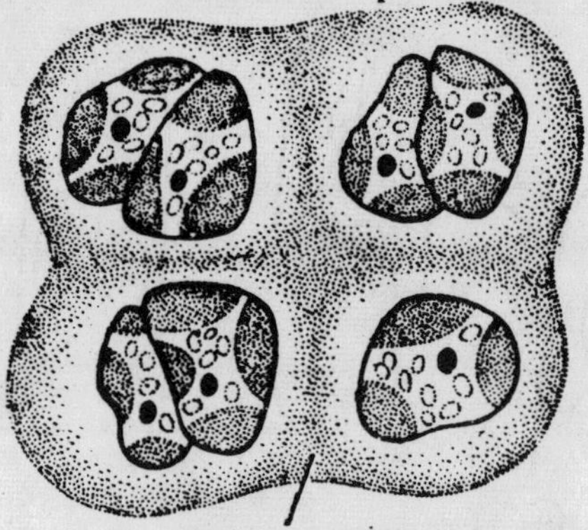

Fig 2.11 *Euglena* showing pallmella stage.

Encystment. This is not a method of reproduction but is a mean to overcome unfavourable conditions. In the process of encystment, the cysts are formed by a special type of yellowish-brown carbohydrates. Their shape is generally rounded and their walls are formed of three concentric layers. During encystment binary fission may occur one or more times. On return of favourable conditions cyst wall breaks, the animal becomes active and emerge from the cyst to lead a normal life

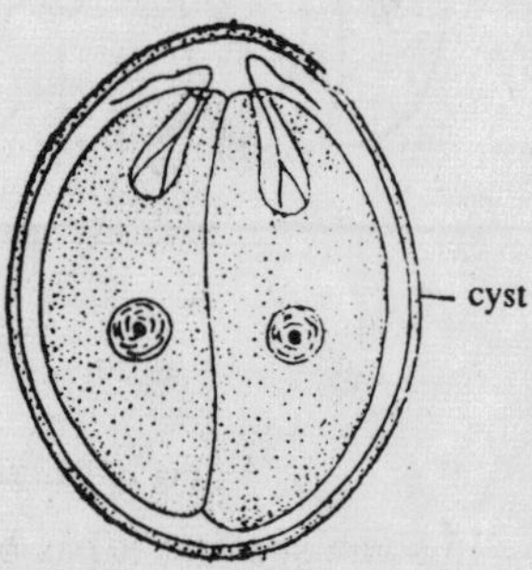

Fig 2.12 Binary fission in a cyst.

In different species of *Euglena*, the cyst may be thick, stalked or operculated and the organism lying centrally or eccentrically.

POSITION OF EUGLENA

Euglena is considered to be an animal by zodogists and plant by botanists.

POSITION OF EUGLENA

Euglena is considered to be an animal by zoologists and plant by botanists.

Animal Characters. The animal characters of *Euglena* include:

1. Pellicle is made up of protein and not of cellulose as in plants.
2. Presence of cytostome and gullet which resembles similar structures in the relatives of *Euglena* that are distinctly animals.
3. Presence of contractile vacuole.
4. Nutrition is heterotrophic, partial or complete.
5. Utalizaation of amino acids, pentoses or polypeptides as sources of nitrogen.
6. Reproduction by binnary fission

Plant Characters The plant characters of Englena are

1. Presence of chromatophore and chlorophyll.
2. Pyrenoid bodies are present
3. Mode of nutrition is holophytic
4. Existence of palmella stage.
5. Photosynthesis takes place in sun-light.

The main palnt like structure is presence of chromatophores and pyrenoids. But these are present only in some species and these can survice even without them. Moreover, the chlorophyll of this organism contains fewer pigments then the chlorophyll of plants. Only few species of *Englena* are autotrophic and they too only partly. In the absence of light the mode of nutrition becomes saprozoic. During sun light the carbohydrate formed is paramylon which differs from the palnt starch as it does not give violet colour with iodine solution. Palmella stage is an adaptation rather than affinity with plants.

Thus it appears that some of the species of *Euglena* are true animal like while few others behave as animal at one time and plant at other time. Its position suggest that we should regard *Euglena* as the intermediate stage in the evolution of plants and animals.

3

TRYPANOSOMA

Ford and *Dutton* (1902) were first to report that *Trypanosoma* was a human parasite and also the causative of 'Gambia' fever: *Kleine* (1909) asserted that tsetse fly (*Glossina)* was the intermediate host in the life history of *Trypanosoma.*

Trypanosoma belongs to the Order Kinetoplastida which comprises relatively simple forms possessing 1-2 flagella provided with a few accessory kinetic structures. *Trypanosoma* and its relatives constitute a closely knit family of pleomorphic parasites which are of great medical and economic importance. Members of the genus are found in all the classes of vertebrates but are pathogenic only to man and domestic mammals, probably representing recently acquired hosts. The cause of their extreme harmfulness to man and domestic mammals is obscure but may consist in the liberation of toxins. The pathogenic trypanosomes are confined to the tropical contries. They are the transmitted from one host to another by blood sucking invertebrates such as insects, mites, ticks and leeches, in whose intestine they undergo a definite cycle of development, requiring a number of days, before they become again infective to vertebrates.

There are two species of *Trypanosoma* found in man. The African human *trypanosomiasis,* known as sleeping sickness is caused by *T. brucei gambiense* and *T. brucei rhodesiense* are found. The first subspecies is confined to the central and West Africa particularly in Congo and Nigeria. The second subspecies in Rhodesia. Tne victors for these trypanosomes are species of blood sucking tsetse fly-*Glossina palpalis* or *Glossina tachenoides.* Another species is *Trypanosoma cruzi,* it is found in South and Central Africa in man and a diversity of mammals, between which it is transmitted by blood sucking bugs, *Rhodnius* or *Triatoma megista.*

The following describtion is based on the biology of *T. brucei gambiense.*

SYSTEMIC POSITION

Phylum	-	Protozoa
Sub-plylum	-	Sarcomastigophora
Super-class	-	Mastigophora
Class	-	Zoomastigiphorea
Order	-	Kinetoplastida.
Genus	-	*Trypanosoma*
Species	-	*brucei*
Sub species	-	*gambiense.*

Morphology

Shape and Size. Trypanosoma is acellular, microscopis animal. The body of adult animal is fusiform or spindle-shaped and is pointed at both the ends. It measures 10-60μ in length and 1-3μ in width. The anterior end tapers gradually and is more pointed than the posterior end and it is well marked by the emergence of the free flagellum.

Pellicle. The body of the organism is surrounded by a thin, elastic, firm and protective pellicle. Due to the presence of pellicle the shape of the body remains constant. The pellicle itself is made up of fine microtubules running spirally around the body of the animal.

Flagellum. Trypanosoma bears a single flagellum (uniflagellate) which arises from a minute granule, the *blepharoplast* or *basal granule,* situated at the posterior end of the body. Just posterior to the blepharoplast, is a small rod-shaped body, the *Kinetoplast* (sometimes erroneously called the parabasal body), which is connected to the blepharoplast by fine fibrils. These two structures usually stain as one. The flagellum extends throughout the length of the body. It remains attached to the body all along its length by means of a thin membrane called *undulating membrane* which is considered as an extension of the cytoplasm in the form of a membrane and then emerges as a free flagellum for a short

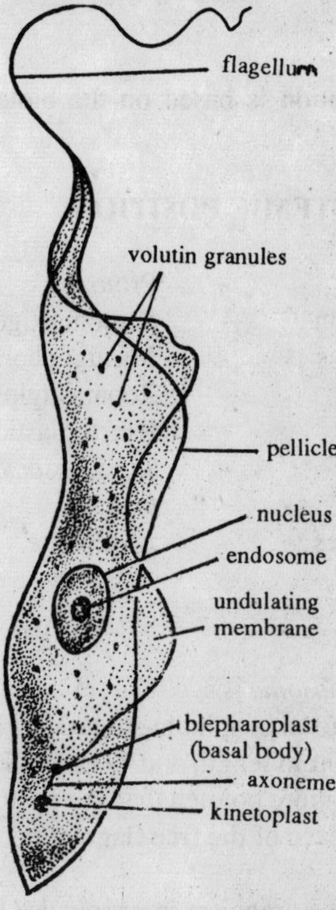

Fig 3.1 Structure of *Trypanosoma*

istances. In life, waves pass into the undulating membrane from the base of the flagellum to its fip. The undulating membrane is said to be an adaptation to movement in a viscous medium i.e. blood. The flagellum and accessory structures are called the *kinetic apparatus*.

In a transverse section the flagellum shows the pattern of 9+2 internal fibres.

Cytoplasm. The cytoplasm is not distinguished into ectoplasm and endoplasm under electronmicroscope. In the cytoplasm there lies a large, rounded or disc-shaped vesicular *nucleus*. It is usually centrally placed but its position varies under different circumstances. In the nucleus lies a large *endosome* or *nucleolus* surrounded by a clear space. Chromatin is either embedded in the endosome or adherent to the nuclear membrane. In the cytoplasm are scattered, greenish refractile particles called *volutin granules* under light microscope. The volutin granules appar

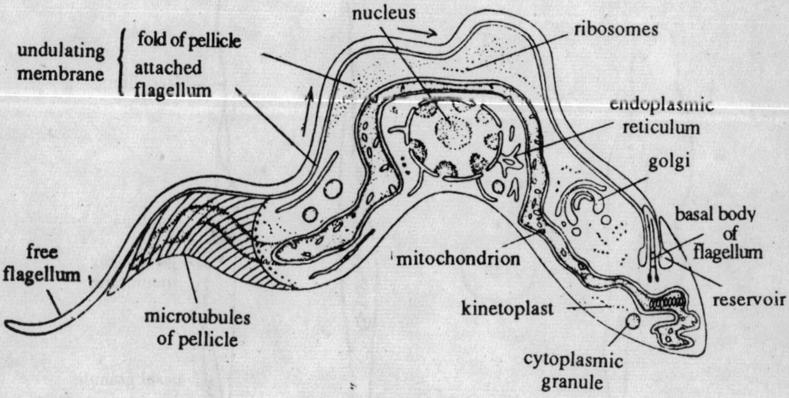

Fig 3.2 *Trypanosoms* seen under electron microscope

ently represent reserve nucleic acid. The contractile vacuoles and food vacuoles are altogether absent. Fine structure of Trypansoma have been studies by *Vickerman* (1969). Under electron-microscope, throughout the cytoplasm of the body is a tubular network of rough endoplasmic reticulum. A small pocket or *reservoir* is present at the posterior end of the body. These is no cytostone. A single Golgi-complex is present between the nucleus and the reservoir. A single elongated giant mitochondrion running the whole length of the body. Its cristate may be present or absent depending upon the form and state of the parrasite. near the basal granule and within the mitochondrion is a small mass of DNA, till recently called kinetosome and parabased body.

Polymorphism in *Trypanosoma*

Four forms of trypanosomes have beens recognised on the basis of the position of kinetic apparatus. These are as follows:

1. *Leishmania* or *Amastigote*. These are round or oval forms with a nucleus, blepharoplast and kinetoplast. The flagellum remains embedded in cytoplasm i.e. it does not come out of cytoplasm.

2. *Leptomonad* or *Promastigote*. Body is elongated The kinetic apparatus is anteriorly placed. The flagellum is short and unattached.
3. *Crithidia* or *Epimastigote*. Body is elongated and the blepharoplast

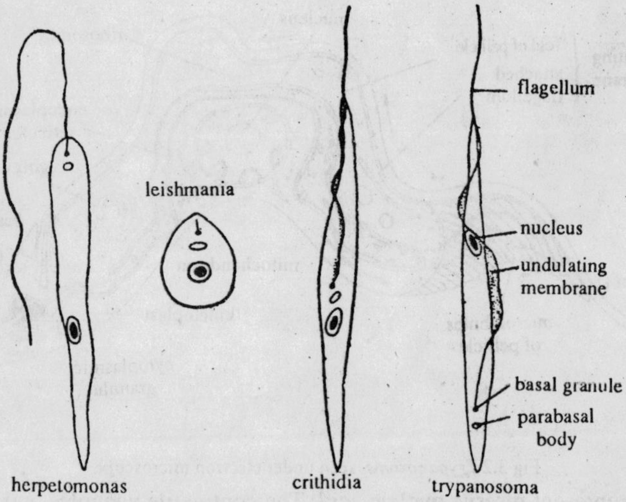

Fig 3.3 Trypanosoma showing polymorphism.

and kinetoplast are situated anterior to the nucleus. Free flagellum and inconspicuous undulating membrane are present.

4. *Trypanosome* or *Trypomastigote*. It has the structure with blepharoplast and kinetoplast situated at the posterior part of body. Flagellum and undulating membrane are conspicuous.

The crithidial form occurs only in the invertebrate host ie. tsetse fly.

PHYSIOLOGY

Locomotion

The locomotion is performed by the wavy movements of the undulating membrane and by the flagellum. The whole body also produces waves for locomotion.

Nutrition

Food must be absorbed from the blood since no mouth is present. The nutritional requirements have been studied in bacteria-free culture media and are quite complex probably as a result of adaptation to the special habitat. Ascorbic acid and hematin have been shown to be essential growth factors. Pinocytosis has been observed in the reservoir.

Respiration

The gaseous exchange takes place by diffusion through the pellicle. The animal is *aerobic* takes oxygen from surroundisng medium. Oxygen consumption increases with an increase in temperature and glucose content of blood. With the increase of antibodies in blood, the rate of oxygen cousumption is reduced As a metabolic activities CO_2 is formed which goes out by diffusion.

Excretion

The chief nitrogenous waste product is ammonia. It is eleminated by diffusion.

Osmoregulation.

Since the cytoplasm of *Trypanosoma* is isotonic with the blood plasma, hence less fluid is absorbed and the animal has no problem of osmoregulation. Thus the contractile vacuole is absent.

Reproduction.

In *Trypanosong* only asexual reproduction occurs by *binary fission,* and most commonly it divides longitudinally. Division is thus said to by *symmetrogenic,* that is, producing mirror-image daughter cells. During binary fission, the blepharoplast divides first and then the kinetoplast. A new flagellum begins to grow out along the margin of undaulating membrane. The nucleus divides mitotically into two daughter nuclei. Finally the body splits mid-longitudinally, from anterior to the posterior end, forming two daughter trypanosomes. Thus one daughter individual retains the old flagellum white the other a new one.

The sexual reproduction is unknown in *Trypanosoma.*

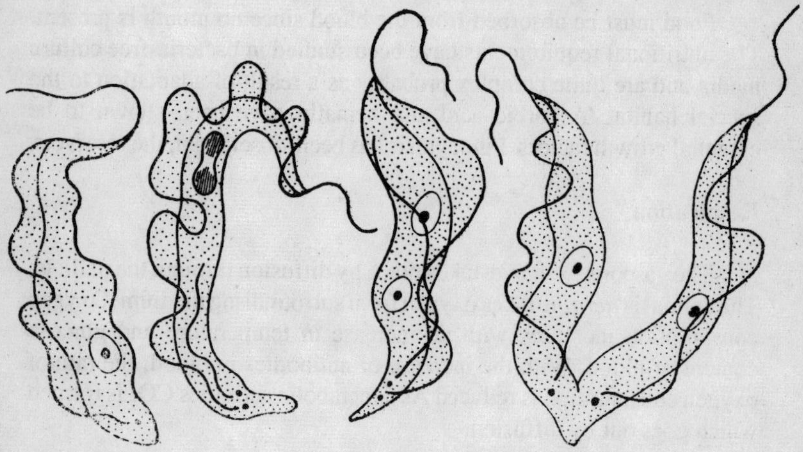

Fig 3.4 *Trypanosoma* stages in binary fission

LIFE CYCLE

Trypanosmonas is a *digenetic* parasite which completes its life-cycle in two hosts, man and tsetse fly. The man serves as primary host, man and tsetse fly. The man serves as primary host and tse-tse fly as secondary host or victor host. Besides the mammals, such as cattles, antelopes, pigs etc. act as reservoir host where the parasite lives until the appropriate vertebrate host is reached.

Life-cycle in man.

The parasite get entry into the blood of man with the saliva of an infected tsetse fly *Glossina palpalis* when the fly sucks the blood of man The saliva contains anticoagulant to prevent the clotting of blood. The trypanosomes, which initiate infection in man, are in the *metacyclic forms* which devoid of free flagella. Soon the metacyclic forms become transformed into the *long slender forms*. They swim freely in the blood and multiply actively by longitudinal binary fission. As their number is increased the host produces antibodies which prevernts glycolysis so that the inactive mitochondrion becomes active (by the appearence of cristae and production of enzymes). Now the parasite stop multipleation and shrink to a *short-stumby form* via an *intermediate form*. Ovoid *leishmania forms* are produced in liver, spleen and lungs of the host. These are

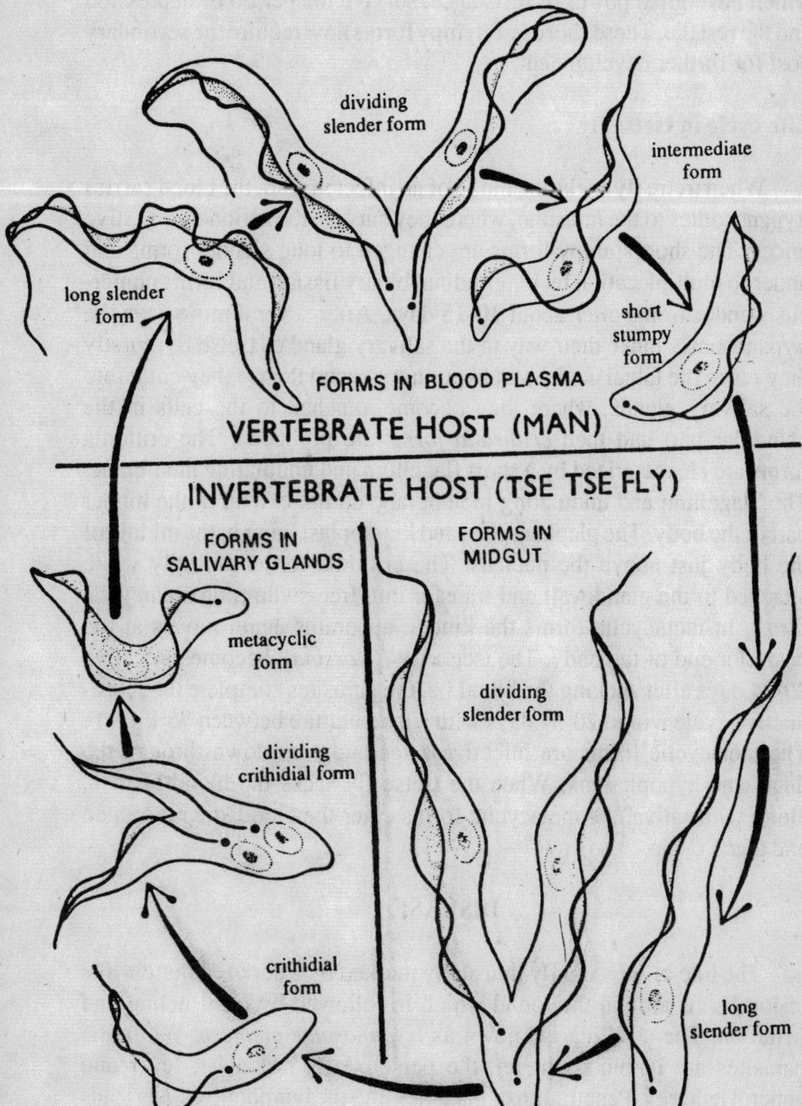

Fig 3.5 *Trypanosoma*. Life-cycle

probably latent forms ment for replacing the losses which occur in the blood of host. During the period of decrease the short and stumpy forms, which have great power of resistance, survive the period of depression and the rest die. These short and stympy forms now require the secondary host for further development.

Life cycle in tsetse fly

When tretre fly sucks the blood of an infected man, the blood carries trypanosomes to the intestine, where they survive the action of digestive juices. The short stumpy forms are changed to long slender forms and undergo multiplecation by longitudinal binary fission and forms numerous slender froms after about 10-15 days. After several more days the trypanosomes make their way to the salivary gland of tsetse fly. Firstly they reach the labial cavity and hypopharynx and thence they enter into the salivary glands, where they become attached to the cells in the glandular part and then *crithridia forms* are produced. The crithidia forms are characterized by a short flagellum and undulating membrane. The blagellum and undulating memberane do not extend in the hinder part of the body. The plepharoplast and kinetoplast lying in the middle of the body just above the nucleus. The crithidia forms multiply while attached to the gland wall and transfer into free-swimming *metacyclic forms*. In metacyclic forms the kinetic apparatus again moves at the posterior end of the body. The tsetse fly *(Glossina)* becomes infective 20-34 days after sucking the blood i.e. the parasites complete the reproductive cycle within 20-34 days with a temperature between 75°F-85°F. The metacyclic forms are infective stage and pass down through the ducts and hypopharynx. When the tsetse fly sucks the blood of man, along with saliva, the metacyclic forms enter the blood stream of man and starts cycle.

DISEASE

The bite of infected fly in usually marked by a dark red button-like lesion forms around the round which in followed by local itcling and irritation. The swelling is known as *trypanosome chancre*. While the parasites are in blood stream, the person feels headache, fever and general lethargy. Penetration of parasites into the lymphatic glands leads to swelling of spleen, liver, glands lymph etc. With the entery of parasites into cerebrospinal fluid, the patient becomes progressively more lethargic, anemic and weak finally the patient loses consciousness and sleeping sickness goes to a fatal end.

TREATMENT

Two drugs, *Bayer 205* (also called *Antrypol, Germanin* or *Suramin*) and *Pentamide* or *Lomidine* are now widely used in the early stages of infection (*Trypanosomiasis*). Previously arsenic and antimony compounds were used, now a days they are rarely used except for late stages when the parasite have invaded the nervous system.

CONTROL MEASURES

The following measures are suggested to control the African human trypanosomiasis:

1. Kill the tsetse fly with suitable insecticides such as DDT.
2. Clear forest and bushes around towns and villages. This will eradicate tsetse flies as the flies take the shelter of trees or bushes during the hot periods.
3. Kill the wild animals that serve as as inexhaustible reservoir of trypanosomes. But it seems inpracticable at present.
4. Human population from the endemic areas may be isolated.
5. A single intramuscular injection of 4 mg / kg of pentamidine remains effective for about six months against the disease.

4

LEISHMANIA

Leishmania first seen by *Cunningham* (1885), *Leishman* and *Donovan* (1903) described similar organism from the case of Dum Dum fever. *Laveran* and *Mesnil* (1903) supposed these bodies to be *Piroplasma*, and *Ross* (1903) believed them to be sporozoan. *Rogers* (1904) notice the Leptomonas stage in its development by placing Leishmania bodies in culture solution. The connection of *L. donoveni* to kala-azar was discovered by *Bautley* (1904). The organism of oriental zone was discovered by *Cunningham* (1885) and rediscovered by *Wright* (1903) in an American child in Boston. The flagellate stage of this species was obtained in culture by *Nicolle* (1908).

SYSTEMATIC POSITION

Phylum	-	Protozoa
Subphylum	-	Sarcomastigophora
Superclass	-	Mastigophora
Class	-	Zoomastigophorea
order	-	Kinetoplastida
Genus	-	*Leishmania*

Leishmania is a haemoflagellate like *Trypanosoma*. It occurs as amastigote form in the vertebrate host and premastigote forrm in the invertebrate host. This protozoan is a *digenetic* as its life-cycle is passed in two hosts. The primary host in man and the secondary lost is the sand fly (*Phlebotomus* sp). There are three very closely similar species of *Leishmania* and also are pathogenic to man. These are:

1. *Leishmania donovani.* It lives in the reticulo-endothelial cells of the visceral organs specially the spleen and cause *visceral leishmaniasis* or *kala-azar*. It was discovered by *Leishman* and *Donovan* (1903).

2. *Leishmania tropica.* It harbours the blood and reticulo-endothelial cells causing *Oriental Sore*.

3. *Leishmania brasiliensis.* It occurs in cutaneous and mucocutaneous parts causing *mucosal leishmaniasis* or *Espundia*.

Morphology. L. donovani occurs as amistigote form in man, measuring 2-4μ in diameter. Under light microscope a nucleus is seen. A rod-shaped kinetoplast and sometimes a basel granule is visible. Under electron microscope it possesses a single mitochondrion, Golgi-complex and endoplasmic reticulum. The mitochondrion with few cristae. The kinetoplast is composed of DNA and is situated within the mitochondrion.

In the secondary host i.e. the sand fly, *Phlebotomus argentipes* it appears as promastigote or leptomonad form. Here the body is elongated and measures 14-20μ in length and 1.5-3.5μ in width. A round nucleus

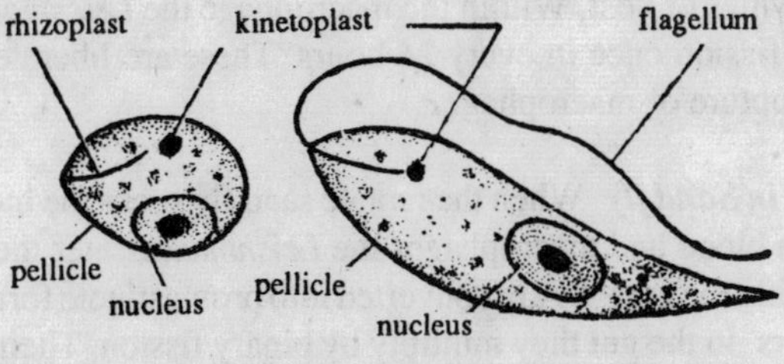

Fig. 4.1 Leishmania

is placed in the centre of the body. An oval kinetoplast lies transversely near the anterior end of body, ahead of which is a small basal granule from which a flagellum is given out.

The animal takes its nourishment in saprozoic manner. It absorbs glucose, fructore etc. from the blood of man and from the alimentary canal of sand fly.

The mode of respiration is aerobic. Oxygen is taken from the blood of host through general body susface by diffusion. Carbon dioxide is eleminated out by diffusion through general body surface.

Nitrogenous waste products are also eleminated by diffusion.

Reproduction

Both amastigote and promastigote forms multiply by binary fission which is no doubt longitudinal. In side the human cell the animal multiplies by repeated binary fission and about 50-200 parasites may be found inside the single cell. These are released by the rupture of the cell and they invade fresh host cell.

The promastigote or leptomonads may also divide by repated binary fission forming flafellate parasites.

Life-Cycle

In man. When an infected sand fly sucks the blood of man along with its saliva a number of promastigote forms enter the blood stream of man. Here they become amastigote forms by losing their flagella. It attacks the macrophare cell of liver or spleen. It appears to be well adapted to a life inside these cells. They remain uneffected by the proteolytic enzymes of host. Within the macrophage the *Leishmania* divides by binary fission once in every 24 hours. These are liberated in blood after the rupture of macrophage.

Cycle in Sand fly When the female sand fly sucks the blood of man along with blood and macrophages the *Leishmania* enter the alimentry canal. Within the gut they are converted into promastigote forms in about 24-36 hours. In the gut they multiply by binary fission. Their number is increased during 4-5 days of blood meal. Soon the parasites migrate forward and the foreget is occupied by them. It gradually moves into the salivary glands by 7-8 days. The development is confined to fore-gut only.

Pathogenicity

Leishmania donorani causes *visceral leishmaniasis* or *Kala-azar* of man throughout the warmer parts of Asia, South America, N. and E. Africa and the Mediterranean coast. In some areas, like China, Central Asia and S. America, the dogs are often infected and serve as reservoirs of the parasites.

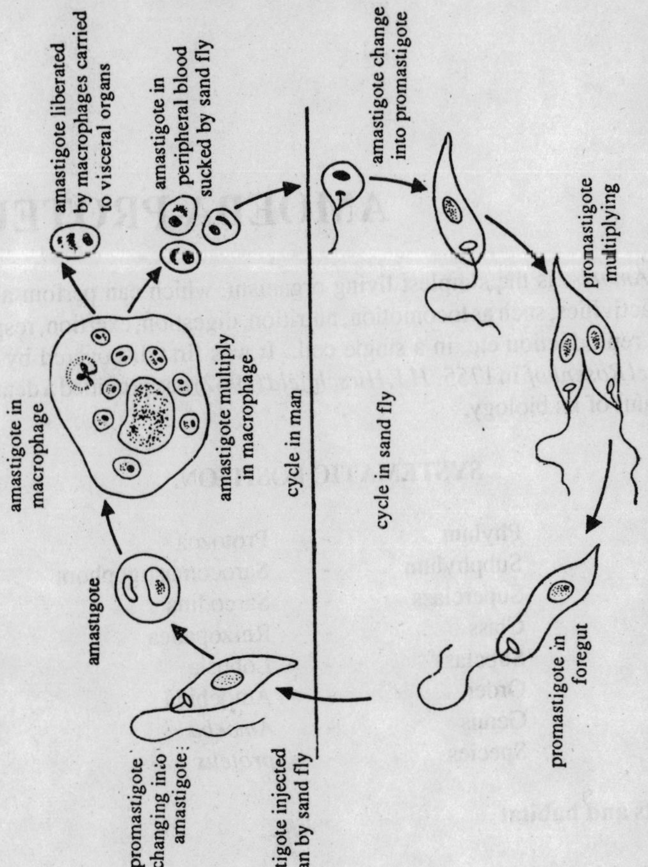

Fig. 4.2 Life cycle of *Leishmania donorani*

In man the kala-azar starts as a primary leison in the skin. After 4-6 months the parasites spread to the viscera with a preference for liver, spleen and bone-marrow infecting the reticulo-endothelial system. Anemia results froms the invasion of bone marrow. Leishmaneasis is slow disease and is usually fatal if not treated. It is some times followed by a nodular eruption of the skin which may be allergic in origin (*post kala-azar dermal leishmoid*), parasites are found in nodules. Treatment with compounds of antimony is often, but not always, effective.

5

AMOEBA PROTEUS

Amoeba is the simplest living organism, which can perfom all its life-activities, such as locomotion, nutrition, digestion, exertion, respiration, reproduction etc. in a single cell. It was first discovered by *Von Rossel Rosenhof* in 1755. *H.I. Hirschfield* (1962) has provided a detailed account of its biology.

SYSTEMATIC POSITION.

Phylum	-	Protozoa
Subphylum	-	Sarocomastigophora
Superclass	-	Sarcodina
Class	-	Rhizopadea
Subclass	-	Lobosia
Order	-	Amoebida
Genus	-	*Amoeba*
Species	-	*proteus*

Habits and habitat

Amoeba proteus lives in fresh water ponds, pools, ditches and streams having green plants. It is also found in damp soil. *Amoeba proteus* name is derived from two Greek words (Gr. *amoeba* = change ; *proteus* = a mythological sea-god, who could change its shape) because it always changes its shape.

Culture

Amoeba may be obtained for laboratory purpose from a variety of places such as organic ooze from decaying vegetation or the lower surface of the lily pads.

Culture of *Amoeba* can be prepared in laboratory. Place some pond water, mud and leaves in 100 ml of water containing a few wheat grains. Amoebae will appear after a few days. To make a pure culture, boil 4-5 grains of wheat in 100 ml of distilled water for about 10 minutes and cool for a few days. To this solution add some amoebae from the previous culture and cover with glass plate. In about 10 days many amoebae will be formed in this culture.

Morphology.

Shape and size : Amoeba is unicellular and can be seen clearly with the help of a microscope. It appears as a colourless, transparent and gelatinous mass. Its shape keeps on changing because of presence of many finger like process, the pseudopodia. The size varies from 0.2 - 0.3 mass in diameter. Some larger forms measure 0.5 mm in diameter, being just visible to the naked eyes as tiny white speaks.

Plasmalemma. The pleasmalemma is a very thin, invisible, elastic external membrance which bounds the cytoplasm. It measures 1-3 μ in thickness. The membrane is selective permeable as water and some small molecules can pass freely cross it in both the directions, but larger molecules can not pass through. Under electron microscope it is a trilaminar structure, consisting of two, darkly stained, layers separated by a clear layer. Each dark layer is composed of mucoproteins and the light layer of lipid (bimolecular layer). The plasmalemme externally posses numerous, ridge-like extensions, the *microvilli*. Zoologists believe that these have adhesive properties and serve to bind the animal to the substratum.

Cytoplasm. The cytoplasm is differentiated into an outer thin cortical layer called *ectoplasm* and an inner modullary mass called *endoplasm.*

(i) *Ectoplasm.* The ectoplasm forms the outer and relatively firm layer lying just beneath the plasmalemma. It is thin, non-granular and almost transparent. It is slightly thicker on the advancing side of the body and at the tips of pseudopodia where it forms the *hyaline cap.*

(ii) *Endoplasm.* It is less fibrous and more fluid zone of granular cytoplasm. It occurs in two colloidal states (*Mast,* 1926). The peripheral viscid or *get* state called *plasmagel* and the central flow

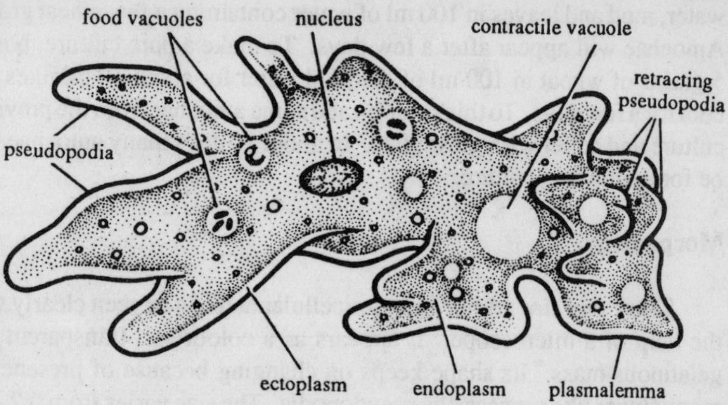

Fig. 5.1 *Ameba proteus.*

ing or *sol* state called *plasmasol*. Modern studies have proved that the ectoplasm is gel and the endoplasm is sol state of cytoplasm. The endoplasm shows distinct streaming movements, called *cyclosis*.

Nucleus. *Amoeba* has a single, disc-like or biconcave nucleus. It is usually located in the central part of plasmasol. It is highly granular and, therefore, refractive to light. The nucleus has a firm nuclear membrane which is double and intercepted by pores. Electron microscope studies show a honey comb layer under bilayered nuclear membrance. The nucleoplasm contains few nucleoli and about 500 small spherical chromosomes.

Contractile Vaculole. Lying near the nucleus is a bubble-like spherical body, which is known as contractile vacuole. At regular intervals, it moves to the surface, where it contract and discharge the water and wate products into the surrounding water. Thus it *is osmo-regulatory* in function. The contractile vacuole in bounded by a delicate elastic *condensation membrane*. It is surrounded by many vesicles and mitochondria. Its position is not fixed due to streaming movement of the cytoplasm.

Food vacuoles A number of spherical food vaculoes small and large, containing food and water in various phase of digestion occur in the endoplasm. These are not permanent structures, but are formed when *Amoeba* ingest food, They disappear with the egestion of undigestible food from the body.

Water globules or vacuoles. Innumerable water vacuoles are scattered in the endoplasm. They are colourless, transparent and non-contractile. Their significance is not well known.

Other organelles. Under election-microscope the endoplasm shows various organelles such as *endoplasmic reticulum, mitochondira, Golgi-complex, lysosomes and ribosomes.* Thin filamentous microfibrils are also present in endoplasm. Besides above organelles numerous minute, regularly shaped crystals caller. *biurets* and *triurets* are *present.*
Polarity.

Although the animal is shapeless but *Amoeba* is considered to have a definite polarity i.e. it has definite anterior end and posterior end. The anterior end possesses pseudopodia while the posterior end in marked by a wrinkled region, called *uroid.*

Physiology

Location. Amoeba moves from one place to another to capture its food-material and other organisms by forming the temporary finger-like protections called pseudopodia (Gr.*pseudo* = foot). These pseudopodia are formed at any part of the body. The characteristic irregular movements found in *Amoeba* are termed as *amoeboid movements.* The pseudopodium is formed by pushing out an enlargement of ectoplasm so as to form a blunt projection. The granular endoplasm than flows in it. As a result of which *Amoeba* moves forward in the direction of pseudopodium. This method of pseudopodial formation is called *profluent type of Hyman.* When several pseudopodia are formed by this method it is called *lobose type* and pseudopodia as *lobopodia.* In some small forms, the lobopodia are formed by *eruptive method.* Here the pseudopodia are formed by bursting out of ectoplasm and endoplasm. Sometimes several pseudopodia are formed at the start but usually one survive, which becomes larger and effective, while others disappear. Several theories have been put forward about the formation of pseudopodia. These theories are as follow:

1. *Adhesion Theory.* According to this theory, amoeboid locomotion is brought about by the force of adhesion between the cytoplasm and the substratum. Just as a drop of water, which spreads irregularly on an uneven glass plate. It was thought that *Amoeba* moved irregu-

larly. Due to adhesive properties pseudopadia generally grow in the path of adhesion. When the surface is clear, adhesion is more perfect, and when it is greasy, adhesion is less suitable.

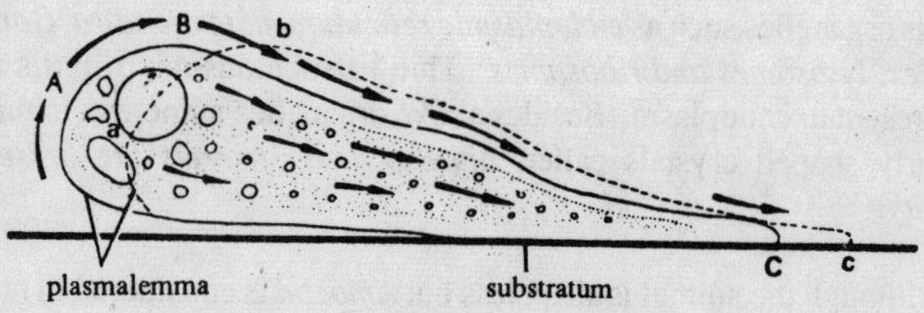

Fig. 5.2 Amoeba verrucosa showing rolling movement

This theory is not satisfactory and does not hold good as the pseudopodia are sometimes given out independently even without any contact with any surface.

2. *Rolling movement Theory.* This theory was proposed by *Jennings* (1904). If a particle of carmine is placed on the surface of *Amoeba verrucosa,* it is seen that the particle flows forward, rolls over the anterior edge, then it stops on substratum until the entire animal has passed over it, then the particle move upwards at the posterior end and comes on the upper surface and moves forwards. *Jenmnings* though that the movement of particle was due to streaming movement of cytoplasm which made the body to roll and move ahead. This theory is also called *Jemnning's contraction theory.*

   However, there are two objections to this theory :

   (a) *Jennings* performed this experiment on *A.verrucosa* only in which pseudopodia are absent. So that this theory could not be apply to *A. proteus.*

   (b) He was unable to explain, with certainty, the locomotion of contractile substances, which are responsible for forward movement.

3. *Surface - tension Theory. Berthold* (1886) proposed that constant internal tension forces the cytoplasm of *Amoeba* to shoot out from its surface in the form of pseudopodium whenever surface tension suddenly lowers due to local effects. This theory has been supported

by *Rhumbler* and *Butschli* (1898). Surface tension does exists on a fluid surface, but its existence upon the surface of plasmalemma of *Amoeba* is doubtful. Thus this theory is not supported now-a-days.

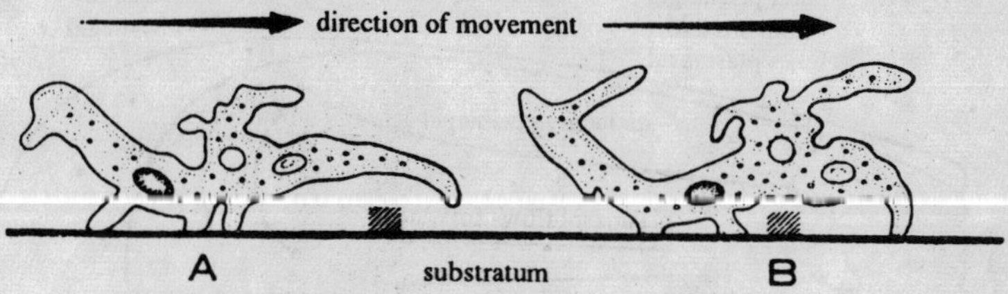

Fig. 5.3 Walking Movement

4. *Walking movement Theory.* This theory was proposed by *Dellinger* (1906). He observed *Amoeba* from the side and found that locomotion was a walking process. These movements start by putting out one pseudopodium and then followed by another. On making forward movements pseudopodium is projected out, which takes hold and after wards contracts. During locomotion, the main part of body remains lifted from the substratum and supported on tips of pseudopodia. Thus *Amoeba* virtually walk on pseudopodia just as the animals move on their legs. *Dellinger* pointed out that *Amoeba* can walk on ceiling as well as floor.

   This theory is also called *Dellinger's contraction Theory.* This theory is unable to explain how the pseudopodia are formed.

5. *Sol-gel Theory.* This theory was first proposed by *Hyman (1917).* Later it was supported by *Pantin* (1923-26) and *Mast* (1925). This is the most acceptable theory now-a-days. According to this theory, the pseudopodia are formed by change of cytoplasm from get to sol and sol to gel.

   According to *Mast* amoeboid movement is brought about four process :

   (i) Plasmalemma attaches to the sbustratum.

   (ii) Plasmasol of moves in the direction of locomotion and comes out by rupturning the plasmalemma at the weakest point. In this newly formed pseudopodium, the plasmasol now changes into the plasmagel due to which a strong gelatinous tube is formed in pseudopodium.

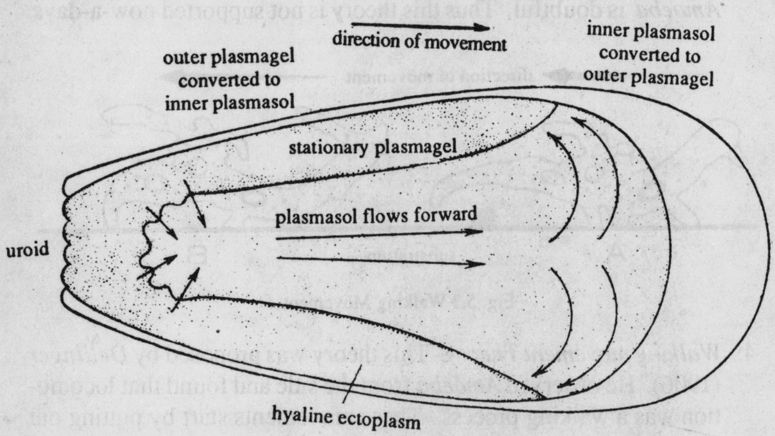

Fig. 5.4 Amoeboid movement after 'sol and get' theory

(iii) At the same time, the plasmagel of hinder end keeps on changing into the plasmasol. This sol flows towards the newly formed pseudopodium so much so that the whole of body cytoplasm comes into it.

(iv) Now the plasmagel tube contracts and the body moves forward. Soon after this, a new pseudopodium is again formed in this direction and all above processes are performed repeatedly due to which *Amoeba* moves forward.

6. *Folding and unfolding Theory.* This theory is proposed by *Goldacre and Lorch* (1950). According to them, "These is no doubt that gelation and solution of cytoplasm occur, but it is unlikely that contraction upon gelation would supply the force capable to moving an *Amoeba.*" They state that all proteins gelate when their molecules unfold and the solate when their molecules fold. In the fluid endoplasm the protein molecules fold. In the fluid endoplasm the protein molecules lie folded compactly, these moleculem unfold at the tip of the advancing pseudopodia to form a layer of straightened and attached molecules. Posteriorty the protein molecules contraction force. In *Amoeba* the contraction is confined towards the posterior side which forces the contracted proteins towards the anterior end. They compared contractibility in amoeboid in amoeboid

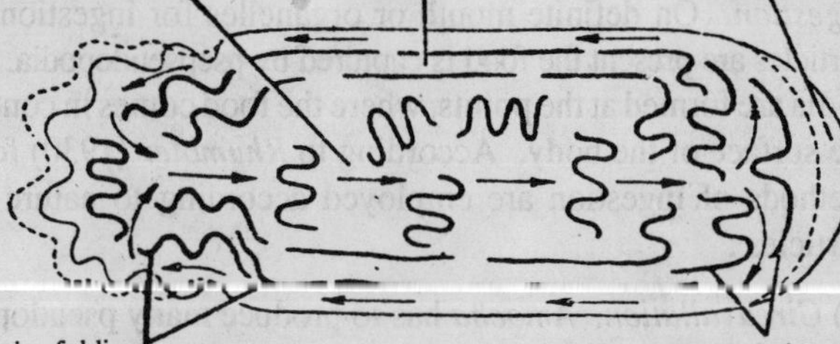

Fig. 5.5 Amoeboid movement after 'folding and unfolding' theory.

movement with that found in muscle-fibres. It is supposed that the energy for the movement of *Amoeba*, folding and unfolding of protein moleules is provided by *adenosine triphosphate* (ATP).

7. *Fountain-zone contraction Theory. Allen* (1962) found that an molecular level, amoenboid movement is actually a type of slow contraction simila in many ways to muscle contraction. Endoplasm is regarded to contain long chains of protein which contract at the anterior end. The endoplasm (sol) is chains are in contracted state. The protein chains become extended at the posterior end, where the ectoplasm is liquified during its conversion to endoplasm. Thus according to this theory, the body of *Amoeba* is virtually pulled forward by the contraction at the anterior end.

8. *Contraction - hydraulic Theory.Rinaldi* and *Jahn* (1963) has supported the theory proposed by *Mast* in an improved form. According to them the contraction of posterior Hasmage is responsible for hydralic resistance to the pressure in least anteriorly. The plasmongel always remains under-continuous tension and localized reduction in the elastic strength is responsible for the formation of pseudopodium.

Nutrition. Amoeba can not prepare its own food it requires ready-made organic substance for food. Such a mode of nutrition in which solid food particles are ingested is called *Hotozoic* or *Zootrophic*. Its food consists of micro-organisisms such as bacteria, diatoms, other flagellate or ciliate, protozoans, minute aquatic plants and mosses.

Amoeba is, thus *Omnivorous* is diet. The process of nutrition is studied under following steps :

(1) *Ingestion.* On definite mouth or organelles for ingestion of food particles are present the food is captured by psepseudopodia. Pseudopodia are formed at the points, where the food comes in contact with the surface of the body. According to *Rhumbler* (1930) following methods of ingestion are employed according to nature of food particles :

(a) *Circuvallation. Amoeba* has to produce many pseudopodia for ingesting a motile food particles like ciliates and flagellates. When it approaches near food-particle the part immediately in line with it stops moving, and pseudopodia form a cup-like structure, the *food-up* around the food-particle. The rim of food-cup narrows and soon closes on the other side of food-particle. The food-particle, thus, gets into the body. Along with food-particle a droplet of water is also engulfed. This water droplet surrounds the food-particle and forms the *food vacuole.* The food vaduole is bounded by plasmalemma.

(b) *Circumfluence.* This method is applied by *Amoeba* when the prey is less-active or motionless. It extends its pseudopodia around the organism and envelops it completely with cytoplasm. The enclosing of the captured organism results in the formation of food vacuole which gets slowly into the cytoplasm.

(c) *Invagination.* In this method the ectoplasm and plasmalemma invaginates into the endoplams to form a small canal as soon as it comes in contact with the food. Later the food particles slips into the tube along with a little water and the mouth or opening of the tube is then closed.

(d) *Import.* In this method, the food itself embeds into the cytoplasm of *Amoeba* as soon as it comes in its contact. Thus *Amoeba* has not to make any special effort to ingest the food in this method.

(2) *Digestion.* The digestion is intra-cellular and takes place in the food vacuoles. The lysosomes, which contain digestive enzymen fuse with the food vacuoles and digestion of contained food organism

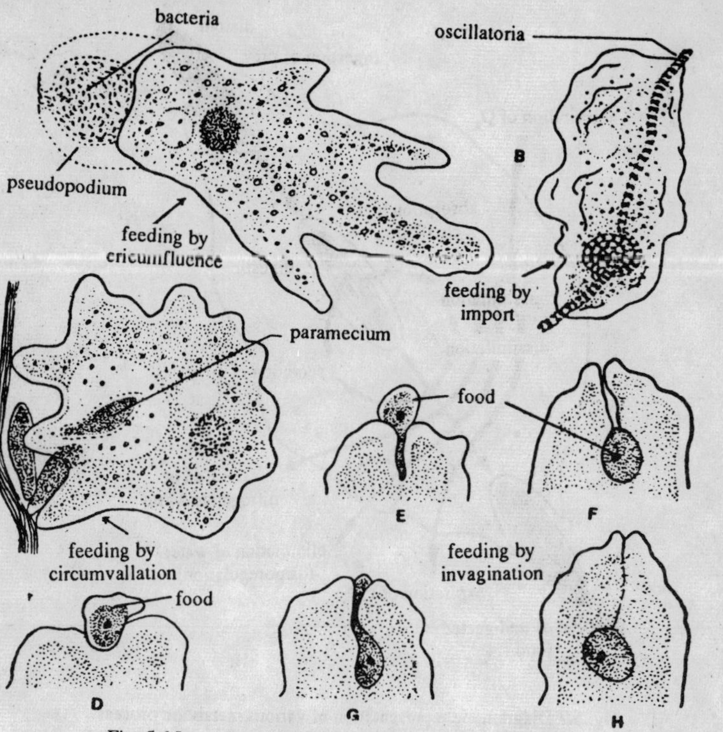

Fig. 5.6 Ingestion of food by A--Circumfluence B--Import C--Circumvallation; and D--H- invagination.

slowly proceeds. The digestions is first carried in acidic medium and lalter in alkaline medium. The digestion of starch, protein and fats takes place respectively with the help of *amylase, protease* and *lipase* in the alkaline medium.

(3) *Absorption and assimilation.* The digested food, minerals and water are absorbed by the protoplasm and circulate in it. The amino acids are built up to form living protoplasm. Sugars, fatty acids and glycerols provide energy and are also synthesized into storage products, like glycogen and fats.

(4) *Egestion.* After digestion, the undigested food can be expelled out from any part of body because there is no definite aperture or *cytopyge* for this purpose. The food vacuoles are shifted backwards

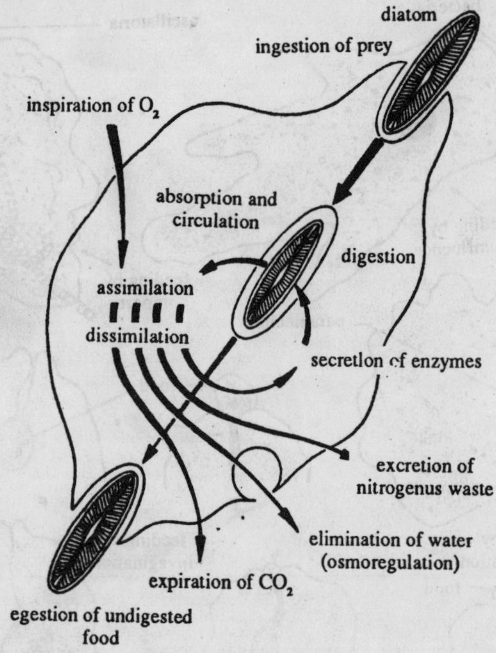

Fig. 5.7 Diagramatic representation of various metabolic process goint on in a living *Amoeba*.

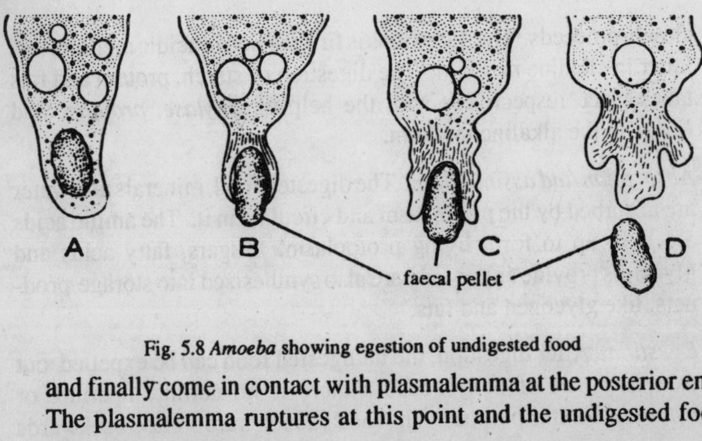

Fig. 5.8 *Amoeba* showing egestion of undigested food

and finally come in contact with plasmalemma at the posterior end. The plasmalemma ruptures at this point and the undigested food goes out as the animal moves ahead. The plasmalemma soon gets repaired to prevent the outflow of cytoplasm.

Respiration Respiration in *Amoeba* is aerobic and takes place through general body surface by diffusion, as there is no special respiratory organ or pigments. Oxygen constantly diffuses into the cytoplasm for its concentration in water is always higher than in the body. This oxygen is used in the oxidation of food-stuffs like carbohydrates, fats and even protein to obtain energy. As a result water, carbon dioxide etc. are formed. The carbon dioxide is diffused out in the surrounding water.

Excretion. The nitrogenous waste products consist of compounds of ammonia such as urea and occasionally urates. In the dissolved state, they diffused out from the general body surface. It has been recently claimed that the cystals called *biurets* or *bipyramidal* and *triurets* or *tripyramidal,* found in cytoplasm. These crystals are formed of an excretory substance, *carbonyl diurea*. Some of these crystals are removed with undigested material and some are redissolve and diffused out. Contractile vacuole also helps removing liquid exertion to some extent.

Osmoregulation. Since the cytoplasm of *Amoeba* is hypertonic to the surrounding medium, water continuously enters the body by endosmosis. Some of the water also comes in cytoplasm from food vacuoles. If all this water is allowed to accumulate, the body will soon swell and burst. So to regulate the amont of water in cytoplasm the excessive water

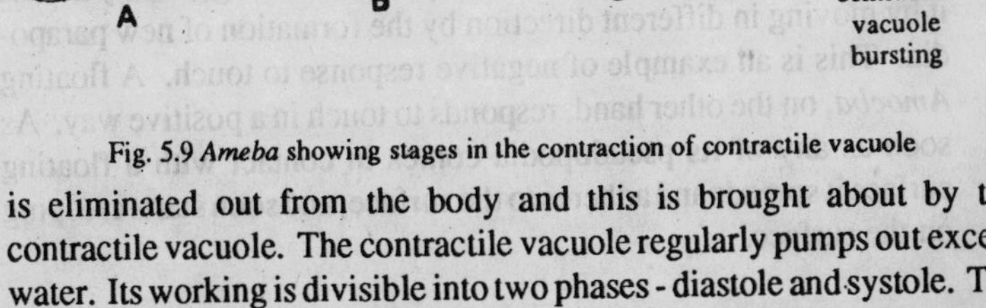

Fig. 5.9 *Ameba* showing stages in the contraction of contractile vacuole

is eliminated out from the body and this is brought about by the contractile vacuole. The contractile vacuole regularly pumps out excess water. Its working is divisible into two phases - diastole and systole. The growing phase up to the maximum size reached is called *diasotle* and the collapsed condition is referred to as *systole*. How the water is forced out of the protoplasm into the vacuole is not known. There are three theories for the growth of contractile vacuole.

(i) *Osmosis Theory*. According to this theory, the water passes into the growing vacuole by the simple process of osmosis.

(ii) *Filteration Theory*. According to this view the water fitters through the vacuolar membrane into the vacuole because of the hydrostatic pressure of the endoplasm.

(iii) *Secretion Theory*. According to this theory the water is first absorbed in the vacuolar membrane and then secreted into the vacuole.

It has been found that the growth of the vacuole takes place in spurts and not steadily. From this it is concluded that the fluid enters the vacuole by the fusion of similar smaller vacuoles. On attaining the maximum size (diastole) the contractile vaccole migrates to the surface until it reaches the plasmalemma which subsequently bursts discharging the contents (systole).

It is possible that the expelled water may contain some excretory products, but there is no experimental evidence to show such function.

Behaviour. *Amoeba* responds to both external and internal stimuli. Responses of *Amoeba* to the stimuli constitute *behaviour or irritability*. *Amoeba* responds to stimuli mainly by locomotary behaviour. When stimulated, it either continues advancing towards the source of stimulation (*positive response*), or moves away from it (*negative response*). Its responses are as follows:

1. *Reaction to mechanical stimuli (Thigmotaxis or Thigmotropism)*. On coming in contact with a solid object. *Amoeba* generally avoids it by moving in different direction by the formation of new parapodia. This is an example of negative response to touch. A floating *Amoeba*, on the other hand, responds to touch in a positive way. As soon as any of its pseudopodia comes in contact with a floating surface it spreads and adheres to the surface, and soon starts creeping on the surface.

2. *Reaction to light (Phototaxis or Phototropism). Amoeba* avoids both direct sunlight as well as total darkness (negative response). It responds positively to diffuse or normal light.

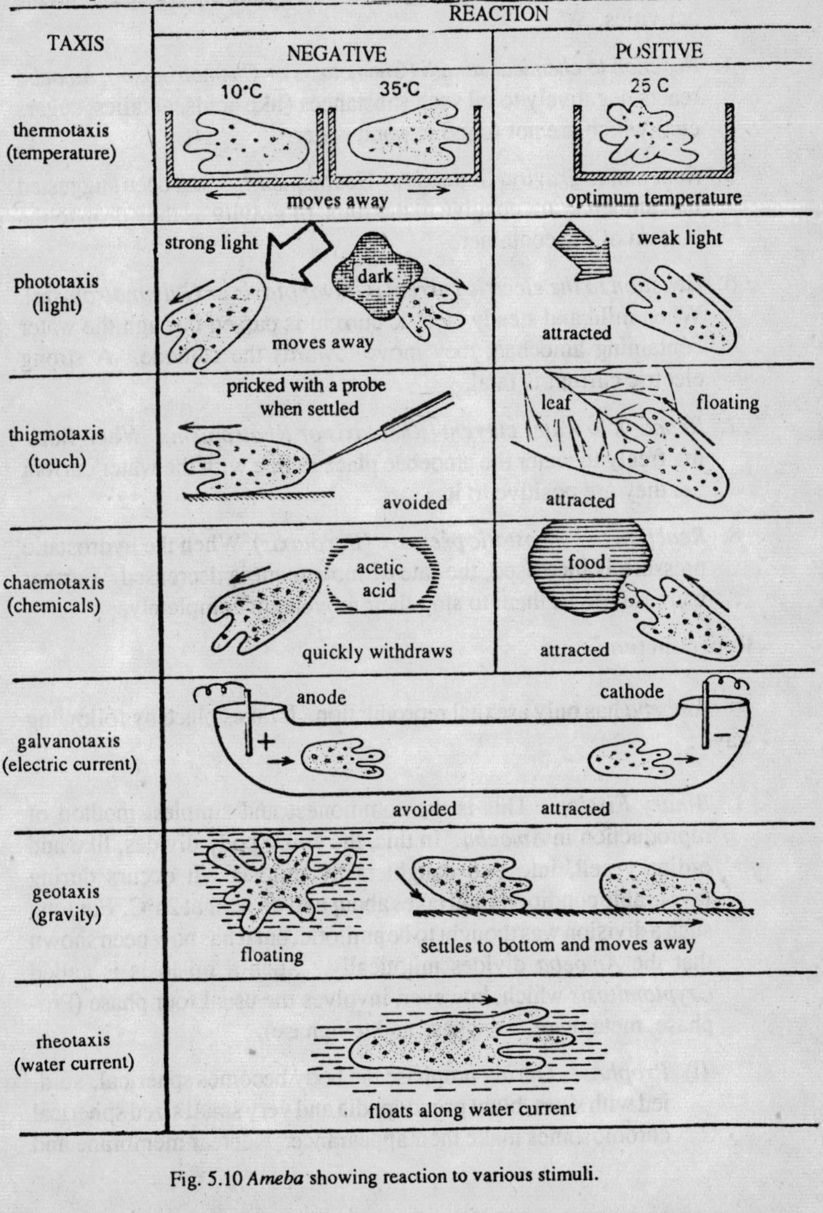

Fig. 5.10 *Ameba* showing reaction to various stimuli.

3. *Reaction to temperature (Thermotaxis or Thermotropism).* It avoids higher temperature as well as low temperature. The optimum temperature ranges between 20-25°C, Beyond 30°C it ceases all the activities.

4. *Reaction to chemical stimuli (Chemotaxis or Chemotropism), Amoeba* reacts negatively to all such substances (like acids, alkalies, sugars etc.) which are not found in pond water.

5. *Reaction to gravity (Geotaxis or Geotropism).* It has been suggested that amoebae are positive to the force of gravity as they drop to the bottom of the container.

6. *Reaction to the electric current (Galvanotaxis or Galvanotropism).* When mild and steady electric current is passed through the water containing amoebae, they move towards the cathode. A strong electric current is fatal.

7. *Reaction to water current (Rheotaxis or Rhotropism).* When floating freely in water the amoebae place in line with the water current i.e. they are positive to it.

8. *Reaction to hydrostatic pressure (Barotaxis),* When the hydrostatic pressure is increased, the rate of movement is decreased. Further increase forces them to stop their movement completely.

Reproduction

Amoeba has only asexual reproduction. It takes place by following ways :

1. *Binary Fission.* This is the commonest and simplest method of reproduction in *Amoeba.* In this, the entire body divides, like and ordinary cell, into two daughters by mitosis. It occurs during favourable conditions and takes about half an hour at 24°C. For long such a division was thought to be amitotic, but it has now been shown that the *Amoeba* divides mitotically. Such a mistosis is called *Cryptomitosis* which, however, involves the usual four phase (Prophase, metaphase, anaphase and telophase).

   (i) *Prophase,* During prophase the body becomes spherical, studied with short, blunt pseudopodia and very small sized spherical chromosomes make their appearance. Nuclear membrane and

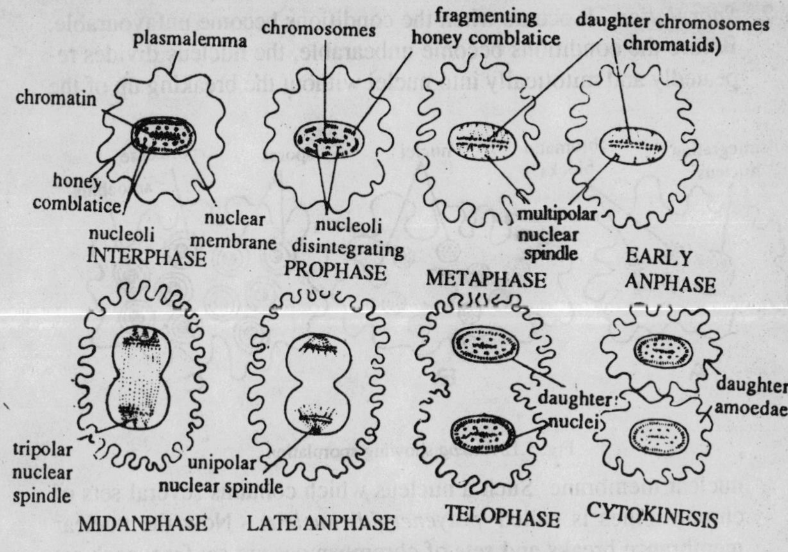

Fig. 5.11 *Ameba* showing binary fission

endosome intact, working of contractile vacuole slows down. The prophase lasts for 10 minutes,

(ii) *Metaphase.* The chromosomes arrange at the equator of the spindle-shaped nucleus to form a *metaphase plate*. The nucleus becomes larger endosome breaks up into pieces and the nuclear membrane breaks down. The chromosomes segregate into two sets. The metaphase lasts for about 4-6 minutes.

(iii) *Anaphase* . It is marked by a change in the external form, the pseudopodia become course and thick. The two sets of chromosomes move apart almost entirely by the enormous elongation of the inter-chromosomal region. Anaphase lasts for about 10 minutes.

(iv) *Telephone* In telophase, pseudopodia assume normal shape; the body first elongates, then constricts in the middle. Two sets of chromosomes reach the potes where each set in enveloped by a nuclear membrane. The construction finally divides the *Amoeba* into two daughters each having a daughter nucleus. The contractile vacuole is retained by one of the daughter, the other one acquires a new vacuole. The telophase takes 8 minutes. Daughter amoebae may begin to divide after 24 hours.

2. *Sporulation.* It occurs when the conditions become unfavourable. Before the conditions become unbearable, the nucleus divides repeatedly and mitotically into nuclei without the breaking up of the

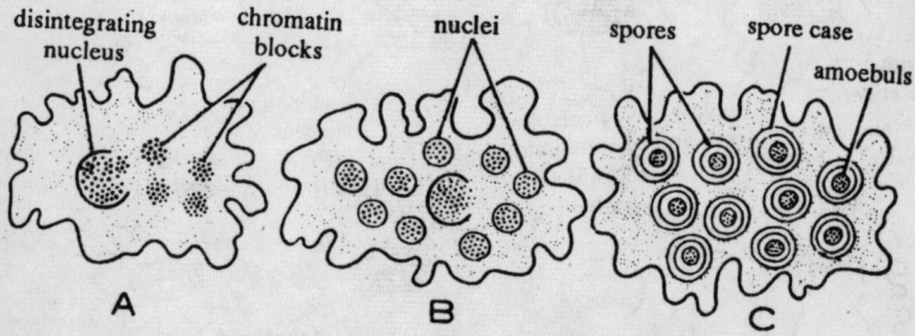

Fig. 5.12 *Ameba* showing sporulation

nuclear membrane Such a nucleus which contains several sets of chromosomes is called *polyenergid nucleus.* Now the nuclear membrance breaks and sets of chromosmoes are set free, each set forming a daughter nucleus. Now, the cytoplasm of parent body segregates into small masses, each around a daughter nucleus. A spore case is formed around this cytoplasm. Thus, about 200 spores are formed. *Amoeba* sinks and disintigrates liberating the spores. On return of favourable conditions spore case bursts open by absorbing water, and a small amoeba is liberated from each spore.

3. *Encystment and multiple fission.* In unfavourable conditions, *Amoeba* becomes rounded and secretes three layered chitinous cyst around itself. This cyst acts as a protective armour within which *Amoeba* passes an inactive life. It comes out of cyst during favourable conditions. According to some workers the nucleus undergoes repeated mitotic divisions forming about 500 minute daughter nuclei. These nuclei arrange themselves along the periphery of the cyst. A small amount of cytoplasm collects round each nucleus. Forming a daughter *Amoeba* called *amoebula or pseudopodiospore.* When favourable conditions arrive, the cyst breaks off liberating amoebulae, each with fine pseudopodia. They lead an active life and grow into adults in due course of time.

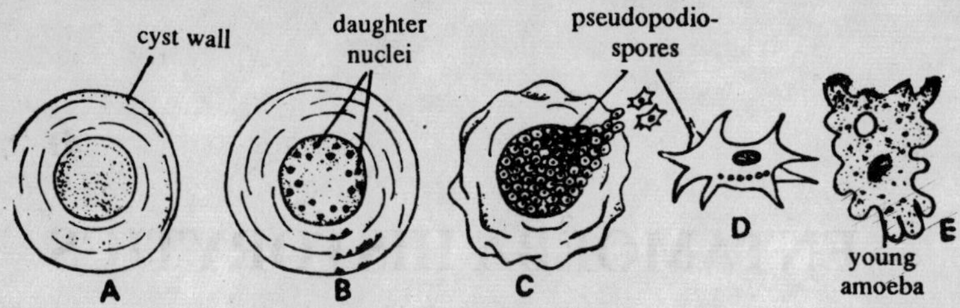

Fig. 5.13 *Ameba* showing encystments & multiple fission

According to modern workers, multiple fission in encysted *Amoeba* is doubtful.

4. *Conjugation.* According to some workers, sometimes two amoebae conjugate for some time and them separate to lead their independent lives. Although there is no definite evidence for conjugation.

Regeneration.

Amoeba has a great power of regeneration. If it is cut into small piece, each piece, containing a part of nucleus, rapidly regenerate into a complete *Amoeba*. A piece without nuclear part, does not regenerate and soon dies.

6

ENTAMOEBA HISTORYTICS

Most amoebae are free-living, but some are endoparasites. They have adapted themselves to live actively in the intestine of many invertebrates and vertebrates. Besides weakening the host body by obtaining nourishment from it, these many cause diseases and, hence, called *pathogenic parasites*. One of such pathogenic amoeabea is *Entamoeba histolytica*. It was discovered by *Lable (1859)*. *Losch* (1875) discovered it pathogenic nature. *Koch* (1887) first saw it in the intestinal section. *Kortulis* (1887) in the liver abscess. The name *E. histolytica* was given by *Schaudin* (1903). The structure of this parasite has been studied by *Noller* (1922) and *Kofoid* (1927). *Elmassion* (1909) and, *Walker* and *Sellards* (1913) studied its precytic stage while *Hartmann* and *Prowazek* (1907) studied it cystic stages. *Dobell* (1919-28) studied its binary fission. *Cleveland* and *Sanders* (1939) studied the excystment. The ultrastructure has been studied by *Miller et al* (1961) and *Feltcher et al* (1962). It causes 'amoebic dystentry or 'amoebiasis' that is why it is commonly called 'dysentery amoeba'. *E. histolytica* also occur in other mammals, lower vertebrates and some invertebrates, often without evident damage to these hosts. Laboratory tutles, snakes and frogs are common sources for study. The contents of lower intestine may be examined by diluting with physiological saline solution on slide. Iodine solutions disclose the nuclei in temporary stain, while iron hemotoxylin in preferred for permanent preparations.

SYSTEMATIC POSITION

Phylum	-	Protosoa
Sub-phylum	-	Sarocmastigophora
Superclass	-	Sarcodina
Class	-	Rhizopodea
Sub-class	-	Lobosia

Order	-	Amoebida
Genus	-	*Entamoeba.*
Species	-	*histolytica.*

Distribution.

E. histolytica is world - wide in distribution. It is known to be widely distributed in tropical, subtropical, temperate regions. The incidence is high in India, China, Maxico and parts of S. America. *Craig* (1962) estimated that about 10% of the human population is infected by this parasite.

Habits and habitat

As mentioned *E. histolytica* occurs in man and other mammals like dogs, cats, rats, monkeys, baboon, chimpanzee, gorilla etc. It lives in the upper part of large intestine (colon) of man. It lives in the mucous and submucous layers of colon. It secretes a toxic substance with dissolves and destroys the lining mucous. From here, certain individuals are carried by blood stream to liver, lungs, kidneys, gonads and even the brain. The parasite feeds on the dissolved tissues, bacteria, and blood corpuscles. It causes a serious and often fatel disease known as *amoebic dysentery*. Patient of this disease discharge mucous and blood in their stool.

Morphology.

The active motile form of *E. histolytica* is called the *trophozoite* which usually ranges in long diameter from 15 - 25 M. It resembles *Amoeba* in its structural details. It is bounded by a thin, flexible, transparent semipermeable membrane, the *plasmalemma*. The cytoplasm in differentiated into enter ectoplasm and inner endoplasm. The ectoplasm forms a thin, clear, relatively firm outer layer. The endoplasm is granuclar and more fluid part. A single pseudopodium is formed of ectoplasm only.

In the endoplasm a nucleus and several vacuoules are present. The nucleus is vesicular and rounded. It is bounded by a delicate nuclear

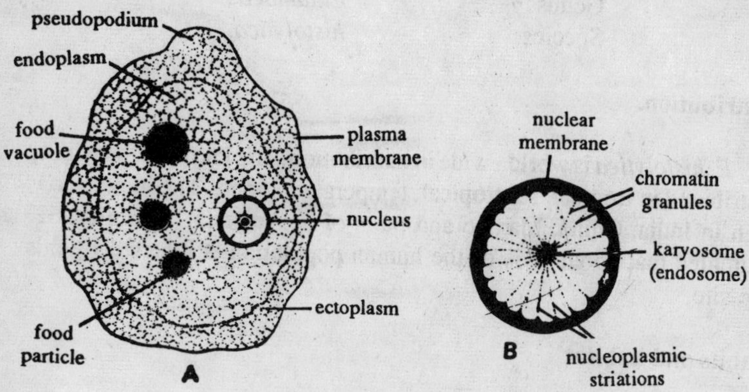

Fig. 6.1 A= Trophozoite of *Entamoeba histolytica*, B= nucleus

membrane, whose inner surface is encrusted with a fine peripheral layer of *chromatin granules*. The *karyosome* or *endosome* is small and centrally placed in the nucleus. It is often surrounded by a clear area or *halo*. The size of nucleus is about 4-6μ in diameter and it contains six chromosomes.

In the endoplasm, food vaculoes very in size and number are present. They enclosing red blood corpuscles, white blood corpuscles and debris of epithetial cells and bacteria. The contractile vacuole is entirely absent.

Physiology.

Locomotion. E. histolytica is a monopodial as it gives out a single, large and broad pseudopodium in the direction of movement. The pseudopodium is called *lobopodium*. In the lobopodium, the outer clear ectoplasm remains sharply differentiated from the inner granular endoplasm. The movement is slow and sometimes called *limax type* movement as it resembles the slow crawling of slug (*Limax*).

Nutrition. The mode of nutrition is *holozoic*. It feeds mainly upon blood corpuscles, fragments of epithetial cells and bacteria. It is believed to secrete *Cytolysin*, a protein digesting enzume that dissolves the epithial mucous membrane and liberates the blood. It than capture blood corpuscles with pseudopodium. It also absorbs substances saprozoically from the surrounding medium.

Respiration. E. histolytica is aerobic when oxygen is available but anaerobic when oxygen is not available. It requires very little oxygen hence *microaerobi.*

Exertion. The nitrogenous waste product formed in ammonia which is diffused out in host's tissue.

Reproduction and life-cycle.

In *E. histolytica* only asexual reproduction occurs. It occurs in two ways : *binary fission* and *multiple fission.* The binary fission takes place in the active trophozoite in gut of host. Multiple fission starts in the encysted stage in one host and completed in the new host after excystation.

The life cycle includes two phases of development i.e. trophozoite and cyst, with a transitory precystic stage.

Binary fission. The trophozoite multiplies asexually within the wall of intestine of host. The nucleus divides mitoically into two. With this division, the body of the organism is also divided into two by a constriction. They rapidly grow in size, feeding upon the bacteria and the host tissue elements. They again multiply by binary fission. Some of them attack and penetrate fresh host cells while others, under certain circumstances which are not well known, remain small (7-10μ in diameter) and retreat into the lumen of the intestine. These small forms are rounded and sluggish and are known as *precystic* or *minuta forms.* The minutes forms do not eat tissue cells and blood corpuscles but feed only on the bacteria. It is the non-motile and non feeding stage thus it is non-pathogenic to man.

Encystment. The precystic forms undergo encystation only in the human of the intestine. They round up and secrete a thin, flexible, refractile and resistant cystwall around themselves. The cyst of *E. histolytica* is spherical body measuring 10-15μ in diameter. It cytoplasm is clear and hydine containing two *chromatoid bodies* and many *glycogen vacuoles.* According to *Pitelka* (1963), the chromatoid bodies are composed of ribo-nucleoproteins. These bodies disappear as the cyst matures. At this stage of cyst is *uninucleate.* Soon the single nucleus

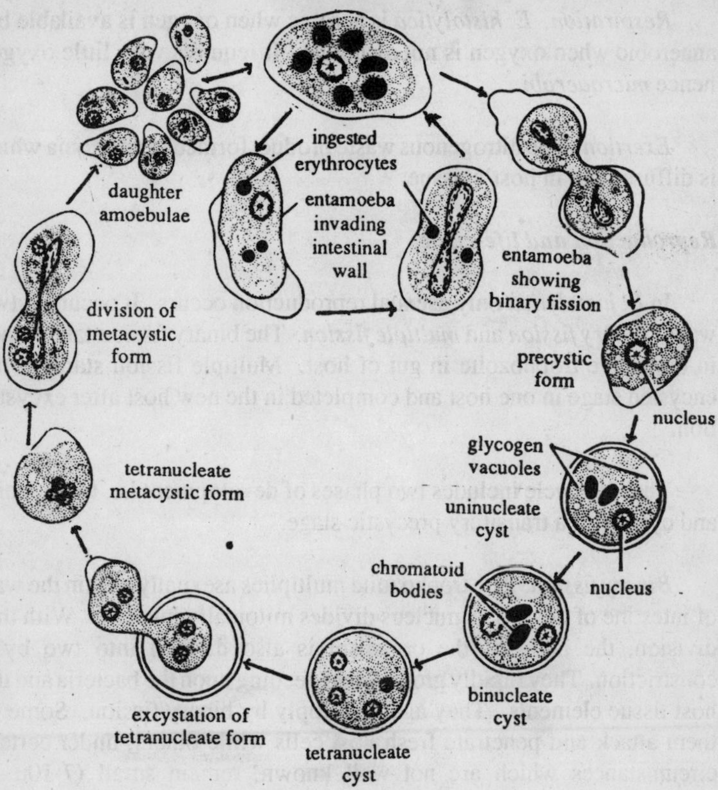

Fig. 6.2 Life - Cycle of *Entamoeba histolytica*

divides mitotically into two and the cyst becomes *binucleate*. Next division result in four nuclei and the cyst now becomes *tetranucleate* or *quadirnucleate*. Now the cysts come out of the body of host with the faces. The cysts can remain alive for many weeks outside the body of the host till they reach in the new host.

Transfer to new host. The cysts reach the aliminatary canal of new host with contamination of food and water. The food can be contaminated by any of the following ways:

(1) By dirty feeding habit.

(2) By coming in contact with contaminated water.

(3) By collection of cysts on food substances by insects like house-fly, cockroach etc.

(4) By eating uncooked vegetables and fruits,

Excystation and metacystic forms. The excystation of cyst and metacystic development have been observed and studied by *Dobell* (1928) and, *Cleveland* and *Sanders* (1930) in cultures. Excystation in the process of transformation of cysts to the trophozoites. The cysts reach to aliminetary canal of host. There is no effect of gastric juice in stomach on them. On reaching the intestine their cyst wall in dissolved by the action of trypsin and a tetranucleated amoeba (the *metacystic form*) comes out in the lumen of intestine. A metacystic form undergoes a series of nuclear and cytoplasmic division producing eight uninucleate daughter amoebae. The young amoebae being actively motile, make their way into the large intestine, invade the mucous and submucous lining and grow into mature trophozoites.

Pathogenic effects

Although about 10% of human population is infected, yet most of them are carriers and no harm is caused by *Entamoeba histolytica.* According to *Hoarse* (1962), *E. histolytica* is characteristically a great commensal and, only in a few cases symptoms of disease.

1. *Amoebic dysentery* . The trophozoites penetrates the mucosa and submucosa of intestine (colon) and causes its necrosis and form small wounds or abscesses which later become flask-shaped bleeding *ulcers.* The cavity of ulcers is generally filled with mucous, bacteria, amoeba and cell-debris. The abscesses pour their contents into the lumen of intestine. In amoebic dysentery, the stools are acidic and contain pure blood and mucous in which swarms of amoebae and blood corpuscles, are usually present. The patient is largely discomforted owing to intense gripping pains with the passing of blood and mucous with stools after every few minutes.
2. *Abscesses in liwer, lungs and brain.* Sometimes, *E. histolytica* pass into the portal circulation and then to liver where the parasites settle and form abscesses in the liver by attacking the liver tissue. The patient has pain in liver and fever may develop. Lung abscesses may also occur. The trophozoites through blood circulation may also

enter spleen, gonads and brain where also they destroy the tissue causing specific abscesses. Abscesses elsewhere are rare. In acute cases, they might prove to be fatal.

Treatment. The permanent cure is sometimes difficult. For prompt relief of acute dysentery the injections of *Emetin* are given. Certain antibiotics such as *Fumgillin, Aureomycin, Erythromycin and Terramycin* are more effective for oral treatment. In chronic cases, certain arsenic compound like *Vioform* and *Carbarsone* and iodine compounds such as *Yatran, Diclogiun* are effective. For amoebiasis of liver and lungs *Chloquine* in quite effective *Prevention.* Following measures are essential in the prevention of the disease :

1. washing of hand with antiseptic soaps and water before taking food and after toilet.
2. Sanitary disposal of faecal matter.
3. Perfect sanitation and protection of water and vegetable from pollution.
4. Protection of foods and drinks from contamination by houseflies, cockroaches etc.
5. Use of boiled water and properly washing the vegetable and fruits.
6. Avoiding open field and road sides for passing out stools and chemicals treatment of human faeces if used as fertilizers.

SPECIES OF ENTAMOEBA

Entamoeba coli.

It is the most common species to occur in about 50% of the human population world over. It lives in the lumen of colon and does not enter the tissues of the wall. It is a nonpathogenic species as it does not cause any disease to man. It feeds on bacteria, undigested food and other debris but never on the tissue. It is, therefore, called endocommensel.

Like *E. histolytica* its trophoites measure 15-40μ in diameter. The cytoplasm is less distinct in ectoplasm and endoplasm. the nucleus has a comparatively larger nucleus or endosome not placed in the centre. The nucleus measure 5-8μ in diameter. Numerous food vacuoles are seen but they contain bacteria, faecal debris etc. The cysts are spherical or often avoid. Immature cysts contain 1, 2, 4 nuclei while the mature cyst has 8 nuclei. In young cyst besides 4 nuclei one or more glycogen bodies and small number of chromatoid bodies are present. The infection takes place through cysts in stool of infected man passed out and taken by new host by contaminated water or food. Its life. History is not known. According to *Hegner*, the cysts hatch as 8 nucleated amoebae.

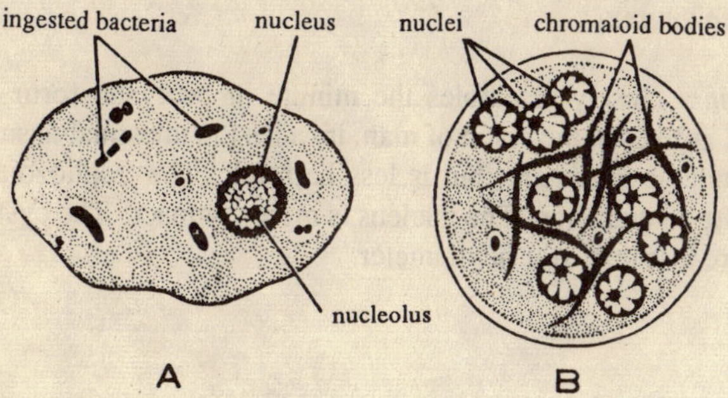

Fig. 6.3. *E. coli*, A= Trophozoite B= Cyst

Entamoeba gingivalis

The mouth amoeba, *E.* ginglivalis lives in the mouth usually between the teeth and gums. It is also found in abscesses of gums, tonsils etc. According to *Kofold* 75% people over 40 years of age harbour this amoeba. the trophozoite measures 10-20μ in diameter. The cytoplasm in differentiated into an outer clear ectoplasm and a hyaline and vacuolated endoplasm. The nucleus is present in endoplasm measuring 2-4μ in diameter. It contains a centrally placed endosome or nucleus. The trophozoite feeds on bacteria, cellular debries and white blood corpuscles. There are many pseudopodia all are with blunt ends.

In *E. gingivalis* the cyst have not been seen since it is directly transmitted from one human to another by contact during kissing or in feeding.

It was believed earlier that *E.gingivalis* caused *pyorrhoea.* Fortunately, they are not pathogenic but probably they aggravate the disease of pyorrhoea by destroying the tissues in gums.

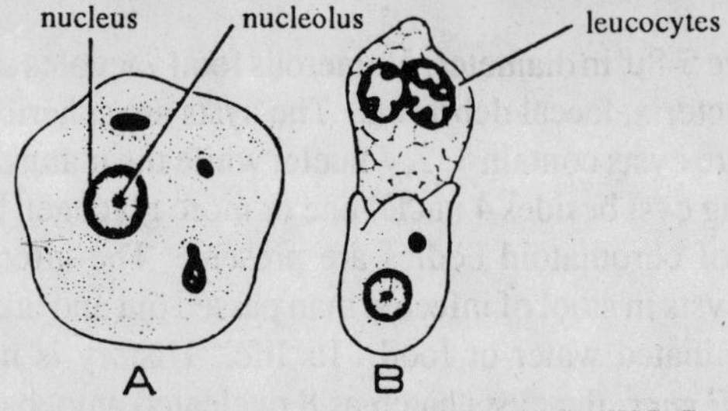

Fig. 6.4. *E. gingivalis.* A=Trophozoite B=Engulfing a W.B.C.

Entamoeba hartmanni.

E.hartmanni closely resembles the minute or precystic form of *E.histolytica*. it lives in the colon of man, invades the intestinal tissues and cause amoebic dysentery, but is less harmful. The trophozoites measure 9-14μ in diameter. The nucleus is more compact. The cystic forms measure less than 10μ in diameter.

7

ELPHIDIUM OR POLYSTOMELLA

Elphidium is a shelled protozoan belongs to the order-Foraminifida, class-Rhizopodea. Its shall is perforated by numerous minute pores hence the name foraminifera (L. *forare*-pores; *feree*-to bear) is given to this order.

SYSTEMATIC POSITION

Phylum	-	Protozoa
Sub phylum	-	Sarcomastigophora
Super-Class	-	Sarcodina
Class	-	Rhizopodea
Sub-class	-	Granuloreticulosia
Order	-	Foramaniferida
Genus	-	*Elphidium*
Species	-	*Crispum*

Habits and habitat

Elphidium is a free-living and marine protozoan. It is a bottom dwelling form found at the sea floor to depth of 540-600 meters. It slowly creeps along the sand and sea weeds. It is omnivorous. It shows dimorphism and the life-cycle shows alternation of generations.

Morphology

Body of *Elphidium* is covered with a hard and translucent; biconvex and oval or spherical shell made up of calcium carbonate and silica. The shall is perforated with several chambers (about 45-50). All the chambers are filled with cytoplasm (*Polythalamous*). The shall is pale-yellow in colour and measures about 1 mm in diameter. The surface of the shall is beautifully chiselled bearing tiny tubercles. The central part of the shall forms the rounded umbo while peripheral part the keel. The chambers of the shall are V-shaped, laid down serially and arranged in a flat spiral in which each whorl of chambers overlaps the previous one i.e. *equitant*. The overlapping portions are called *alar processes*. Due to overlapping of the chambers only the last chamber is visible from outside. The hinder margin of each chamber bears a row of projections, the *retral processes*. The adjacent chambers remain separated from each other by perforated septa through which cytoplasmic continuity is maintained.

The first or initial chamber formed by the animal is known as *proloculum*. As the young grows, its cytoplasm overflows through the aperture of proloculum. The extruded cytoplasm rounds itself off and secretes a new chamber. By the repetition of this process a multichambered shall is formed. Each new chamber is larger than the predecessor.

In *Elphidium* there are two types of shells differisng in relative sizes of the proculum. If the proculum is small and is fromed by the individual developing from zygote, the shall is called *microspheric*. In others the initial chamber or proculum is proportionately larger and they have produced by individuals resulting from schizogony. Such shells are called *megalospheric*. Thus the animals show dimorphism. The differences between two forms given in the table.

Cytoplasm. The shall-chambers are filled with cytoplasm. The cytoplasm is viscous and granular in nature. There is no marked distinction of ectoplasm and endoplasm. The cytoplasm contains single nucleus in megalospheric forms and many nuclei in microspheric forms. The nucleus is vericular in form and contains several nucleoli in its nucleoplasm. The contractile vacuoles are absent. Coloured granules or globules called *xanthosomes*, commonly occur in cytoplasm. These representexcretory matter.

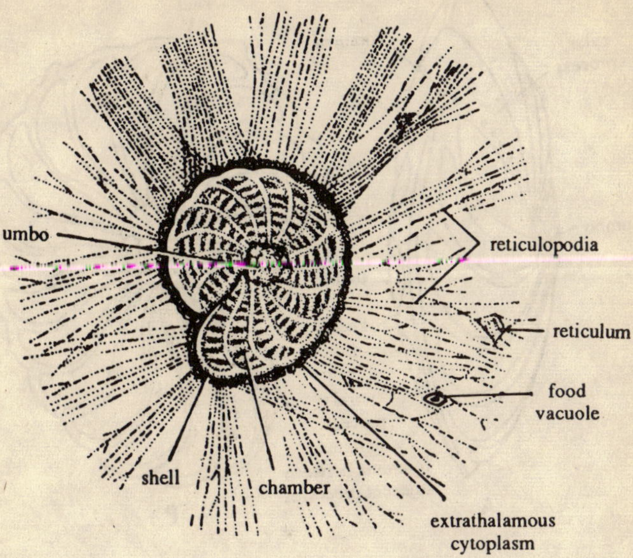

Fig. 7.1 *Elphidium Polytomella*

Megalospheric	Microspheric
1. Proloculum large but thin walled-diameter 60-100μ.	1. Proloculum small but thick-walled-diameter about 10μ.
2. Second chamber horn shaped.	2. Second chamber spherical.
3. Uninucleate.	3. Multinucleate.
4. Gamont-haploid.	4. Schizont-diploid.
5. More common.	5. Less Common.
6. Retral processes present in all the chamber.	6. Retral process found in later chambers only.
7. Rotation of chamber counter clockwise.	7. Rotation of chambers clockwise.

Under electron-microscope ribosomes, endoplasmic reticulum, Golgi-complex, mitochondria are visible.

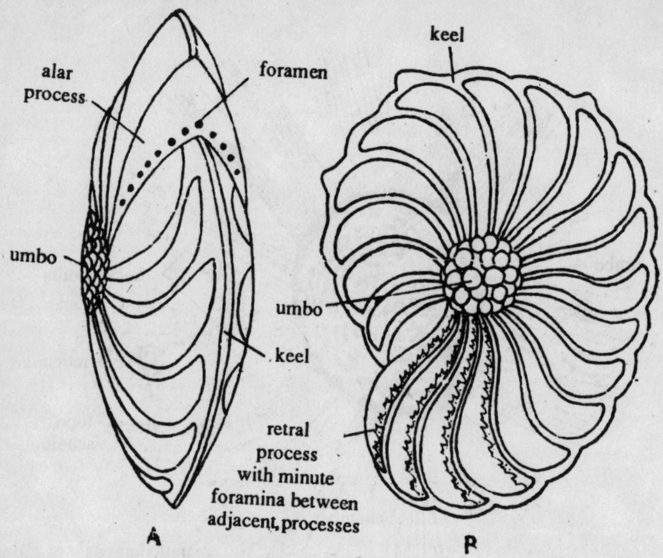

Fig. 7.2 Shell of *Elphidium* A= End on view, B= Lateral view

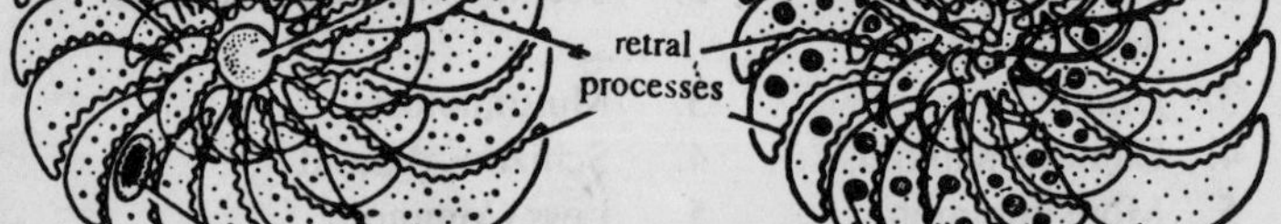

Fig. 7.3 *Elphidium.* A= Macropheric form, B--Micropheric form.

Preudopodia Preudopodia are temporary protrusion of the cytoplasm by means of which these animals crawl about. They are long, slender, branching and anastomosing. These are known as *reticulopodia* or *myxopodia*. Each reticulopodium consists of an inner fibril-like *axis* and on outer fluid-like *cortex*. They are very active. They are withdrawn and put out very quickly, and freely wave about in water. The reticulopodia are arranged in bundles and help the organisms in locomotion as well as in nutrition.

Physiology

Locomotion. The animal slowly creeps along the sand or sea-weeds at the sea floor. The locomotion is performed by reticulopodia, which help the animal to drag its shell along. The movement is very show i.e. 1 6 mm. per hour.

Nutrition The mode of nutrition is *holozoic*. The animal is omnivorous as it feeds on other protozoans, algae and minute crustancean larvae. The food is traped by pseudopodial net, and is promptly surrounded by cytoplasm. Digestion occurs in food vacuole and is completed before the food reaches the shell.

Respiration and Excretion The respiration takes place through general body surface by diffusion. Oxygen is taken in and carbon dioxide is eleminited out. The nitrogenous waste products also got rid off by diffusion. The xanthosomes are excretory granules or globules which are passed out by the retracting pseudopodia, as the animal creeps along.

Osmoregulation As the animal is marine so that the osmotic pressure of cytoplasm and its surrounding medium is almost equal. No water enters in the body in excess. Thus there is no necessity of osmoregulation hence the contractile vacuole is absent.

Reproduction and Life-Cycle

The first attempt at the study of life-cycle was made by *Schauddin* (1900). Later *Lister* (1903) added some information on the reproduction in *Polystomella. Myers* (1930-40) made further observations and most of the earlier findings were confirmed by *Le Calvez* (1938, 1950) who provided the first complete account of the life cycle of this genus. *Jepps* (1962) gave a detailed account on the life-cycle. *Grell* (1967) found that gametes are heploid and the amoebulae produced asexually by the microspheric forms are diploid.

Typically there are two generations in the life-cycle of *Elphidium*-a sexual one (*gamont*) which alternaties with an asexual one (*schizont*). This alternation of generations is further combined with dimorphism in the adult condition.

1. Asexual phase or microspheric forms.

The microspheric form reproduces asexually which is therefore known as the *agamont* or *schizont*. The microspheric are smallest forms as described by *Schaudinn*. It has nine chambers and 28 nuclei. The nuclei are irregularly scattered, although they are absent in terminal chambers. Nuclei multiply by simple division. Nuclei give off irregular strands of darkly staining substances, which are called *chromidia* by some authors.

The first indication of reproduction starts with the great increase in the number of pseudopodia the protoplasm streams out and forms a *halo* around the brown shall. Within a short time, coarse brown granules, the *chromidia* pass out. A large number of new nuclei are recognized from the chromidia. A bit of cytoplasm geathered around each nucleus. These give rise to small amoeboid bodies called *amoebulae* or *pseudopodiospores*. Each amoebula secretes a shall that later becomes proloculum of macrospheric form. New chambers are added as the animal grows in size. Thus a macrospheric form is formed from the microspheric form by multiple fission.

2. Sexual phase or macrospheric forms.

The sexual reproduction occurs in macrospheric forms so that they are called *gamonts*. The nucleus of gamont divides into numerous nuclei and also the cytoplasm. Each nucleus is surrounded by a bit of cytoplasm, which acquires a pair of flagella. This minute biflagellated gamete is called *flagellula* or *zoospore*. The formation of gametes occur within the parental cell. The gametes are released in the surrounding water through the foramina of parent shell, They swim about freely for sometime and then become somewhat elongated. Tthe gametes from different macrospheric forms unit in pairs to form the *zygote*. The zygote lose its flagella. Each zygote secretes a shall around itself and becomes a young microspheric form. The first chamber is small and called proloculum. New chambers are laid down. Nucleus divides several times to form several nuclei. The first division is meiotic so all the nuclei are haploid. With the division of nucleus, the cytoplasm does not divides so a multinucleated condition is found. The microspheridc form on becoming full-grown, undergoes asexual reproduction i.e. multiple fission.

According to *Jepps,* it takes about two years for a complete dimorphic cycle in *Elphidium crispum.* The slow pace of the cycle is related to

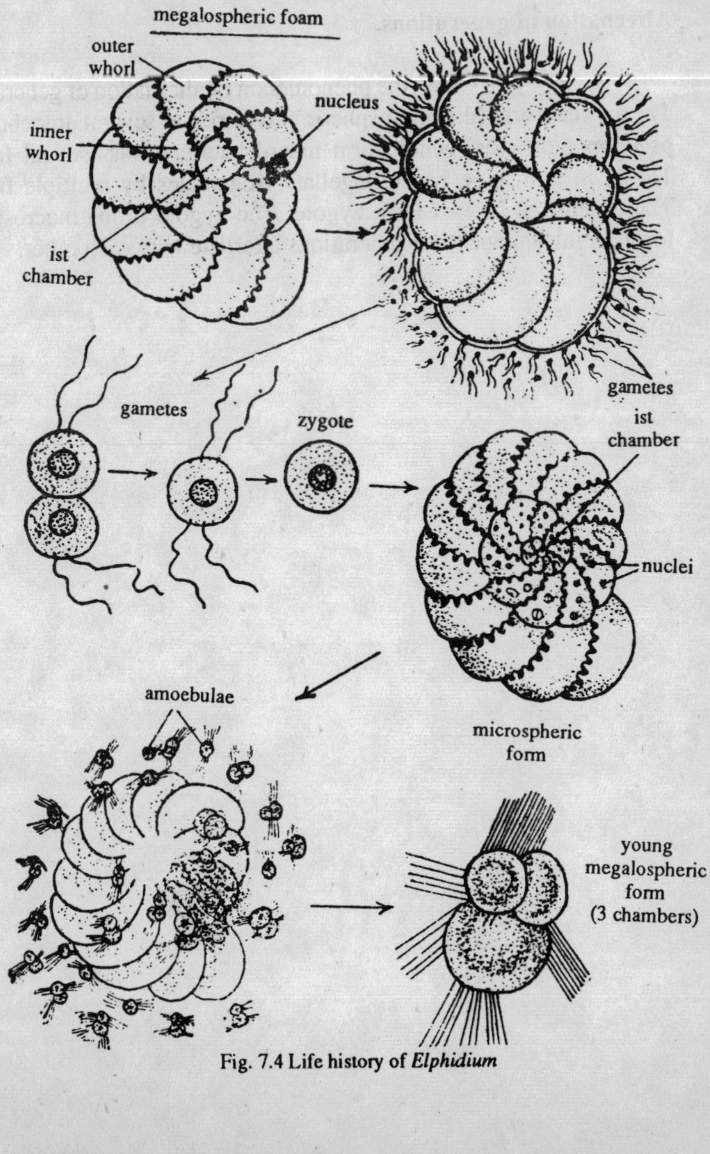

Fig. 7.4 Life history of *Elphidium*

the amount of growth, the young gamont must undergo before maturity. The gamont develops a test with about 4 or 5 chambers in about eight days. The 6th chamber is complited after about eleven days, the 15th chamber in about one month and usually 40th almost four months. From the above description it is evident that each phase takes a sufficiently long time to mature, and that is why the life-cycle takes a long period.

Alternation of generations.

The life-cycle shows the phenomenon of alternation of generations. The asexual form (the microspheric) gives rise to several amoebulae or pseudopodiospores, which form macrospheric forms (sexual forms). These, in turn, give rise biflagellate isogametes by multiple fission. These gametes fuse to form zygote. The zygote forms macrospheric form. In this manner two generations alternate with each other.

8

RADIOLARIA

Radiolaria are exclusively marine and chiefly plagic. They are found equal well in warm and cold water; littoral or oceanic; surface dweller or inhabiting deep down upto 500 meters, West areas of oceanic floor are covered by the ooze of radiolarians. Many of the radiolarians are bioluminescece. Their zoological history gets back from the Salurian Period and they are one of the earliest known animals.

The body is generally spherical and ranges from 50μ to several mm., and the colonial forms to several cms. The prevailing symmetry is radial or bilateral although asymmetrical form are also found.

The cytosome is divided into two regions: an *extra-capsular* and an *intra-capsular* by the central *capsule*. The capsule is a thin, delicate structure of a single or double membrane of pseudo-chitinous or mucoid nature. In form it is spherical, ovoid, lobate or branched. The capsule may increase in diameter with the growth of animal and it dissolves during fission in simple forms. Perforations either distributed uniformly or concentrated in one or more groups. The perforations permit continuity between the extra-capsular and intra-capsular cytosome. The extra-capsular cystosome is primarily considered with floatation and digestion of food. Three zones are distinguished in it sarcomatrix, calymma and sarcodictyum.

The sarcomatrix or assimilative layer immediately surround the central capsule. It contains pigments, excretory substances and undigested food material. In some Tripyleans aggregation of the excretory substances and food material produces greenish or brownish mass surround the aperture of capsule called phoedium. The calymma or the vacuolated layer lies in a wide zone next to sarcomatrix. It is filled with vacuoles containing water, saturated with carbon dioxide which gives buoyancy. When the vacuoles burst the specific gravity of individual increases and it sinks; when the vacuole reappears the specific gravity decreases and the animal rises. In the calymma are usually symbiotic zooxanthellae, which are modified dinoflagellate. Zooxanthellae are absent in deep sea individuals. Fat drops, oil spheres and pigments are also present in the calymma. The sarcodictyum is in a thin film on the out side of calymma, from it extends slender filamentous pseudopodia.

The typical pseudopodia are *axopodia* which arise from inside the central capsule and radiate in all directions, some pseudopodia are provided with myoneme, which produce circular group of short, rod like body clustered round each radial spine.

The intra-capsular part is the seat of reproductive activity, it is granular often vacuolated. It contains one or more nuclei, pigments, oil-droplets. fat-globules and crystals. Zooxanthellae are also found in it sometimes.

The skeleton is absent in certain forms as *Thalassicola*. In others the skeleton is present in various stages of development. It is formed of strontium sulphate while other consists of siliceous substances. In Actipylina the skeleton is composed of 20 radiating from centre and emerging from body in five circle comparable one equitoreal, two tropical and two circumpolar region. This arrangements after *Muller* is known as Muller's Law.

The siliceous skeleton may be in the form of rods and spines which usually lies outside the capsule. The rods and spines may be intracapsular or inter nuclear in some Forms like *Acantheria*. The skeleton may be in the form of one or more layers of lattic work, peripheral and concentric with central capsule. In the later case the lattice work may lie in the nucleus, in the intra-capsular cytosome and extra-capsular cytosome.

Radiolarian are not swimmers. Some of them rise and sink in response to changes in the environmental conditions by regeneration or collapse of vacuoles, such a mechanism allows animals living near the surface to sink when the sea is rough or temperature in the upper region of water is unfavourable. The filamentous, cylinder pseudopodia which

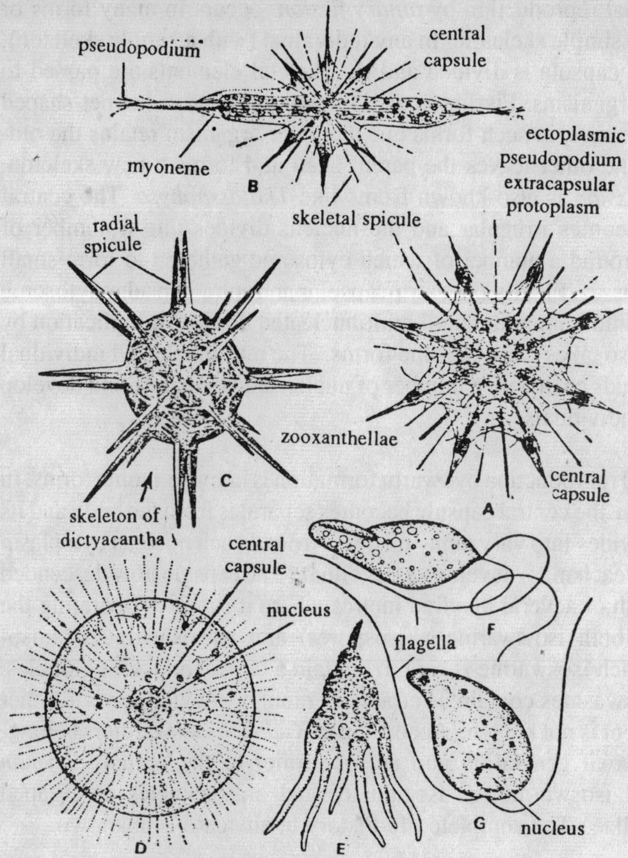

Fig. 8.1 Some Radiolarians. A=*Acanthometra*, B=*Amphilonche*, C=*Dictyacantha*, D=*Thalassicola*,

come out from extra-capsular cytosome help in creeping along hard surface.

Radiolarians feed on microplankton like diatoms and various protozoa. The food is captured by pseudopodia and by their contraction is drown in sarcomatrix where it is digisted in food vacuole. Radiolarian, can live without food if kept in sun-light due to zooxanthellae which occur in calymma except in Actipylina where it is found in intra-capsular cytosome. In the absence of organic food the radiolarian lives on the reserve made by zooxanthellae through holophytic nutrition.

Anexual reproduction by *binary fission* occurs in many forms or forms with simple skeleton. In any individual (with a simple skeleton), the central capsule is divided and the skeletal elements are passed to daughter organisms. Fission is also known to occur in helmet-shaped skeleton bearers, in such forms one daughter organism retains the old-shell and the other leaves the parent shall and forms a new skeleton. *Multiple fission* is also known froms like *Thalassophysa*. The central capsule becomes irregular and the nucleus divides into a number of globules around a number of which cytosome gathered to form small ovoidal masses. Finally the extra-capsular and intra-capsular cytosome breaks up into numerous small multinucleated body. Multiplication by budding also takes place in some forms. The multinucleated individual gives off buds each with a number of nuclei. Buds separate and develop into new individuals.

Sexual reproduction by swarm formation is known in some forms. In *Thalasicola* the central capsule becomes separates from the body and its nucleus divides into vary many nuclei. Around each nucleus cytoplasm gather and each mass develops a flagellum. The capsule now descended to the depth of several hundred metres where it ruptures liberating the swarmess both iso-swarmess (isospores) and anisoswarmers (anisospores). Each isoswarme is said to contain a crystal and a fat globules. The anisoswasmes contain a refractile granules in cytoplasm. Further development is not known. According to *Kudo* the isoswarmers belonging to asexual generation and anisoswarmer to the sexual. *Hyman* considered isoswarmer as isogametes and anisogametes as escaped zooxanthellae. The complete life-history in any case is unknown.

9

MONOCYSTIS

One of the largest and best-known accphaline gregarines is *Monocystis*, which almost always can be found in the seminal vesicle of the earthworms, *Pheretima posthuma, Lumbricus terrestris, L. rubellus and L. Castaneus.*

SYSTEMATIC POSITION

Phylum	-	Protozoa
Sub-phylum	-	Sporozora
Class	-	Telosporea
Sub-class	-	Gregarinia
Family	-	Monocystidae
Genus	-	Monocystic

Habits and habitat

Monocystis is an inhabitant of a part of reproductive apparatus of the earthworm. Various stages in the life-history are invariably found in the seminal vesicles of earthworm (endoparasites). The common species of *Monocystis* are: *Monocystis lumbrici, M. pheretimi* (*Bhatia* and *Chater-jee*), *M. beddardi (Ghosh), M.lloidi (Ghosh), M.magna, M. agilis* and *M. bengalensis.*

To obtain living parasites, an infected worm can be anesthetized by submerging it in 7% alcohol or 5% chloretone. After such treatment, the anterior third of the animal should be slit open along the mid-dorsal line and the body wall reflected. If small pieces of seminal vesicles, large lobed organs located approximately in segments 9-16, are pinched off with forceps and gently teased in 0.75% NaCl solution, large ciliated - appearing trophozoites (200 by 70μ) and cysts (160μ in diameter) will readily be found.

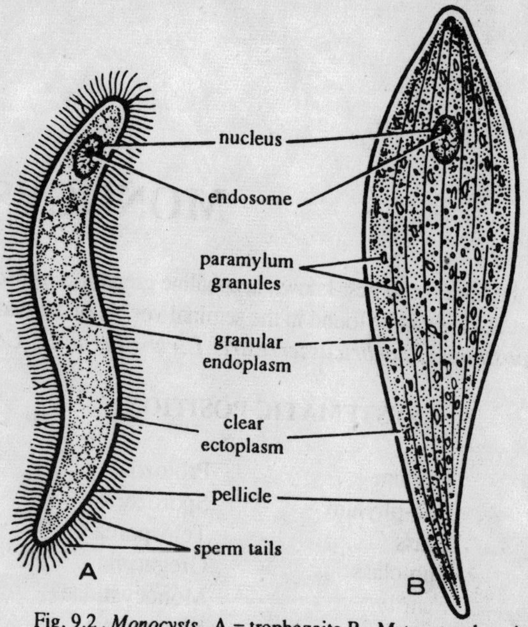

Fig. 9.2 . *Monocysts*. A = trophozoite B= Mature trophozoite

MORPHOLOGY

Shape and Size. The *trophozoite* is the adult and feeding stage of the *Mynocystis*. It is an elongated spindle-shaped and flatenned animal. It is about 200μ in length and 70μ in width.

Pellicle. The body of the trophozoite is bounded externally by the thick, smooth and porous pellicle (also called *ectocyte)*. It may be modified in various ways i.e., it may be striated or thrown into elevation and furrows or hair-like processes.

In *M.agilis*, at the anterior end and on one side is a small some what pointed rose thorn-like process called *mucron*. It is supposed to help in locomotion.

Cytoplasm. The cytoplasm can be differentiated into outer *ectoplasm* and inner *endoplasm*. The ectoplasm in further differentiated into an outer homogeneous *sarcocyte* and an inner *myocyte*. In the myocyte are found longitudinally running tracts of cytoplasm called *myonemes*, especially developed contractile structures. In some species circular myonemes are also present.

The endoplasm contains granules of *paraglycogen* in large number. Besides *volutin* granules and fat globuled are also present.

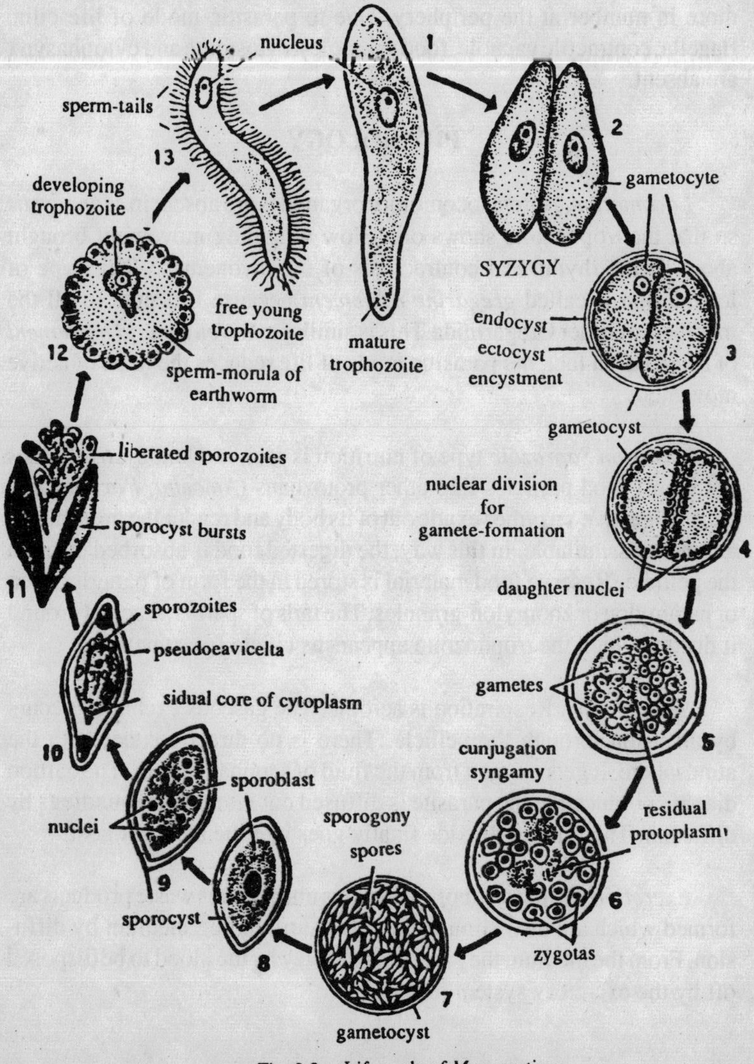

Fig. 9.2 Life cycle of *Monocystis*

Nucleus. Within the endoplasm there lies a single vesicular nucleus. It is surrounded by a delicate nuclear membrane and contains a clear liquid in which are suspended denuser bodies called *nucleoli or karyosomes.* Chromatin is distributed throughout the nucleoplasm.

Under electronmicroscope the endoplasm is seen to contains other organelles like *Golgi complex , mitochondria* etc. The mitochondria are more in number at the periphery. Due to parasitic mode of life cilia, flagella, contractile vacuole, food vaculoles, cytostome and cytophasynx are absent.

PHYSIOLOGY

Locomotion. Active locomotory organelles are absent in *Monocystis* so that the trophozoite shows only slow wriggling movement brought about be the rhythmical contractions of the myonemes. This type of locomotion is called *gregarine movement* because it occurs in all the members of order Gregarinida.This is similar to the *euglenoid movement* of *Euglena.* In fact, the parasitic mode of life reduces the need of active movement.

Nutrition Saprozoic type of nutrition is seen in *Monocystis.* It does not ingest food particles like other protozoans (*Amoeba, Paramecium* etc.). Digestive enzymes exude out of its body and render the protoplasm around is assimilable. In this way, the digested food is absorbed through the pellicle. Reserve food material is stored in the form of paraglycogen or paramylon or zoomylon granules, The tails of sperms attached around it due to which the trophozoite appears as ciliate for sometimes.

Respiration. Respiration is aerobic. The gaseous exchange occurs by diffusion through the pellicle. There is no direct contact with the atmosphere. It gets oxygen from the fluid of seminal vesicle. The carbon dioxide produced in the parasite is diffused out into the surroundings by diffusion. This carbon dioxide finally goes into the blood of host.

Excretion. By the catabolic activities nitrogenous waste products are formed which are also eliminated in the surrounding medium by diffusion. From the medium they are drained away by the blood to be disposed off by the excretory system of host.

Behaviour Very little is known about the irritability is *Monocystis* as it lives in a remarkably uniform environment so there is no need of any active response by the organism.

REPRODUCTION AND LIFE CYCLE

The life-cycle of *Monocystic* is completed in one host i.e. the earthworm. Thus *Monocystis* is monogenetic. The tropohozoite lives freely for sometimes in the sperm morula (it is a group of developing sperms or spermatozoa of host) of host. It feeds on the sperm photoplasm and grows in size. Soon the fully adult trophozoite enters the reproductive phase which is divided into three stages:

1. Gametogony or gamontogamy.
2. Syngamy or Conjugation
3. Sporogony.

1. Gametogony.

The gametogony is a method of sexual reproduction which involves pairing of gamonts and formation of gametes.

(a) *Syzygy*. After feeding stage each trophozoite changes into *gametocyte* or *gamont*. Gametocytes come together in pairs. The two then secrete a common resistant and protective cyst wall called *gametocyst* or *gamontocyst* or *association cyst*. The latter word has recently been given *by Marshall* and *Williams* (1972). The cyst wall is bilayered, an outer thick and rigid *ectocyst* or *epicyst* and an inner thin *endocyst*. Fusion or conjugation between two gametocytes does not take place in cyst. This type of pairing of gametocytes is called *syzygy*.

(b) *Gamete Formation*. Within the cyst-wall the gametocytes shrink further and exude some fluid that fills the narrow space between them and endocyst and is called the cystic fluid.

Each gametecyte produce a number of nuclei by the mitotic division of its nucleus. The nuclei thus formed, move to the periphery of gametocytes and each gets surrounded by a small amount of cytoplasm to form gamete. The gametes project from the surface of the gametecyte, giving it the appearance of mulberry. A small amount of cytoplasm is left as *residual cytoplasm* in the gametocyte. This cytoplasm is vaculated and contains some paraglycogen granules and serves as a medium in which the gametes live. The gametes are *isogametes* in *M. pheretimi* as they are similar in shape and size. They are haploid and each contains 5 chromosomes. Recently it has been suggested, on the basis of new findings of some workers, that sex differentiation occurs between gametes of two gametocytes. These may be in the form of cytoplasmic inclusions or in shape (*anisogametes)*. For instance, it has been pointed out that the male gametes have a pointed tail or are flagellated. Flagel

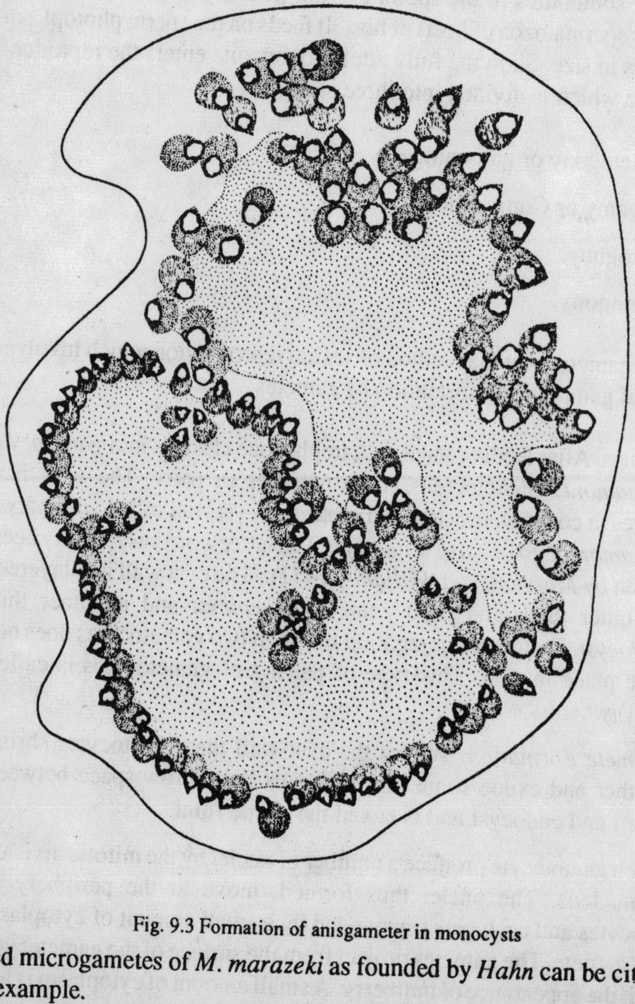

Fig. 9.3 Formation of anisgameter in monocysts

lated microgametes of *M. marazeki* as founded by *Hahn* can be cited as an example.

The gametes produced in one gametocyte are of the same sex and are equal in number to those formed in its partner.

2. Syngamy

It involved the fusion of gametes sooner or later, the middle wall separating the gametocytes break down and their contents intermingled. The gametes now fuse in pairs to produce dipoloid zygotes. It is as a rule that two gametes uniting in a pair come from different gametocyte. Many zygotes are formed within the gametocyst by syngamy.

3. Sporogony.

The developing zygote is also called *sporblast*. It is spherical in the beginning but transforms itself into a single-walled boat-shaped body. It secretes around itself singled-layered, transparent chitinous wall called *sporcyst*. The sporoblast thus changes into the *spore*. It possesses a mucoid plug at either end. The spore now acquires the characteristic form of a boat resembling a diatom, *Nevicella* and therefore, it was termed by some earlier Zoologist as Pseudonevicella, but it is now called zygocyst or spore.

After a brief resting period, the spore undergoes multiple-fission called *sporogony*. In this process, its nucleus is divided three times, the first division is meiotic, resulting in eight nuclei, Each nucleus soon gets surrounded by a bit of cytoplasm. The uninucleated masses of cytoplasm, thus formed, are known as *sporozoites* or *zoites*. Each sporzoite is slender, elongated, sickle-shaped body and, all the sporozoites are arranged like the loculi of an orange around a core of residual cytoplasm. All the sporzoites are hoploid and occur within the gametocyst formed at the time of syzygy.

The spores are liberated into the cavity of seminal vesicle of earthworm by the rupture of the gametocyst. In those species of *Monocystis* where reproduction takes place in coelomic fluid, the spores are found in the posterior region of the body.

The sporozoites can develop further only if the spore containing them are transferred to new host which is no doubt an earthworm.

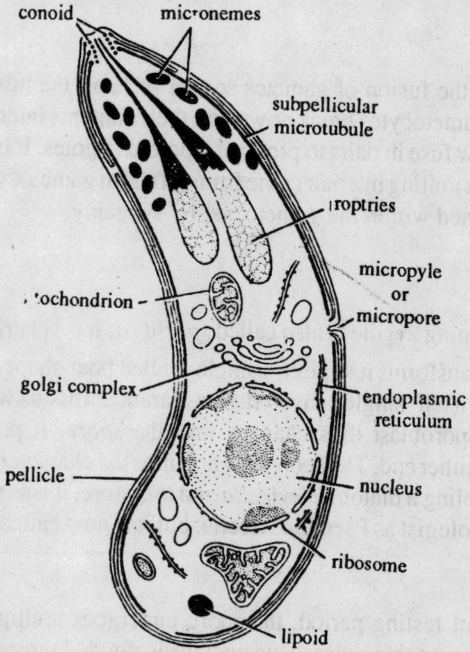

Fig. 9.4 *Monocystis*. A sporozoite seen under electron microscope.

MODE OF TRANSFERENCE

It is not definitely known as to how the spores of *Monocystis* are transmitted from one host to other. However, their transmission is believed to take place by any of the following methods:

1. *By the death of host.* When the host dies and decays, the spores are scattered in the soil. When such soil is eaten by another worm. they reach its alimentary canal.

2. *During copulation.* Some zoologist are of the opinion that the spores are usually transferred from one host to another during copulation, when spermatozoa and seminal fluid of one worm are passed into the spermatheca of another, but spores are never seen in cocoons.

3. *By birds. Pfiffer* found spores in the gut contents of various birds. In

such cases sporocysts are passed outside with the excreta of birds into the soil. If such a soil is swallowed up by other worms they start their life cycle again.

4. *Autotomization.* In certain species of *Monocystis* which reproduce in coelomie fluid their spores are found in the posterior segments of the host body, when the posterior body is autotomized and the spores are liberated in the soil. When such soil is swallowed up by other earthworm, The sporozoites start their life-cycle again.

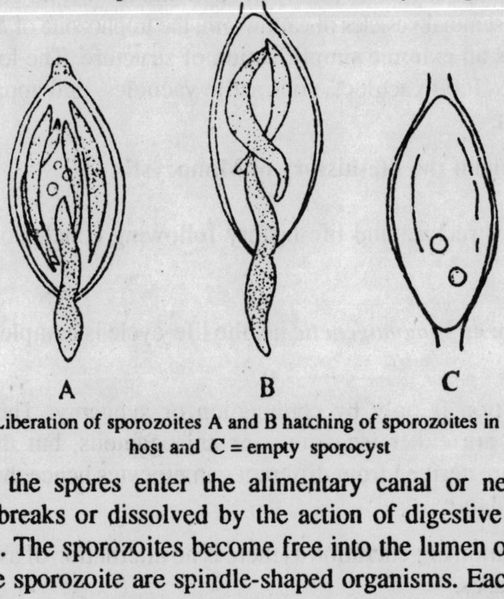

Fig. 9.5 Liberation of sporozoites A and B hatching of sporozoites in intestine of host and C = empty sporocyst

When the spores enter the alimentary canal or new host, the sporocyst breaks or dissolved by the action of digestive enzymes of earthworm. The sporozoites become free into the lumen of alimentary canal. The sporozoite are spindle-shaped organisms. Each contains a nucleus, one or two mitochondria, Golgi complex. The pellicle forms the shape of sporozote. At the anterior most region there is a pair of secretary organelles or *roptries*. These structure secrete a secretion which helps in penetration through the tissues. At the anterior end besides the roptries *conoid* and *micronemes* of unknown function are present.

Being actively motile the sporozoites pass out through the alimentary canal into the coelom and than migrate into the seminal vesicles. On reaching the seminal vesicles of a new host each sporozoite penetrates into a new sperm morula, becomes a trophozoite, and the life cycle begins again.

Effect of Parasitism

1. *On the Earthworm. Monocystis* seems to cause no appreciable harm to the host. No doubt it feeds on the spermatozoa, but this does not affect the fertility of the host because the earhworm produces a large number of sperms. The worm is able to combat the parasites by forming a resistant envelope around the trophozites and by killing the cysts through phagocytosis.

2. *On the Parasite.* Due to planty of food and almost no competition inside the seminal vesicles of earthworm, the trophozoite of *Monocystis* undergoes an extreme simplification of structure. The locomotory organelles, food vaculoes, contractile vacuoles, oral apparatus etc. are absent.

Conclusions from the life-history of Monocystis.

From the structure and life-history following conclusions can be derived:

1. The organism is *monogenetic* i.e. the life-cycle is completed in one host.

2. Reproduction is only by conjugation or syngamy. The gametes produced are either isogamous or anisogamous, but the pairing gametes are derived from different gametocytes hence the gametes are *exogamous*.

3. No alternation of generation as there is no alternation of asexual and sexual phases.

4. Neither the trophozoite nor the sporozoite is free-living.

5. As the sporozoites are subjected to pass through adverse condition outside the body of host, hence, are protected by sporocyst.

6. Due to parasitic mode of life, the locomotory organelles, food vacuoles, contractile vacuoles, etc. are absent.

7. Respiration and excretion by simple diffusion.

10

PLASMODIUM (MALARIAL PARASITE)

Plasmodium is a member of class-Telosporea of phylum Protozoa. It is a parasite and cause malaria fever in man and other animals. Malaria (Gr. *mal* bad; *aria* - air)meaning bad air, was so named as its was previously thought that damp night-air of marshes and swamps caused this disease. Malarial parasite was first discovered by *Charls Laveran* (1880), who was, a doctor in the French Army stationed Algiers. He observed *Plasmodium* in the blood of a malaria - patient. *King* (1883) reported that *Plasmodium* were transmitted by mosquitoes. The observation of *Laveran* were confirmed by *Golgi* (1885) and *Sallie* (1885). *Patric Manson* suggested that the malarial parasite enter the blood of man after the bite of mosquitoes. He placed his suggestion before *Sir Ronal Ross* in 1894. In 1895 *Ross* undertook this problem and after continuous labour of two year in 1897 proved that *Plasmodium* are sucked up by female *Anopheles* mosquito and are later on injected into the human blood by *Anopheles* mosquitoes. *Grassi* described the life cycle of *Plasmodium* in the stomach of *Anopheles* female. *Granham, Bird* and *Baker* (1960) studied ultra structure of the sporozoite in birds *Granham et.al.*(1961, 1963) studied ultrastructure of sporozoites in human.

There are about 60 species of *Plasmodium* which cause malaria in man and other animals.

SYSTEMATIC POSITION

Phylum	-	Protozoa
Sub-phylum	-	Sporozoa

Class	-	Telosporea
Sub-Class	-	Coccidia
Genus	-	*Plasmodium*

Four species of *Plasmodium* are known to cause different types of malaria in man. These are *Plasmodium vivix, P. malariae, P. falciparum, and P. berghei found in rats P. rats* and *P. gallinaceum* in chickens.

Habits and habitat.

Plasmodium is an intracellular parasite in red blood corpuscles of man. It is most widely distributed in both temperate and tropical regions. It has two hosts: (a) man or any other vertebrate (reptiles, birds and mammals) which are called *primary hosts* or *definitive hosts* and (b) female *Anopheles* mosquitoes or other blood sucking insects which are *secondary hosts* or intermediate hosts. The life history is completed in the above two hosts: *asexual* in man and *sexual* in female *Anopheles*. The common species of *Anopheles* which cause malaria in India are *A. stephensi, A. maculatus, A. fluvialitis* and *A. culcifancis*.

Morphology

Structure. The adult parasite is called as *Trophozoite* that lives in red blood corpuscles of man. It is amoeboid, uninucleated with granules and vacuoles in the cytoplasm. The granules are chiefly of *haemozoin* pigments. The trophozoite feeds upon the substance of R.B.C.

Its ultrastructure has been studied by *Garnham* (1966) and *Rudzinska* (1969). It possesses a double membrance, the *plasmalemma* which is closely applied to the cytoplasm. The endoplasmic reticulum is in the form of small vesicles, which are both smooth and rough type. Free ribosomes are present in cytoplasm. Mitochondria are few and posses double membrane, the cristae are peripheral while the central region is structureless. Golgi complex is composed of small vesicles arranged in rows. Double-membrane concentric structures in the trophozoites are believed to work as mitochondria. One or two double membrane vacuoles with structureless matrix are present their function is unknown. The nuclear membrane is double to which are attached ribosomes. An endosomes or nucleolus lies eccentrally in the nucleus. Pinocytic vacuoles are common in cytoplasm.

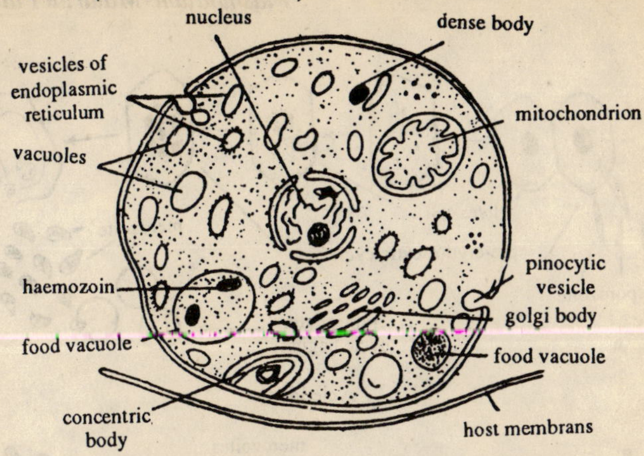

Fig. 10.1 *Plasmodium* Ultrastructure of Trophozoite

The trophozoite feeds upon the haemoglobin of R.B.C. Digestion is intracellular. The residual product called haemozoin is scattered throughout the cytoplasm. Liquid food is taken by the process called pinocytosis.

Life-Cycle

The life - cycle of *Plasmodium vivex* is *digenetic* as its life-cycle is completed in two hosts-man and the *Anopheles* mosquito. The life-cycle is not started by the trophozoite but by the *sporozoite (Marshall* and *Williams,* 1972) The life-cycle may be subdivided into three phases; a phase of growth and asexual multiplication occuring in man and called *Schizogony,* sexual phase *or gamogony* which starts in man and completed in mosquito; and an asexual multiplication phase in mosquito called *sporogony.* Previously it was thought that there were only two phases in the life-cycle.

1. Schizogony

When a female *Anopheles* bites a man to suck the blood, then along with saliva it inject the *sporzoites* of *Plasmodium* into the human blood. Each sporozoite measures 6-15μ in length. These are sickle-shaped single-nucleated sporozoite. *Granham et al* (1961, 1963) studied sporozoite under electron microscope. According to them the pellicel is outer membrane made up of three layers and an intermediate zone is distinguishable. It is continuous over the whole organism except anteriorly and at the base of the micropyle. At the anterior end lies a cup-

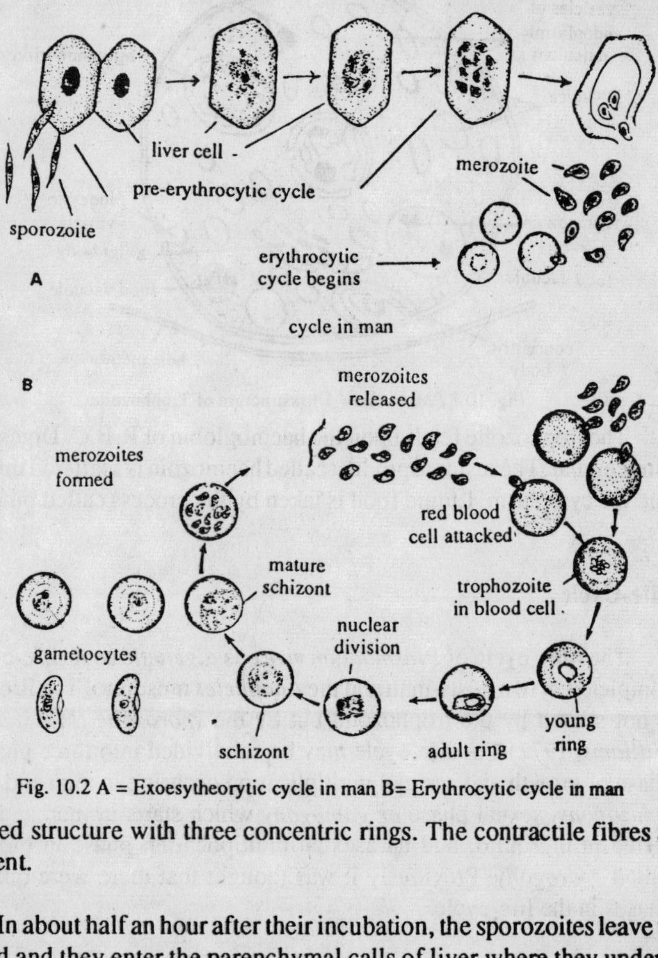

Fig. 10.2 A = Exoesytheorytic cycle in man B= Erythrocytic cycle in man

shaped structure with three concentric rings. The contractile fibres are present.

In about half an hour after their incubation, the sporozoites leave the blood and they enter the parenchymal calls of liver where they undergo at least two schizogonic cycles.

(a) *Pre-erythrocytic cycle.* Sporozoites rapidly grow in size in liver cells by absorbing nourishment. Here they become spherical and are called *schizonts*. They measure 10-40μ in diameter. The nucleus of sporozoite divides into several nuclei. Now the schizont is called *cryptozoite*. Each nucleus is surrounded by a bit of cytoplasm and called *cryptomerzoite*. Their number is about 80-1000. Cryptomerozoite measures 1.2μ in diameter. Due to pressure of cryptomerozoites the body of cryptozoite, as well as, the host liver cell bursts liberating the cryptomerozoites into liver sinusoids.

(b) *Exo-erythrocytic cycle.* The Cryptomerozoites of pre-erythrocytic cycle, which invade fresh liver cell, grow into adult-like form, called

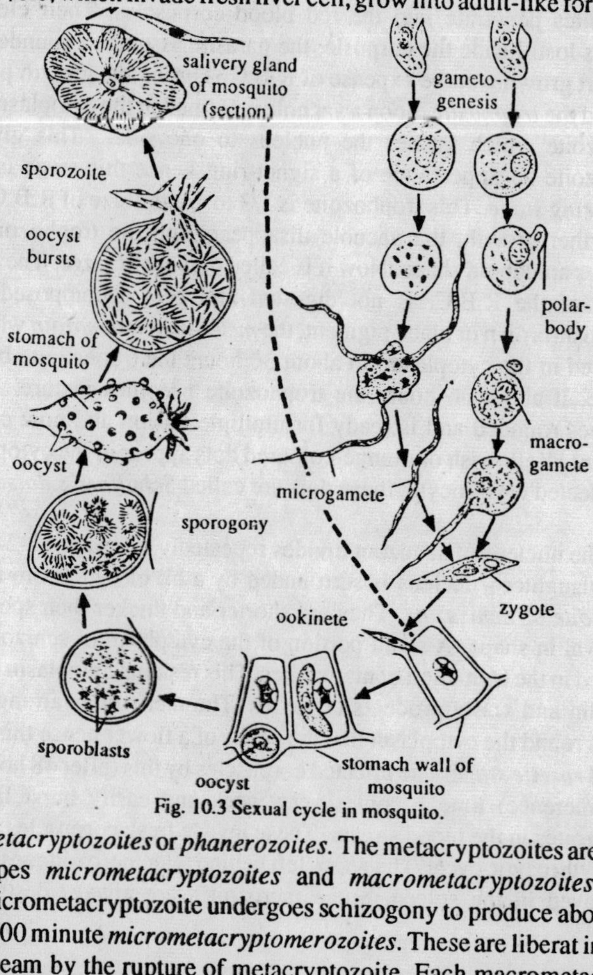

Fig. 10.3 Sexual cycle in mosquito.

metacryptozoites or *phanerozoites*. The metacryptozoites are of two types *micrometacryptozoites* and *macrometacryptozoites*. Each micrometacryptozoite undergoes schizogony to produce about 100-1000 minute *micrometacryptomerozoites*. These are liberat in blood stream by the rupture of metacryptozoite. Each macrometacryptozoite also undergoes schizogony so as to produce 64 *macrometa cryptomerozoites*, which are liberate in blood and invade fresh liver cells to continue exoerythrocytic cycle. There is no effect of various drugs as the parasite is intracellular. Hosts are acquire immunity to the infection. The period of biting of mosquito in man and entry of merozoites in blood is called *pre-patent period.* It takes about 10 days.

Thus, the liver cells provide a sort of reservoir for reolapse of fever. On the oxoeryth rocytic cycle.

Erythrocytic Schizogony. After their release from liver cells the merozoites penetrate into the red blood corpuscles. Their elongated shape is lost. Inside the corpusle; the parasite assumes a rounded form and start growing at the expense of R.B.C. During the growth period it is called the *trophozite.* Soon a vacuole is formed in the cytoplasm of the trophozoite which pushes the nucleus to one side. This gives the trophozoite an appearance of a signet-ring hence this stage is called *signet-ring stage.* This trophozoite is 1/3 to 1/2 the size of R.B.C. With the further growth, the vacuole disappears and the trophozoite now assumes amoeboid shape. Now it is called *amoeboid stage.* The haemoglobin of the R.B.C. is not digested and gets decomposed into a yellowish-brown or black pigment, the *melanin* or *haemozoin* which lies scattered in the cytoplasm. In about 36 hours the trophozoite becomes mature. It about 36 hours the trophozoite becomes mature. It again becomes rounded and is ready for multiple fission and now called as *schizont.* Yellowish or orange-coloured dots appear in the cytoplasm of the infected erythrocyte, these dots are called *Schuffner's dots.*

The nucleus of schozont divides repeatedly, forming 12-24 nuclei. Each daughter - nucleus is surrounded by a bit of cytoplasm to form *merozoite or Schizozote.* They are shorter and thicker then sporozoites and oval in shape. A small portion of the cytoplasm of schizont is not utilized in the formation of merozoites. This residual cytoplasm contains melanin and waste products or toxins. The merozoites arrange themselves round the cytoplasm like the petals of a flower hence this stage is called *rosette stage.* The infected corpuscles by this (after 48 hours from the enterence) time becomes very weak and easily burst liberating merozoites in the blood stream. These invade fresh corpuscles to repeat the erythrocytic cycle. The *ghost,* left behind after merozoites escape, are destroyed in the spleen. Some merozoites are always destroyed by W.B.C.

With the rupturing of the R.B.Cs., the patient suffers an attack of malaria. The fever has been attributed to the release of toxins from the merozoites. The interval between inoculation of sporozoites into human blood and first attack of fever is called *incubation period,*

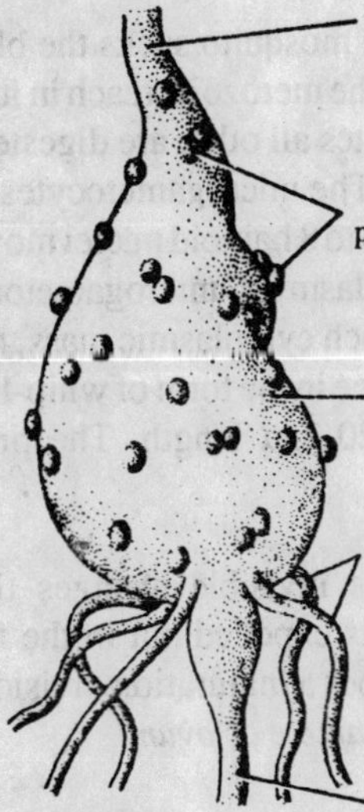

Fig. 10.4 Stomach of infected female *Anopheles* with oocyst, of *plasmodium*.

2. Gametogony.

It starts in man and completed in mosquito. After a number of repeated erythrocytic schizogony,, the mesozoites do not proceed ahead with erythrocytic cycle, but after entering the R.B.Cs., increase in size to become rounded *gametocytes* or *gamonts*. Growth of merozoites into gametocytes is rather slow and characterised by absence of signet-ring stage. The gametocytes have a large, central nucleus and denuser cytoplasm due to the presence of numerous haemozoin granules. The gametocytes are of two type. *male* or *microgametocytes* and *female* or *macrogametocytes*. The microgametocyte is small and contains a large central diffused nucleus. The female gametocyte is larger and contains a small and eccentric nucleus. The cytoplasm contains fewer granules. The gametocytes do not divide, but remain as intracellular parasites within their host R.B.C., until they either die or are ingested by the victor, in which they continue their development.

Sexual cycle in Mosquito

When the female *Anopheles* mosquito sucks the blood of malarial patient the gametocytes and also the merozoite reach in its stomach along with the blood. Except gametocytes all other are digested. The gametocytes now come out of R.B.C.s. The microgametocytes become active. The nucleus of each divides into 6 to 8 haploid nuclei move to the surface of microgametocyte. The cytoplasm of microgametocyte segregates around these nuclei. Suddenly, each cytoplasmic mass, together with its nucleus, shoots out from the surface in the form of whip-like microgametes or sperms each measuring 20-25μ length. The process is called exflagellation.

Macrogametocyte undergoes invisible changes till some of its nuclear material and cytoplasm is expelled out in the form of a *polar body,* or in other words it undergoes a maturation division, the resulting large body is a *female or macrogamete* or *ovum.*

Fertilization. Microgametes make active lashing movements and approach macrogametes. One of the microgametes reaches the protoplasmic projection called cone of reception of macrogamete. The nuclei of microgamete and macrogamete fuse. resulting in the formation of zygote or *synkaryon* with a diploid nucleus.

The zygote remains inactive for sometimes but later on it becomes elongated or worm-like or vermiform and motile. It performs gliding movement and is known as *ookinete.* It measures 15-22μ in length and 3μ in width.

The ultrastructure of ookinete is given by *Granham et. al* (1962) It is enclosed in two layered envelope made up of an outer folded and inner smooth layer. The micropyle is absent. The nucleus is granular with an endosome. At enterior end, inner layer is more dense and is splitted so as to appear like a mouth. It moves into the stomach of mosquito, there it pass on into the epithelium and finally reaches in-between the epithelial and subepithelial tissues. Ookinete becomes enclosed in a cyst called *oocyst.* The occyst is thin, membranous and elastic. It is partly secreted by the ookinet and partly derived from the mid-gut of the host. It grows in size absorbing nourishment from the insect through previous cystwall. The oocysts look as rounded projections on the surface of the stomach. A single infected mosquito may contain 500-5000 such oocysts.

3. Sporogony

The cytoplasm of the oocyst develops a number of vacuoles and its nucleus divides repeatedly to form a very large number of daughter nuclei. The first division is always a meiotic division and subsequent divisions are mitotic (*Bano*, 1959). Thus the nucleui prodoused are haploid. Each daughter nucleus gets surrounded by a mass of cytoplasm. The resulting calls are called the *sporozoites*. Each oocyst contains about 10,000 sporozoites. Each sporozoite has tapering ends and a broad middle part containing a single nucleus. The full-grown sporozoite break off from the residual cytoplasm of the oocyst. When the oocysts rupture, the sporzoites are liberated in the haemocoel of mosquito. Being motile, the sporozoites move to different organs in the body of mosquito but the majority migrate into the salivary glands and wait for their transmission into the blood of man white the mosquito bites. It has been estimated that the salivary glands of a single infected mosquito may contain about 200,000 sporozoites. One bite of such a mosquito may infect upto 1000 sporozoites in man. Cycles in mosquito lasts for 2-4 months.

KIND OF HUMAN MALARIA

The malaria fever is of three type. The names of these thee are according to the period of recurrence of the fever. Mover, these three types of fever are caused by different species of *Plasmodium*. The fever are:

1. *Tertian Fever* The fever comes on every third day in it. It is cause by the presence of *P.vive, P. Falciparum* and *P. ovale*. It is of two types.

   (i) *Benign tertian*. It is due the presence of *P. vive* and *P. ovale*. The temperature of patient may rise up 106°F but also comes down soon. It is not a fatal one.

   (ii) *Malignant tertian*. It is due to the presence of *P.falciparum*. The temperature of the patient does not rise much but other complication take place. The red blood corpuscles, in which the parasites are living, aggregate to block the blood cepillaries due

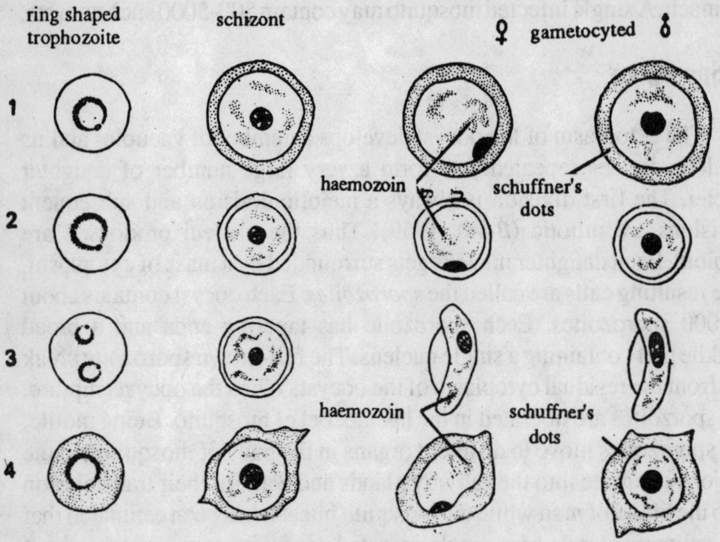

Fig. 10.5 Species of *Plasmodium* 1=*P. vivex*, 2=*P. malaria*, 3 =*P. falciparum*, 4= *p.ovale*

to which the proper blood supply is stopped to different organs of the patient. It is fatal.

2. *Quartan Fever*. The fever comes on every 4th day in this. It is caused by the presence of *P. malarial* It is not fatal.
3. *Mixed Fever*. A patient can suffer both mild tertian and malignant tertian at the same time. Similarly, tertian and quartan can combine to cause mixed fever. In such conditions, the fever comes on every day.

A comperative account of four species of human infecting *Plasmodium*

Character	P. vivex	P.malariae	P.ovale	P. falciparum
1. Distribution	Widely distributed in tropical and subtropical countries	World wide, in tropical and Subtropical regions	Tropical Africa	Very Common in tropical countries

2. Incubation period	10-14 days	18-28 days	10-14 days	9-12 days
3. Duration of erythrocytic Schizogony.	48 hours	72 hours	48 hours	48 hours
4. Duration of exo-erythrocytic cycle.	8 - days	14-15 days	9 days	5 days.
5. No. of Cryptozoite per schizont	10,000	15,000	15,000	30,000
6. No. of merozoites per schizont	12 - 24	6 -12	6 - 12	8 - 20
7. Arrangement of merozoites	In two circles	Rosette-shaped	In one circle	Irregular.
8. Shape of Trophozoites	Amoeboid	Band-shaped and compact	Large and compact	Small and compact.
9. Infected R.B.C.	Large, swollen, distorted and pale with Schuffner's dots	Not enlarged no dots.	Slightly enlarged, irregular, dot numerous	Normal sized, no dots.
10. Haemozoin granules	Light brown, masses.	Dark-brown to abundant	Dark-brown,	Dark-brown in compact mass.
11. Schizont in R.B.C	Occupies more than 2/3 of R.B.C.	Almost entire R.B.C.	About 3/4	Less than 2/3
12. Gameto-cytes.	Rounded or oval	Circular or ovoid.	Rounded	Crescentric.
13. Micro-gametocyte	7-8μ wide with numerous haemozoin granules; 4-8 male gametes produced	Large nucleus with numerous and scattered haemozoin, 4-8 male gametes produced.	Similar to *P. vivex*	Haemozoin golden brown 2-5 gametes are produced.

14.	Macro-gametocyte	8-10μ wide, large, few haemozoin	small nucleus, coarse haemo-zoin.	Same as in p.vivex	Dark, coarse haemozoin around the nucleus
15.	Duration of sexual cycle	10 days	25 days	16 days	10-12 days

Symptoms

Prodromal Symptom. In a typical malarial infection, mild symptoms appear early, even during incubation period. These include loss of appetite, nausea and sometimes sleeplessness. Tongue may be thickly coated. Headache, muscular pain and aches in joints develop.

Paroxym. It is the actual attack of malaria which occurs after a few initial erythrocytic cycles. The period of paroxym varies from person to person. It includes:

(a) *Rigor stage.* That patient feels a terrible chill and shivering, rapid pulse and breathing.

(b) *Febrile stage.* Shivering subside in about one hour or so and the body temperature rises to 104°F to 105°F.

(c) *Defervescent stage.* After a few hours, profuse sweating starts, temperature of body steadily drops and the patient feels healthy again.

Control of Malaria.

The measures for control of malaria may be divided into three categories.

I. Destruction of the victor or *Anopheles* mosquitoes.

II. Presentation of infection or *Prophylaxis.*

III. Treatment of disease

I. Destruction of the victor

This is being done in following ways:

(1) Destruction of adult mosquitoes.

(i) By spraying D.D.T., Flit, melathion or any other insecticide in the houses.

(ii) By fumigating pyrethrum, cresol and other compounds of neptha.

(iii) By killing them after traping them in small boxes made of wire gauze and internally lined by black paper.

(iv) By sterlization of mosquitoes in now being achieved in Japan.

(2) *Destruction of larvae and pupae.* The larvae and pupae live in stagnant water. They can be destroyed.

(i) Dirty water should not be allowed to collect at a place.

(ii) In large bodies of water, kerosine oil, paraffin oil etc. should be spread on the surface.

(iii) By introducing insectivorous predators like ducks, fishes (*Gambusia)* and some plants like *Utricularia.*

II Prevention of infection.

(1) The houses should be made mosquitoes free by building human dwellings on high grounds, keeping houses clean ventilated, and covering doors and windows with meshes and screens.

(2) Sleep in mosquito-net.

(3) Mosquito repellent cream should be applied on the exposed parts of the body before going to sleep.

(4) Anti-malarial drugs should be taken regularly to escape the attack of malaria. These drugs are *quinine, paludrine, deraprin* etc.

III Treatment of Malaria.

Qunine, Atabrine, Comoquin, Chlorquine, Paludrine, Resochin, Pentaquine, Pamaquine etc. are the medicician to check the malarial fever. *Daraprim* is the most effective drug. Its single dose of 25 mg. causes the schizonts of *P. vivex* and *P. falciparum* to disappear.

11

EIMERIA

Eimeria is a parasite like *Monocystis Eimeria schubergi* was discovered by German biologist. *Fritz Schaudinn* in 1900. He called this parasite as *Cocidium schubergi* and worked out its life cycle.

SYSTEMATIC POSITION

Phylum	-	Protozoa
Sub phylum	-	Sporozoa
Class	-	Telosporea.
Sub-Class	-	Coccidia
Family	-	Eimeriidae
Genus	-	*Eimeria*
Species	-	*schubergi.*

Habits and habitat.

E. schubergi is an cytozoic or intracellular parasite in the epithelial cells of alimentary canal of centipede, *Lithobius forficatus*. For some time it lives in the lumen of alimentary canal as a *coelozoic parasite*. The greater past of the life-cycle of *E.schubergi* is spent in the host and a part in open.

Morphology

A full grown trophozoite is somewhat spherical body with a single nucleus. The nucleus contains a single endosome in its centre. Due to parasitc mode of life, the locomotory organelles, mouth, gullet, contractile vacuoles etc. are absent.

Physiology

Locomotion. Although the locomotory structures are absent but it moves about in the alimentary canal, progressing by gliding movements, alternating these by flexions, bending itself like a bow and straightening out again. Inside the epithelial cell, it lives a quiescent life.

Nutrition. It feeds in a saprozoic manner, absorbing food in dissolved form. It also subsists on the contents of the cell of the host in which it is living. When in the open, it subsists on the reserved food materials, which is derived from the host's cells before leaving them.

Respiration and excretion. Both take place by the diffusion.

LIFE CYCLE

E. schubergi is a digenetic and shows an alternation of generation. The life-cycle may be divided into three phases:

1. Schizogony
2. Gamogony
3. Sporogony

1. Schizogony

The parasite enters the host in the form of oocyst. When swallowed the oocyst hatches in the small intetine of host (probably under the influence of mechanical pressure and enzymes). From each oocyst eight motile *sporozoites* are released. Each sporozoite has a tiny sickle-shaped body tapering at both the ends. A nucleus is present in the broad middle part. They measure 15-2 by 4μ. The spotozoite more apart by gliding, bending and Constriction movements till they reach the surface of the epithelium of gut. Each sporozoite penetrates into an epithelial cell of the alimentary canal of the host. When it enters in a cell, it becomes retracted to an ovoid mass of cytoplasm, which measures 16μ in length and about 8μ in width. This stage is known as *trophozoite*. The ovoid body of trophozoite grows quickly, increase in length and thickness. In about 24 hours, the trophozoite becomes full-grown, then it is termed as *schizont*. It is spherical and has a diameter of about 20μ. These is no reserve food material in cytoplasm. There is a large vesicular nucleus with a karyosome. Multiple fission occurs in schizont. In this process, its nucleus devided repeatedly to form 30-40 nuclei, which migrate to the periphery, each each passes into a finger-like projection arising from the surface of schizont. The small uninucleated parasite thus formed is called *merozoite* or *schizozoite*. The portion of cytoplasm in the centre of schizont is termed residual cytoplasm, Merozoites remain attached in a radiating manner. Each merozoite measures about 15μ in breadth. By this time the host cell becomes very week because its contents have been used up by the parasite. The merozoites break from residual cytoplasm of schizont and finally escape into the lumen of gut by the rupture of host cell. The merozoites resemble sporozoites in form, structure and movement but differ from them in being slightly shorter and in having a karysome in the nucleus. The cytoplasm is differentiated into an anterior vacuolated and a posterior denser area instead of being more uniformly vacuolated. The merozoites penetrate fresh epithelial cells, grow into trophozoite and repeat schizogony. The process of multiplication is repeated several times till many parasites are formed. Parasites face difficulty in procuring proper nourishment. Asexual multiplication remains in progress for about 5 days during which 4-5 generations of merozoites are produced and then sexual phase is starts.

2. Gamogony.

For sexual phase, some of the merozoites instead of growing into schozonts develop into two distinct forms known as *male* or *microgametocytes* and *female* or *macrogametocytes*. The male gametocytes are spherical and have clear cytoplasm while the female gametocytes are bean-shaped and have their cytoplasm crowded with reserve food material. Male gametocyte divide to form numerous narrow, biflagellate *microgametes*. Each measures 6-7μ in length and about 1μ in width. The flagella are as long or longer than the body.; It consists almost entirely of nucleus, the cytoplasm being represented by the flagella abone. One of the flagellum is free, while the other runs the hind end.

The macrogametes produces a single *macrogamete*. Its karyosome break up into a number of small pieces, which pass from the nucleus into the cytoplasm and finally leave the gametocyte altogether. With this the macorgametocyte changes into female gamete or macrogamete or *hologamete*. Now the hologamete frees itself from the epithelial cell and

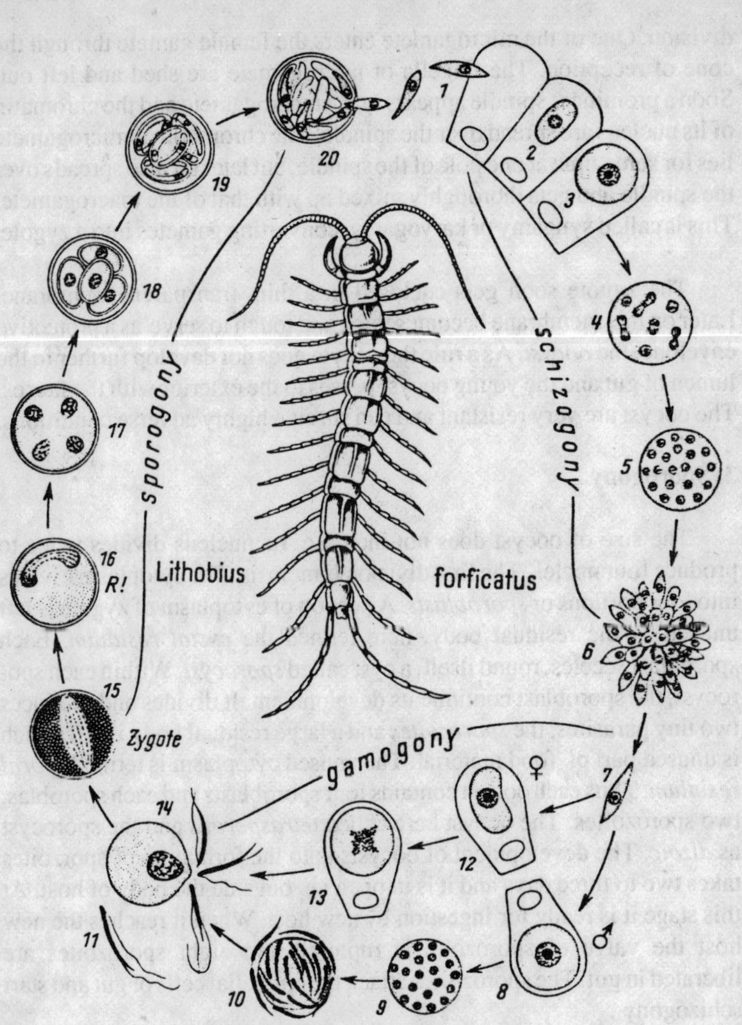

Fig. 11.1 Life-cycle of *Eimeria schubergi*

develops a small projection, the *cone of reception*, on one side. The nucleus also moves towards the cone. This makes the macrogamete ready for fertilization.

The male gamete, also liberated in the lumen of gut. Some male gametes are attracted towards the female gamete only after maturation division. One of the microgamete enters the female gamete through the cone of reception. The flagella of microgamete are shed and left out. Soon a prominent spindle appears in the macrogamete and the chromatin of its nucleus are spread over the spindle. The chromatin of microgamete lies for some time at one pole of the spindle, but later on also spreads over the spindle and gets thoroughly mixed up with that of the macrogamete. This is called syngamy or karyogamy, converting gametes into a zygote.

The zygote soon gets enclosed in a thin, transparent membrane. Later on this membrane becomes thick and tough to serve as a protective envelope, the *oocyst*. As a rule the oocyte does not develop further in the lumen of gut and the young oocyst passes to the exterior with the faeces. The oocyst are very resistant and can survive highly adverse conditions.

3. Sporogony.

The size of oocyst does not increase. Its nucleus divides twice to produce four nuclei. The first division is meiosis. The cytoplasm divides into four portions or *sporoblasts*. A portion of cytoplasm of zygote is left unused as the residual body, here termed the *cystal residium*. Each sporoblast secetes, round itself, a cyst called *sporocyst*. Within each sporocyst, the sporoblast continue its development. It divides and produces two tiny parasites, the *sporozoites* and a large residual body is left which is unused part of food material. The unused cytoplasm is termed *sporal residium*. Thus each oocyst contains four sporoblasts and each sporoblas, two sporozoites. The oocyst here called *tetrasporous* and the sporocyst as *dizoic*. The development of oocysts upto the formation of sporzoites takes two to three days and it is in open i.e. out side the body of host. At this stage it is ready for ingestion by new host. When it reaches the new host the valve of sporozoite is ruptured and eight sporozoites are liberated in gut. The sporozoites attack the epithelial cells of gut and start schizogony .

Alternation of Generation

The life cycle of *Eimeria* shows alternation of generations. The schizogony and sporogony constitute the asexual generations as in these reproduction occurs by asexual method of multiple fission. The gamogony represent the sexual generation as it involves the formation of male and female gametes and their fusion results in the formation of zygote.

Schizogony is repeated several times but it does not continue indefinitely. The sexual generation is followed by asexual generation. The phenomenon in which the sexual and asexual generation are followed by each other regularly is called alternation of generations.

Effect on host.

Eimeria causes an extensive demage to the epithelial cells of the intestine and attached cells. However, these damage is *ephemeral* as the epithelial calls are soon regenerated. Some species of *Eimeria* are very harmful to the host and they render an economic loss the the man. These species are *E. tenella* of chick, *E. stiedae, E. magna, E. sciurorum, E. perforans* of rabbits; *E. debliecki* of pigs and *E. bovis* of cattles.

A disease is called *coccidiosis,* in which the large scale destruction of intestinal epithelium is accompanied by haemorrhage. The symptoms of coccidiosis are loss of appetite, anemia, diarrhoea, bloody stool, weakness and sometimes fever. It often prooves fatal.

12

GREGARINA

Phylum	-	Protozoa
Subphylum	-	Sporozoa
Class	-	Telosporea
Subclass	-	Gregarinia
Genus	-	*Gregarina*
Species	-	*blattarum*

Habits and habitat

G. blattarum, is an intestinal parasite of cockroaches, such as *Blatta orientalis, Periplaneta americana,* and *Blatta germanica.*

Living specimens may be obtained from roaches by removing the intestine to a microscopic slide and splitting it lengthwise. The contents of the tract including parasites, may be teased free into a drop of either distilled water or Ringer's solution.

Morphology & life-cycle

The first stage to be seen wiil probably the large, motile, bipartite *sporadin* or *sporont* or adult *trophozoite*. The trophozoite is composed of an anterior *protomerite* measuring 100-120μ and a posterior *deutomerite,* measuring 130-450μ. The trophozoite stage is found in lumen is dark and denuse in appearance, and quite often a nucleus is present in the deutomerite. The nucleus is spherical. The pellicle is deeply folded in parallel groves and ridges. Myonemes are present below ectoplasm. The endoplasm contains paraglycogen and other inclusions. Locomotion of a

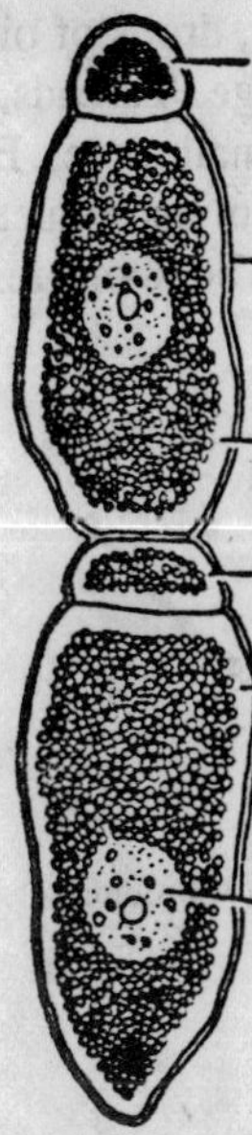

Fig. 12.1. *Gregarina blattarum*

gliding nature on the path of secreted mucous is apparently due to the contraction of longitudinal myonemes. This movement is called *gregarine movements*. It feeds by absorbing carbohydrates, peptones etc.

Often two sporadines will be seen in linear association with the protomerite of the second individual. The anterior one is called a *primite* and the following is called *satellite*. Rarely, one primite may have two satellites of the same or unequal size.

Normally in the host, two associated sporadins will encyst together in a non-sexual union called *syzygy*. Such an encystment may be induced artificially by placing the associated sporadins in egg albumen. The cyst measures 450-900μ. The encysted individuals are called *gamonts*.

The cysts are eleminated from the alimentary canal of the host. If they remains moist for several days. *gametes* develop from the periphery of the gamonts. The gamonts fuse and lose their identity, remaining as a residual mass. The gametes unite in pairs, and spores and formed. After these activities, the cyst produces, on the average, 13 sporoducts which evert through the relations cyst membrance. This, eversion may be due to an enzymic action. When everted, drops of oil first empty from the ducts. This oil lubricates the passage. Threads, measuring 87mm in length and containing about 10,000 small spores. Each spore measures 4-8μ in length. They are liberated from sporoducts. Within the spore a group of 8 sporozoites is formed. The spore resembles the nevicella hence called *pseudonevicella.*

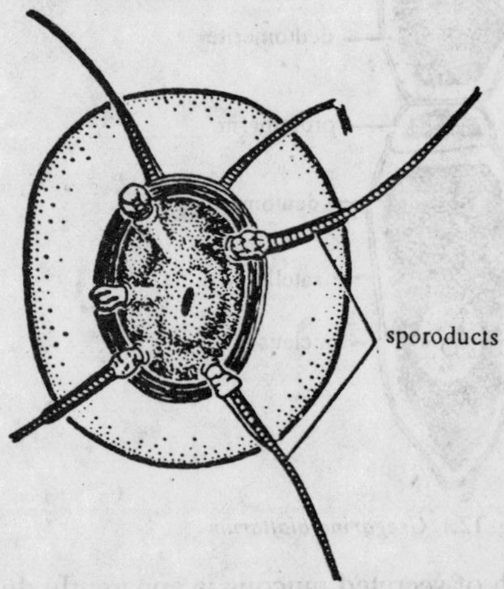

Fig. 12.2 *Gregarina* a cyst with developing sporoducts

A sporozoite is an infective stage. Upon entering the alimentary canal of cockroach the sporozoites are liberated from pseudonevicella.

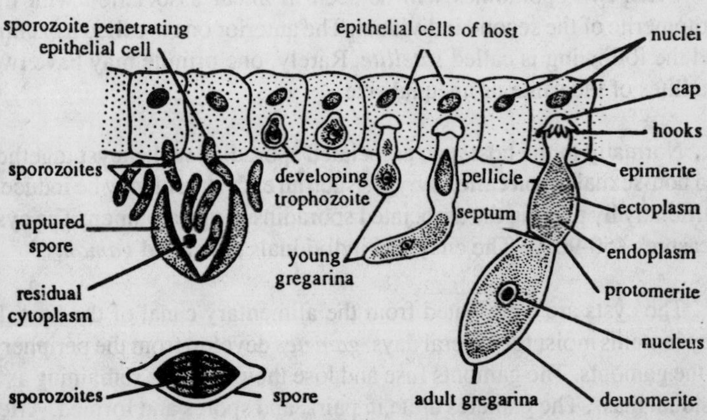

Fig. 12.3 *Gregarina blattarum*, life-cycle.

They penetrate the epithelial cells of gut. Intracellulary it is converted into a *tripartite* young trophozoite: The tripartite trophozoite consisting of an anterior *epimerite*, a middle *protomerite* and a posterior *deutomerite*. The epimerite remains intracellular as a hold fast, presumably ingesting food. The lumen forms with epimerites attaches are *cephalonts*. Cephalonts may be free moving in alimentary tract, but eventually the epimerite is lost and the resulting forms is the already mentioned bipartite sproadin.

13

OPALINA

Opalina are supposed to be intermediate between Ciliata and Mastigophora. They are never amoeboid, do not have a gullet and possess very large number of small equal-sized cilia or flagella. They are endoparasite.

SYSTEMATIC POSITION

Phylum	-	Protozoa
Subphylum	-	Sarcomastigophora
Super-class	-	Opalinata
Order	-	Opalinida
Genus	-	*Opalina*
Species	-	*ranarum*

Habits and habitat.

Opalina ranarum is an endoparsite found in the rectum of frogs and toads.

Morphology.

Its body is much flatenned like a disc and oval in outline. It is almost colourless and transparent measuring about 1mm in size. The body is covered over by tough but thin pellicle. The pellicle interm is covered over by innumerable uniform small sized cilia arranged in rows over the entire surface. He dorsal rows follow sigmoid paths while ventral rows are straight. About half of the rows pass completely around the body. Between the rows of cilia the pellicle is thrown into narrow longitudinal folds, each supported by a simple longitudional ribbon of 12-24 microtubules (*Noirot-Timothee*, 1959 and *Wessenberg* (1966).

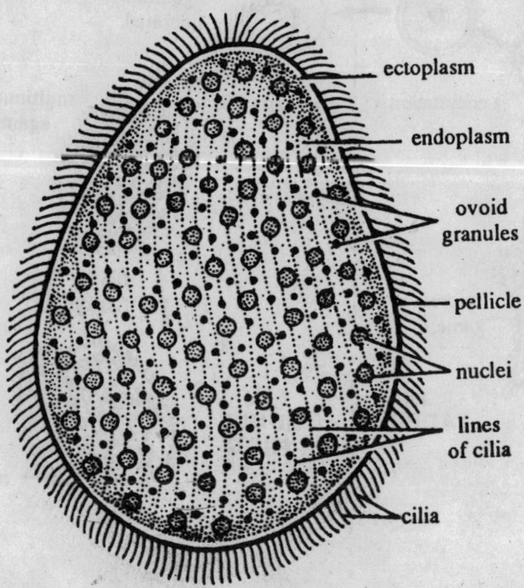

Fig. 13.1 *Opalina*

Mouth, gullet, cytopyge, food vacuoles and contractile vacuoles are absent. The cytoplasm is differentiated into outer clear ectoplasm and inner granular endoplasm. The number of nuclei is very large. These are small, hyaline, spherical and similar (*Monomorphic)*. They are evenly distributed in the endoplasm. Each nucleus has both *trophochromatin* and *idiochromatin*. Endoplasm also contains numerous spherules of chromatin-like material but of unknown function. These are known as *endospheres*.

Physiology

Nutrition. The mode of nutrition is *saprozoic*. It absorbs the liquid food contents of host's intestine through the general body surface by *pincytosis*.

Life - cycle.

Opalina reproduces by binary fission throughout the year. It is most active during the breeding season of its host i.e. frog or toad. The line of cleavage passes between the ciliated lines. This type of fission is known as *plasmotomy*. But in spring, unusually rapid division produces smaller

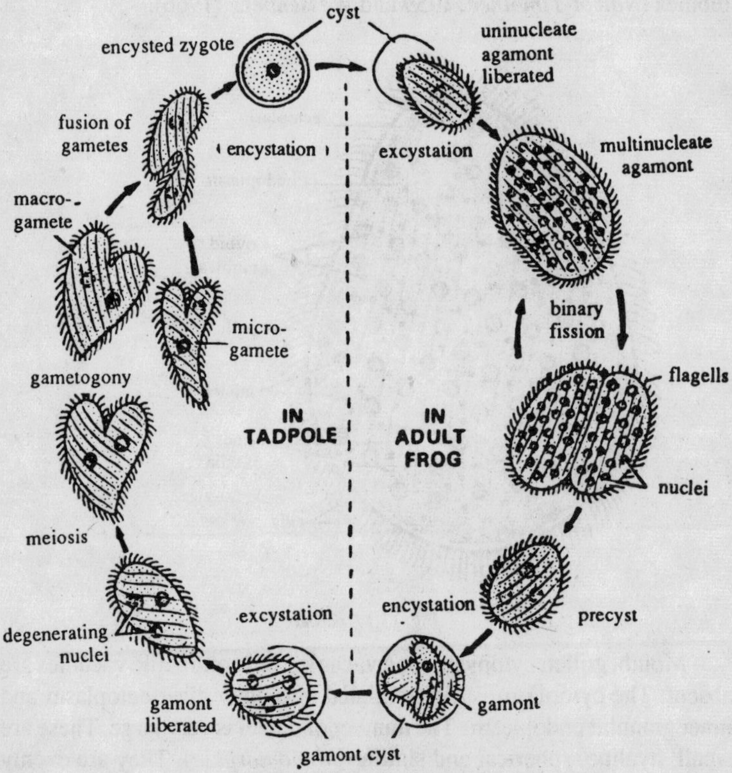

Fig. 13.2 Life cycle of *opalina*

individuals, each with few nuclei (4-6) and few rows of cilia. Each daughter secretes a cyst wall around itself and passes out with the faecal matter of host. The cysts so formed are transparent and spherical to ovoid and usually left over water-plants. Cysts are swallowed by young tadpole as they scrap water-plants. They (cysts) hatch in the rectum as gametocytes. The gametocytes divides by a series of longituditinal fission to form uninucleate gametes. The gametes are anisogamous as they differentiated into microgametes and macrogametes. Gametes fuse and the zygote is formed. The zygotes may encyst for a while as *zygocysts*. With growth and rapid nuclear division each zygote grows into a young *Opalina* called proto-opalina stage.

SYSTEMATIC POSITION OF *OPALINA*

Opalina was once classified as a member of ciliata. They were regarded as primitive ciliates and called *Protociliata* by *Metcalf* (1918). It was based on the basis of numerous nuclei of *opalina* which considered to be macro-and micro-nuclei of Euciliata. In *Opalina* the nuclei are monomosphic i.e. they can not be differentiated into micro and micronuclei (dimorphic). In true ciliate, the nuclei are dimorphic and sexual reproduction is by conjugation, there being neither gametes nor syngamy. In *Opalina* as earlier said the nuclei are monomorphic. The sexual reproduction by fusion of anisogametes (micro-and macro-gametes) and never by conjugation.

Except cilia, most of the characters of *Opalina* are those of flagellates. The behaviour of infraciliature during division is of the flagellate pattern i.e. the cleavage plane is parallel to the kinetia (longitudinal fission). This type of division is called *symmetrigenic*. In *Opalina* the centrioles are absent. They are also absent from some mastigophores like *Euglena*.

Thus *Opalina* shows greater affinities with the flagellates rather than ciliates. Some earlier worker pleased them in class Mastigophora.

The position of *Opalina* is still uncertain. Some workers include them in ciliate and some in Mastigophora. Some authors have put them in a separate class Opalinata, equal in rank with Mastigophora and Sarcodian. Probably they are highly simplified metazoans and in past related to the Turbellaria.

14

PARAMECIUM

Paramecium is the widely studied type of the class-Ciliata. The members of Class Ciliata are much more developed then the other protozoans. Inspite of their being acellular special organelles are found in them for ingestion, excretion etc. Thus division of labour is clrarly seen in ciliates.

SYSTEMATIC POSITION

Phylum	-	Protozoa
Subphylum	-	Ciliophora
Class	-	Ciliata
Sub-class	-	Holotrichia
Order	-	Hymenostomatida.
Family	-	Paramecidae
Genus	-	*Paramecium.*

Habits and Habitat

Paramecium is cosmopolitan in distribution. It is found in fresh water lakes, ponds, puddles, sewage pipes and rice fields. It is abundent in stagnant water containing decaying organic matter. It is omnivorus in diet and reproduces in a variety of ways, which are asexual and sexual both.

There are about 10 species of *Parasmecium*. They differ in shape, size, number of micronuclei etc. The common specis are *Paramecium caudatium P. aurelia, P. bursaria* and *P. multimicronucleatum culture.*

Fore general laboratory purpose, *P. caudatum* will grow satisfactorily on a boiled-hay infusion medium to which wheat or rice grains have been added. It has also been maintained on several complex sterile media and simplest of which the yeast juice. *Paramecia* are easily grown in chalky medium (Nacl-80mg., $NaHCO_3$ - 4mg, $Cacl_2$ - 4 mg, KCL - 4 mg and CaH_4 (PO_4) H_2O 1.6mg. all are dissolved in one litre of distilled water). Add 6 - 8 drops of skimmed milk weekly. The jar is kept away from direct light.

Other less uniform but very satisfactor method of growing paramecia include a monofloral suspension of bacterium, *Pseudomonas ovalis*, and a suspension of pure, living yeart.

MOROPHOLOGY

External Feature

Shape and Size. The first description of *Paramecium* was given by *Christian Huygens* (1678). The term *Paramecium* was coined for the first time by *John Hill* (1752) from the Greek word *paramekos* meaning oblong. *Paramecium,* in general appearence appears like the sole of the slipper, hence it is popularly known as *"Slipper - animalcule."* One species was named *P. caudatum* (*Joblot*) because of its resemblance to an imprint of human food.

They vary in size. The largest species *P. caudatum* measures 180-300μ in lenght; *P. aurelia* is 170-250μ, and *P. trichium* measures only 62μ in length.

Colour. Some species are somewhat translucent and colourless, others are vary from light gray, white to pale yellow. *P. burseria* is green because of the presence of *Zoochlorellae* in the endoplasm.

Structure. The shape of the body is definet because the body of ciliates is typically covered by a complex, living *pellicle*, usually containing a number of different organelles. Under light microscope the pellicle appears homogenous. Under electron-microscope the pellicle has inflated kidney - shaped *alveoli*. The inflated condition and the shape of alveoli produce a polygonal space about the one or two cilia that arise between them. Alternating the alveoli are bottle shaped organelles, the *trichocysts*, which form a second, deeper compact layer of the pellicular

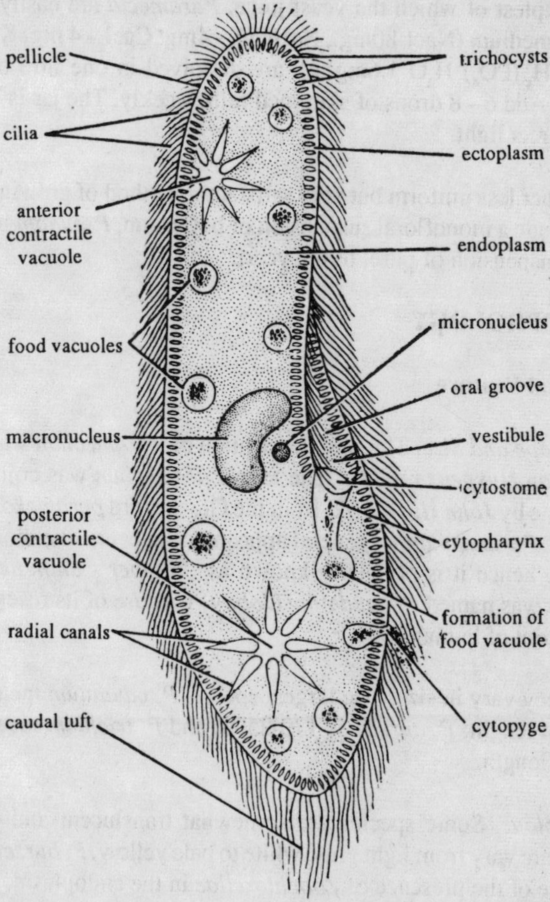

Fig. 14.1 *Paramecium Caudatum*

system. *Ehert* and *Powers* (1959) showed that pellicle is made up of three membranes, the outer most being the plasma membrane which covers the whole organism including the cilia. Beneath it, in most place, there are two further membrance, the *outer* and *inner membranes.* the outer and inner membranes bounding a flattened alveolus would thus form a middle and inner membrane of the pellicle.

The alveoli contribute to the stability of the pellicle and perhaps limit the permeability of the cell-surface (*Pitelka,* 1970).

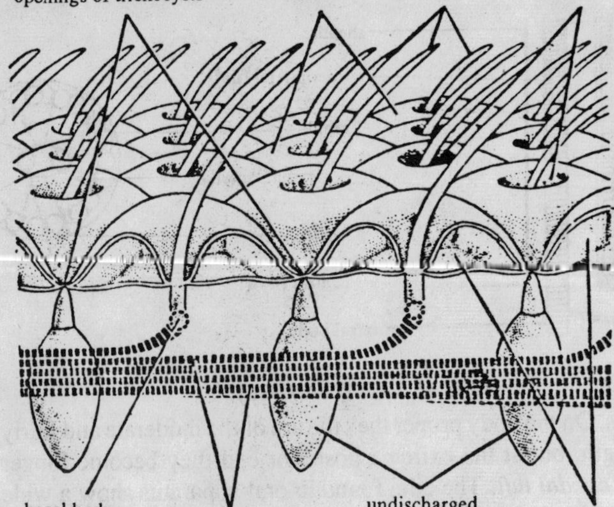

Fig. 14.2 Pellicular system in *Paramecium*

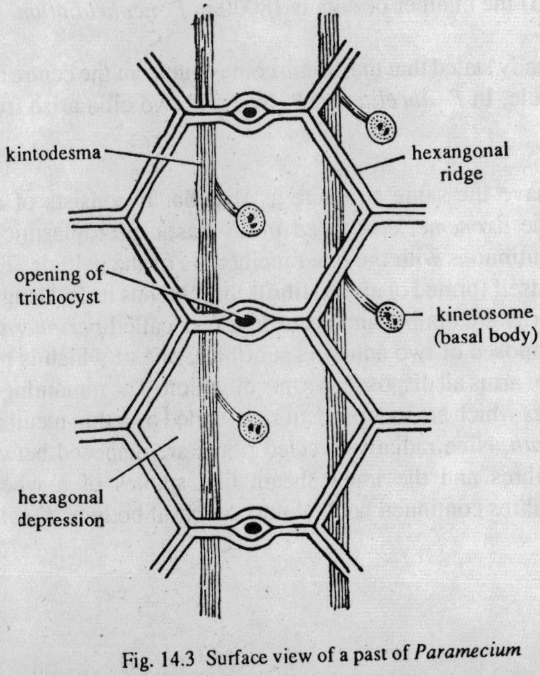

Fig. 14.3 Surface view of a past of *Paramecium*

Ciliature. The entire surface of *Paramecium* is covered by innurmerable fine structures called *cilia*. They are arranged on the body in

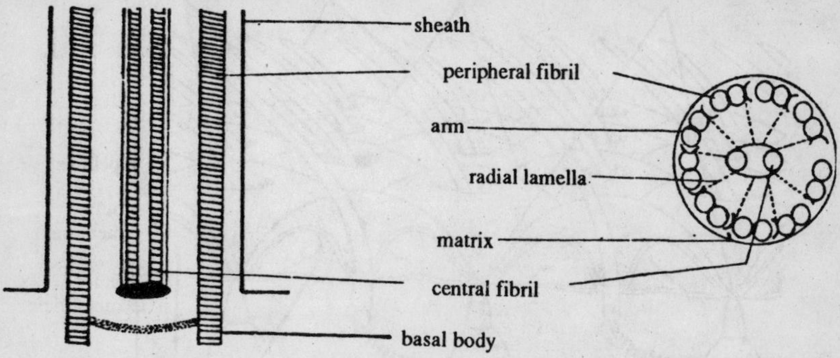

Fig. 14.4 L.S.and T.S. of Cilium

parallel rows. On the body proper the cilia are of the moderate and fairly uniform length, but at the extreme posterior end they become longer forming the *caudal tuft*. The cilia found in oral apparatus show a wide range of variation in lenght. Each cilium measure 10-16μ in lenght and 0.1 - .2μ in width.The number of cilia varies from 350 (*P. trichrium) to 18000. In P. caudatum* their number is 10,000 - 14,000. According ot *Gelei* (1925) the number of cilia is 18000 in *P. nephridiatum.*

As already stated that the cilium comes out from the centre of alveoi of the pellicle. In *P. aurelia* and *P. bursaria* two cilia arise from each area.

Cilia have the same structure as flagella. It consists of an axial filament, the *axoneme*, embedded in an elastic protoplasmic *sheath*, which is continuous with the outer membrane of the pellicle. The axial filament is itself formed of eleven fibrils that extends its full length. Nine of these fibrils are double and occur in a ring called *peripheral fibres*. Each is composed of two adhering subfibres, one of which is provided with a pair of arms all disposed in same direction. The remaining two are *central fibres* which are independent surrounded by a thin membraneous *central sheath*. Nine radially directed fibres are disposed between the peripheal fibres and the inner sheath like spokes of a wheel. The peripheral fibres continued beolow into the basal body or *Kinetosome*.

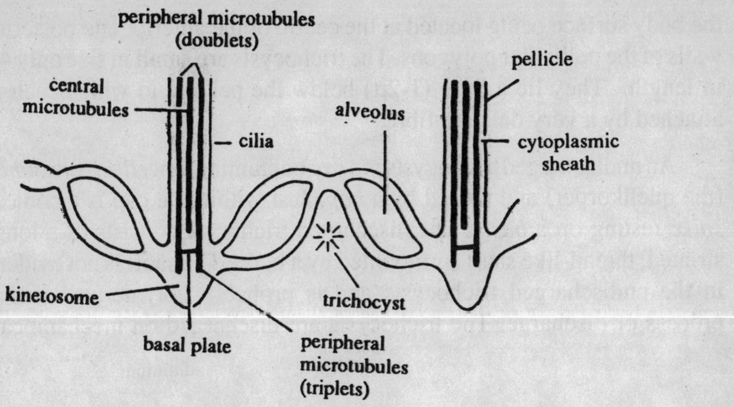

Fig. 14.5 V.S. through the surgace of Paramecium to show position of cilia, kinetosome and Trichocyst

Internal Features

Cytoplasm. The cytoplasm is differentiated into an outer ectoplasm and inner endoplasm.

1. *Ectoplasm.* It is clear and dense part of the cytoplasm which forms its outer layer below the pellicle. In the ectoplasm occurs following three important structures.

(i) *Trichocysts.* Trichocysts are peculiar, rod-like or oval organelles characteristic of many ciliates. They are oriented at right angles to

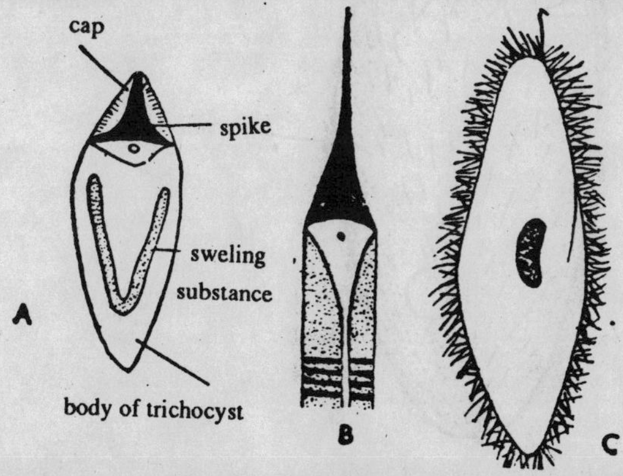

Fig. 14.6 Trichocyst. A= entire, B=tip of discharged Trichocyst, C=Trichocysts discharged after staining with per ink.

the body surface being located at the centre of the anterior and posterior walls of the pellicular polygons. The trichocysts are small in size only 4μ in length. They lie a little (1-2μ) below the pellicle to which each is attached by a very delicate fibre.

An undischarged trichocysts is a *sac* containing a *swelling substance* (the quellkorper) and closed by a *cap*. Just within the cap is a conical *spike* resting on a *basis*. The discharged trichocyst consists of a long, striated, thread-like shaft surmounted by a barb. The shaft is not evident in the undischarged trichocysts and is probably polymerized in the process of discharge. The trichocysts are discharged on mechanincal,

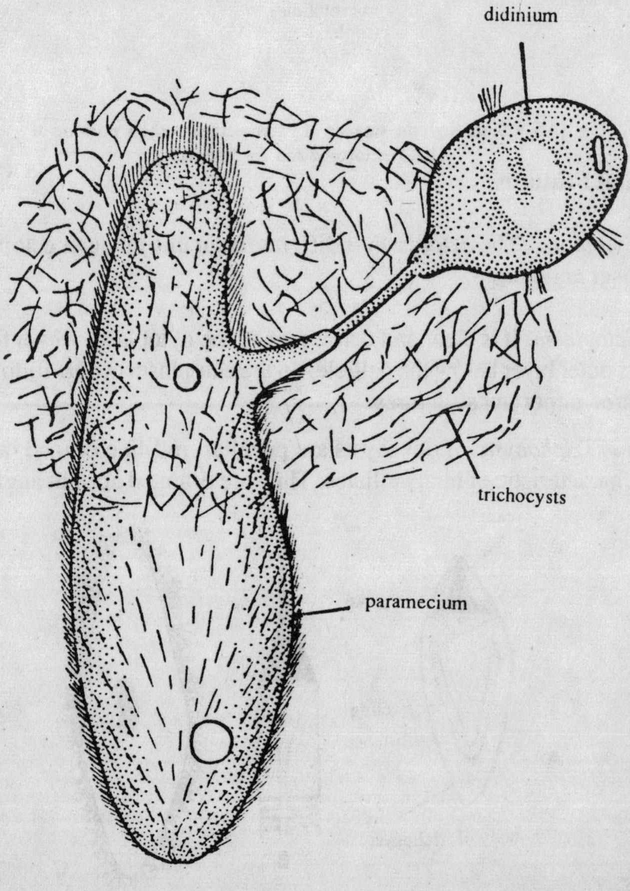

Fig. 14.7 Trychocysts thrown out to protect against *Didinium*.

chemical or electric stimuli. They are discharged very rapidly (within several milliseconds).

The trichocysts are *non-toxic* type in *Paramecium* thus they are not defensive organelle. Probably they may be used in anchoring the animal when feeding.

Area in the neighbouring of the nucleus is believed to the centre of origin of trichocysts. Recently, it has been established that trichocysts are produced by the kinetosomes. A kinetosome first ejects a *trichocysts granule* or *tricystosome* which foms a budding that develops into a trichocyst.

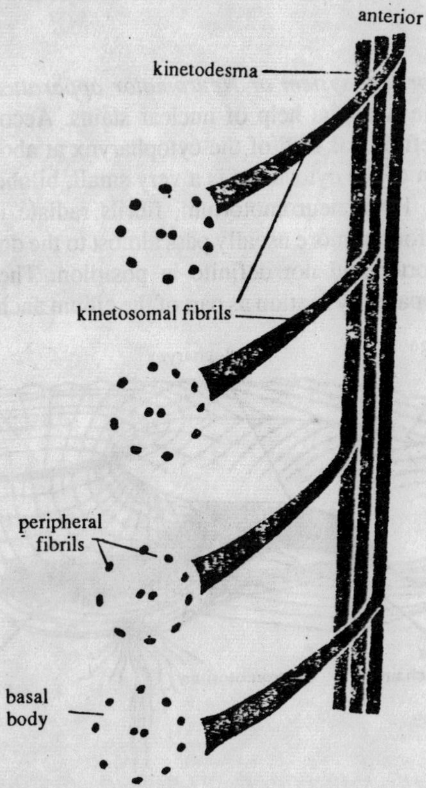

Fig. 14.8 A Ciliary kinety

(ii) *Intra-ciliary system.* As mentioned earlier, each cilium arises from a baseal body or kinetosome, located in tthe alveolar layer. The kinetosomes that form a particular longitudinal row are connected by means of fine. striated fibrils, called *kinetodesma*. The kinetosomes plus the fibrils of that row make up a *kinety*. The longitudinal bundle of fibrils run to the right side of the row of kinetosomes, and each kinetosome gives rise to one kinetodesoms (fibril), which joins the longitudinal bundle and extends anteriorly. A single kinetodesmos is tapered and extends for varying distance, as a part of bundle. At the kinetosome, the Kinetodesmos is connected to certain of the kinetosome triplets.

A kinety system is characteristic of all ciliates although these are variations in details of the pattern. The kineties with their cilia constitute the neurolocomotor organelle. The kineties control and coordinate the beating of cilia.

(iii) *The fibrillar system or Neuromotor apparatus.* The fibrillar system is visible with the help of nuclear stains. According to *Lund* (1933) on the left dorsal wall of the cytopharynx at about the level of posterior margin of the cytostome is a very small, bilobed mass called *neuromotorium.* From neuromotorium, fibrils radiate into the endoplasm. Of these four or more usually pass almost to the dorsal body wall but rest are shorter and not definite in positiion. Their fuuction is uncertain and apparently funtion as part of the cilium anchorage system.

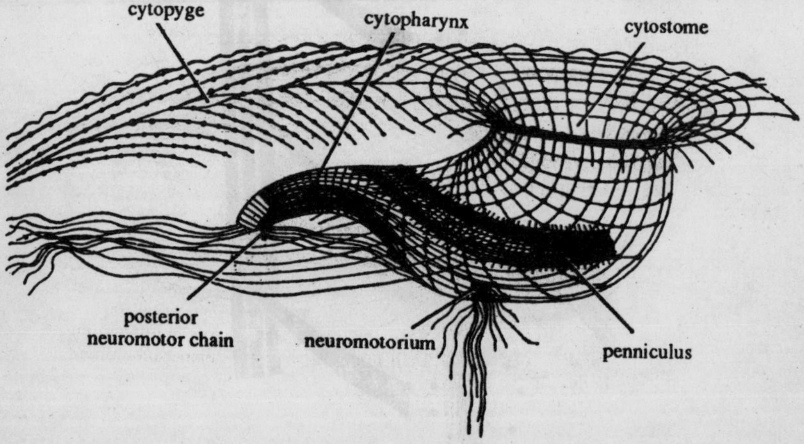

Fig. 14.9 Neuromotor system

But the electron microscopic studies reveral only the presence of the alveoli, kinetosomes and kinetodesmata and there is no evidence regarding the existence of this system.

2. Endoplasm.

The endoplasm of *Paramecium* is semi-fluid and less granular. It contains following structures.

(i) *Nucleus. Paramecium* is *hereokaryotic* having two types of nuclei. The smaller one or the *micronucleus* is lodged in a depression at one side of the larger nucleus called *meganucleus* or *macronucleus*. The mega nucleus is bean or kidney shaped body lying approximately in the centre. It is a compact structure, containing fine threads and tightly packed discrete chromatic granules of variable size. The meganucleus controls the vegetative acttivities of the animal.

The micronuclei vary in number in different species. These are smaller structure and may be compact *(P. caudatum)* or vesicular *(P. aurelia) P. cadatum* and *P.bursaria* eacyh have one micronucleus each. *P. aurelia* has two and *P. multimicronucleatum* has 3-9 micro-nuclei. The micronuclei are concerned with reproduction.

(ii) *Contractile Vacuoles.* In *Paramecium* there are two contractile vacuoles occupying somewhat fixed postions in the endoplasm. One vacuole is located at both the posterior and anterior end of the body. (In *P. multimicronucleatum* there are 2-7 contractile vacuoles). The anterior vacuole is sometimes called the *nuclear vacuole* being near the nucleus and the posterior one is called the *peristomial vacuole* being in vicinity of peristome.Each contractle vacuole is surrounded by 5-12 *radiating canals*. Each radiating canal consists of (a) a tubular terminal part (b) an ampulla and (c) an injector. The injector canal opens into the vacuole. Electron microcopy reveals that the outer or terminal part is surrounded by a network of fine coiled *nephridial tubules* (*Elliot* and *Bak,* 1964). The nephridial tubules collect water from the endoplasm and pass into the terminal part. This fluid is collected in the ampulla which becomes bulb-like when distended, but when the fluid passed out it is of same diameter as the terminal past. When the ampulla is fully distended (diastole) it collapses (systole) and the fluid is passed through the injector canal to form the contractile vacuole. Thus it is apperent that the systole of ampulla becomes the diastole of the vacuole. In *P. trichium* the contractile vacuole is without radial canals. The contractile vacuole opens outside through discharge canel. Bundles of fibres surround the amnpullae and the main vacuoles and concerned in the production or control of the pressure requircd to control the vacuole.

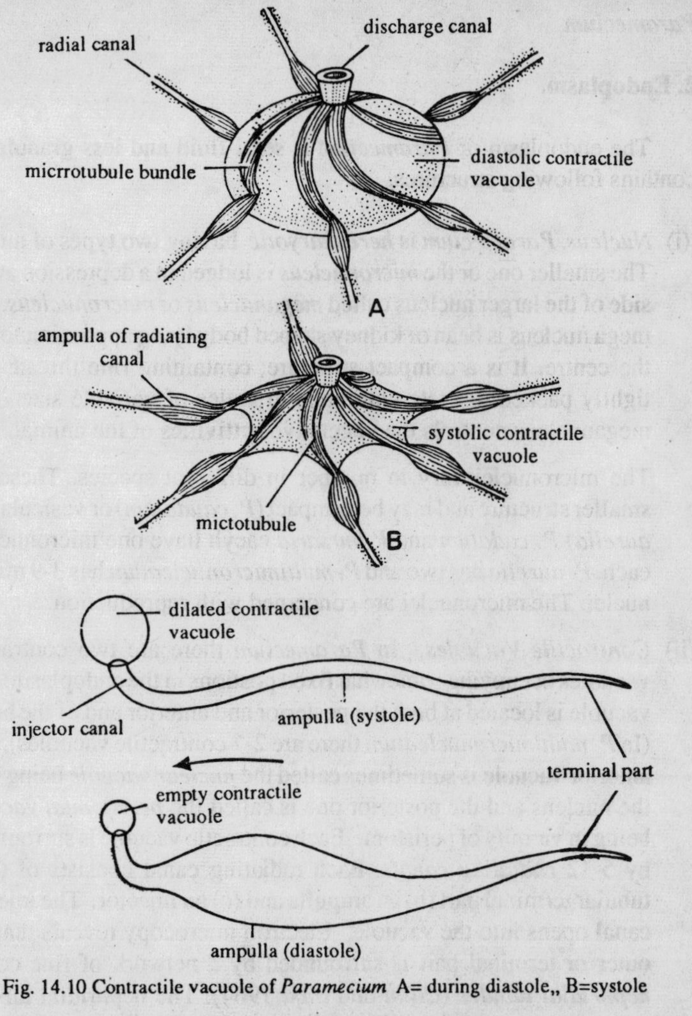

Fig. 14.10 Contractile vacuole of *Paramecium* A= during diastole,, B=systole

The contractile vacuoles pulsate at different rates depending on their position. The posterior vacuole pulsates faster than the anterior vacuole because of large amount of water being delivered into the posterior region by the cytopharynx.

(iii) *Food vacuoles.* These are roughly spherical, non contractile bodies varying in size and number lying in the endoplasm. They contain ingested food materials, mainly bacteria and a small amount of

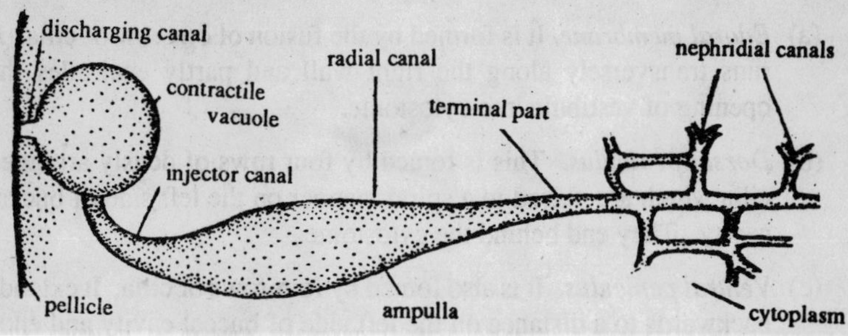

Fig. 14.11 Contractile vacuole of *Paramecium*

water. *Volkovsky* (1934) proposed the name *gastrioles* for these vacuoles.

Oral groove and associated structures. The oral groove is located on the ventral surface extending back-wards into a funnel shaped depression known as *vestibule* consisting of inveginated body pellicle. The vestibule leads into buccal cavity which directly leads into the *cytostome* or *mouth* which is fixed oral opening. The cytostome directly opens into the wide *cytopharynx* or *pharynx* which extends toward the centre of body. The cytopharynx forms a food vacuole at the posterior end.

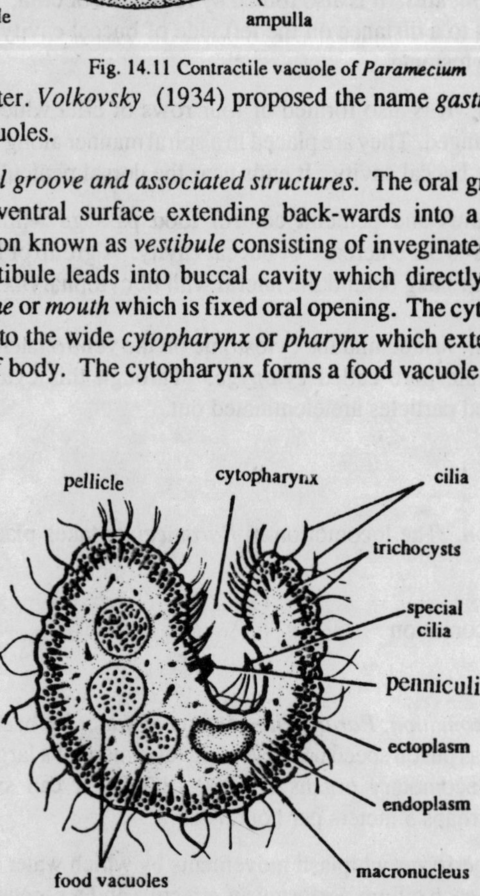

Fig. 14.12 T.S. of *Paramecium* through the region of cyto phorymx to show ciliation.

Large and hard cilia are present in these structures. These are specialized cilia and form four organelles which help in food ingestion.

(a) *Edoral membrane.* It is formed by the fusion of one row of cilia. It runs transversely along the right wall and partly encirciles the opening of vestibule into cytostome.

(b) *Dorsal peniculus.* This is fomed by four rows of densly arranged cilia which are placed in a spiral manner on the left side of buccal cavity. They end behind the cytostome.

(c) *Ventral peniculus.* It is also fomed by four rows of cilia. It extends backwards to a distance on the left side of buccal cavity and ends near the cytostome.

(d) *Quadrulus.* It is also formed of four rows of cilia which are less densty arranged. They are placed in a spiral manner along the dorsal wall of the buccal cavity. It ends near the dorsal peniculus.

The quadulus and peniculi control food passage while endoral membrane guards the enterence of buccal cavity. A group of long cilia, *post buccal cilia* hang behind the lateral wall of cytopharynx.

Cytophyge. Just behind the cytostome on the ventro lateral surface there is a minute pore called *cytopyge.* Through this cytopyge the undigented food particles are eleminated out.

Physiology

Locomotion. The locomotion in *Paramecium* takes place by two methods:

1. Ciliary locomotion
2. Metaboly.

1. *Ciliary locomotion. Paramecium* swims in water by beating of its cilia and has put on speed because of development of a large number of these locomotary organs (10,000-18,000). It can swim very slowly, perhaps 5 meters per hour.

A cilium performs whiplash movements by which water is propolled. In its beat each cilium performs an *effective* and a *recovery stroke.* During the effective stroke the cilium is outstratched and moves from a forward to a backward position. In the recovery stroke the cilium is bent over to the right against the body and is brought back to the forward position in a counterclockwise movement. The recovery position offers less water resistance and is somewhat analagous to feathering an oar. The direction of effective stroke is oblique to the long axis of the body. This causes the animal to swim in a spiral path and at the same time to rotate on its longitudinal axis. The ciliary beat can be reversed, and the animal can move backward. This backward movement is associated with the so-called avoiding reaction (behaviour). The direction and intensity of beat is controlled by levels of Ca^{++} and K^{+} ions (*Eckert*, 1972).

Fig. 14.13 A longitudinal row of cilia showing metachronal rhythm.

All the cilia of body can neither more simultaneously nor are they capable of independent movement. The cilia of a longitudinal row oscillate in such a way that the front cilium moves a little earlier than the one behind it. Thus the cilia more one after the other from front backwards. This movement of cilia is called *metachronal rhythm* . However, the cilia of a transverse row move simultaneously. This movement is called *Synchronous rhythm.* The movement of cilia is controlled by the infra ciliary system and ATP plays an important role (*serawin,* 1967).

Fig. 14.14 A=recovery stroke, B=effective stroke

2. *Metaboly*. The body of *Paramecium* possesses elasticity. It can pass through a passage narrower than its body, after which it regains its normal size. Such type of movement is called *metaboly*. This type of body movement is caused by the cytoplasm.

Nutrition. The free-living ciliates are almost entirely *holozic*. It feeds on bacteria, small protozoan, yeast, algae etc. but its favourit food

Fig. 14.15 Locomotion in *Paramecium*.

is *Bacterium coli*. It can devour 2-5 millians *Bacterium* in 24-hours. Thus *Paramecium* is a selective feeder, showing considerable choice of food, and often rejecting undigestible non-nutritive particles.

Feeding mechanism. It feeds while stationary or swimming slowly. In feeding, the cilia of the oral groove produce a current of water that

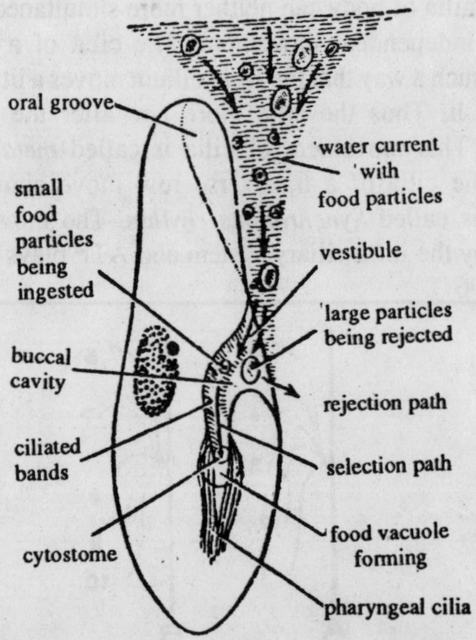

Fig. 14.16 *Paramecium* drawing water current to obtain food.

sweeps in an arclike manner down the side of body and over the oral region. The food particles are sucked with water current into the oral groove and then to the vestibule. Ciliary treat of vestibule direct the food particles into the buccal cavity. Some of these particles are discarded and left out. Quadrulus and peniculi regulate the passage of food into the cytopharynx. At the bottom of cytopharynx the food particles are gathered and form a rounded or ball-like mass, the *food vacuole*. It grows rapidly and when the vacuole attains a maximum size it is pinched off and takes its course in endoplasm. The food vacuole besides food particles also contains a little amount of water. The exect mechanism by which the food vacuole is liberated is not properly known probably post-oral fibres help in this process. The food vacuoles are formed in about 1-5 minutes depending upon the abundance of material and rate of feeding. The size of vacuoles depend upon the quantity and quality of food, viscocity of sorrunding medium and the physiological state of individual.

Digestion and assimilation. The food vacuole thus formed move in the body in a definit path with the streaming movement of endoplasm. The movement of food vacuoles is called *cyclosis*. Soon after it enters the endoplasm the food vacuole moves, posteriorly up to the end where it stops momantarily and them turns dorsally and turning anteriorly it moves forwards. On reaching the anterior end of the body it moves back again along the oral side casting off unabsorbed remains. The food vacuole remains in the body from one to three hours.

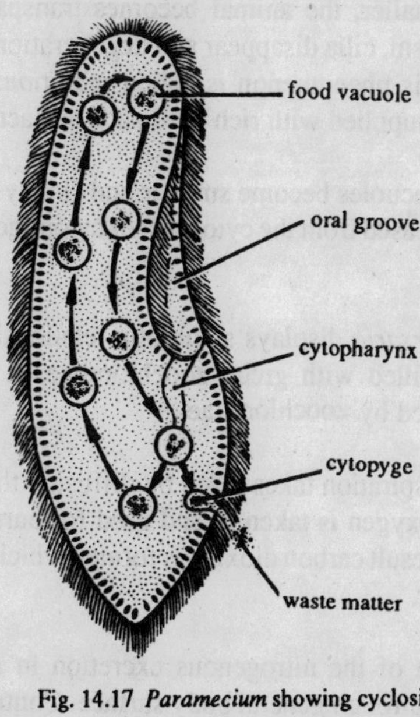

Fig. 14.17 *Paramecium* showing cyclosis

On leaving the cytopharynx the vacuole first decrease in size and finally increases. The decrease is due to osmotic concentration of endoplasm as it is higher than the vacuole, the water is lost from the vacuole and it becomes small.

The digestion of food takes place in the food vacuoles in which the medium is first acidic and later alkaline. As a result of digestion the proteins are hydrolysed into amino acids by *proteases*; Carbohydrates into glucose by *carbolydrases*; and probably fat is also digested. The process of cyclosis can be demonstrated experimentally in *Paramecium.* If milk, coloured with *Congo Red* stain, is given to *Paramecium,* the fat particles of milk first become red due to the acidic reaction of the enzymes in food vacuoles, then they gradually turn purple and finally blue due to alkaline medium in the food vacuole. Complete digestion takes place mainly during the later alkaline phase of food vacuole. The digested food materials are throughly distributed to all parts of body during cyclosis. Reserve food meterial is glycogen and fat-droplets which remain scattered throughout the endoplasm.

If *Paramecium* does not get food, it feeds upon the reserve food materials in the endoplasm. When even these are exbausted the macronucleus becomes smaller, the animal becomes transparent, vacuoles appear in the cytoplasm, cilia disappear and degeneration of specialized structure sets in. This phenomenon is called *inanition.* It restores its normal shape when supplied with rich food such as bacteria.

Egestion. The vacuoles become smaller and finally the undigested food particles are released from the cytoproct or cytopyge into surrounding medium.

Paramecium bursaria displays symbiotic relationship with algae. The endoplasm is filled with green zoochlorellae. It takes its food material manufactured by zoochlorellae.

Respiration. Respiration takes place by diffusion through the general body surface. Oxygen is taken in and used for burning of food to obtain energy. As a result carbon dioxide is formed which is diffused out in sorrounding water.

Excretion. Most of the nitrogenous excretion in *Paramecium* is carried by diffusion through general body surface. Contractile vacuoles also serve for excretion as urea has been detected in them. The chief nitrogenous product is ammonia which is diffused out.

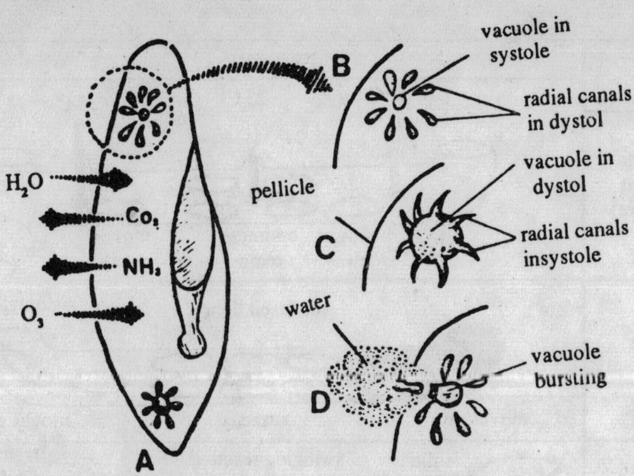

Fig. 14.18. ***Paramecium*** **diagrametic representation of respiration, excretion and osmoregulation.**

*Osmoregulation.*Two contractile vacuoles regulate the water content of the body. They occupy rather fixed position, one being anterior and the other posterior. Each vacuole is surrounded by 5-12 radiating canals. Both the vacuoles derive energy to work from the mitochondria crwoded around them.

Behaviour. Paramecium is extremely sensitive to different types of stimuli and shows its reaction to them. The response is positive when the animal moves towards a stimulus and negative when it moves away. To an adverse stimulus the animal continuous to give the avoiding reaction until it escapes. Various stimuli and their response given by *Paramecium* are:

1. *Reaction to mechanical stimuli (Thigmotaxis or Thigmotropism).* Response to contact is varied. If the anterior end lightly touches a fine point, the animal shows strong avoiding reaction. A show moving individual responds positively to contact with an object by coming to rest upon it.

2. *Reaction to temperature (Thermotaxis or Thermotropism).* The optimum range of temperature for paramecia is 24-28°c, but they can withstand considerably lower and higher temperature than the optimum range. Extreme cool and hot water is avoided.

3. *Reaction to chemical (Chemotaxis* or *Chemotropism).* It shows avoiding reaction to chemical like base. It a drop of weak salt

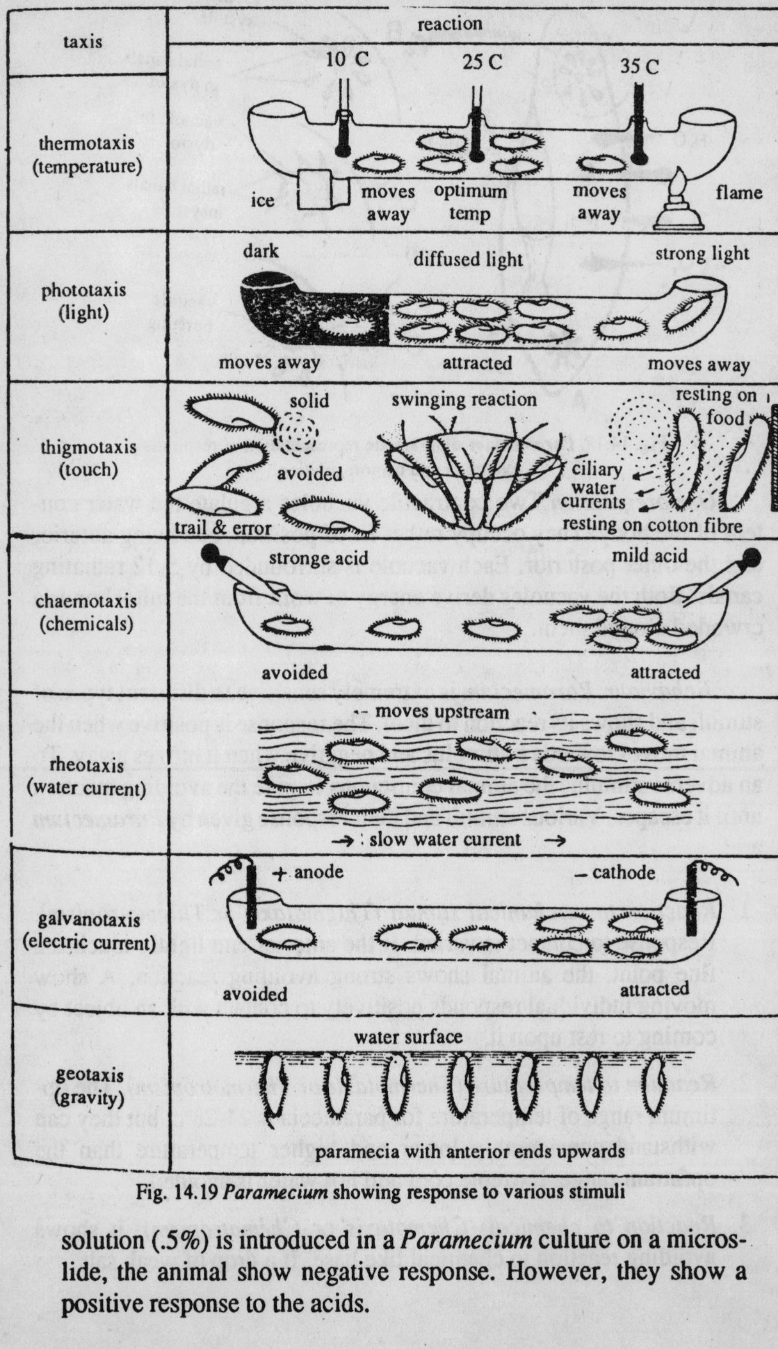

Fig. 14.19 *Paramecium* showing response to various stimuli

solution (.5%) is introduced in a *Paramecium* culture on a microslide, the animal show negative response. However, they show a positive response to the acids.

4. *Reaction to light (Prototaxis or Phototropism).* It shows negative response to bright light as well as dark. It responds to the diffused light. However, *P. bursaria* is positively phototactic.

5. *Reaction to gravity (Geotaxis or Geotropism) Paramecium* generally exhibit a negative respond to gravity as seen in a culture contained in a test tube where they gather close to the surface film with their anterior ends pointed upward. If paramecia are introduced in an inverted water filled'U' shaped tube stoppered at both the ends, they immediately move upward into the horizontal part of the tube.

6. *Reaction to water current (Rheotaxis or Rheotropism).* In a gentle water current the paramecia will mostly move with the flow with their anterior ends upstream thus showing positive response.

7. *Reaction to electric current (Galvenotaxis* or *Galvenotropism).* When two electrodes are placed opposite each other in a shallow dish containing paramecia and a constant current applied all the organism swim towards the cathode.

8. *Reaction to hydrostatic pressure* (*Barotaxis* or *Barotropism*). When the hydrostatic pressure is increased, the rate of movement is decreased.

Reproducttion Ciliates differ from almost all other organisms in possessing two distinct types of nuclei-a usually large macronucleus and one or more small micronucli. They are diploid with little RNA. The micronucleus is a store of genetic material is responsible for genetic exchange and nuclear organization, and also gives rise macronucleus.

In *Paramecium* two types of reproduction is reported.

I Asexual Reproduction

II Sexual Reproduction

I Asexual Reproduction

Asexual reproduction takes place during favourable conditions. Asexual reproduction is always by means of binary fission, which is typiclly transverse. More accurately, fission is described as being homotheogenic, with the division plane cutting across the kineties-the longitudinal rows of cilia.

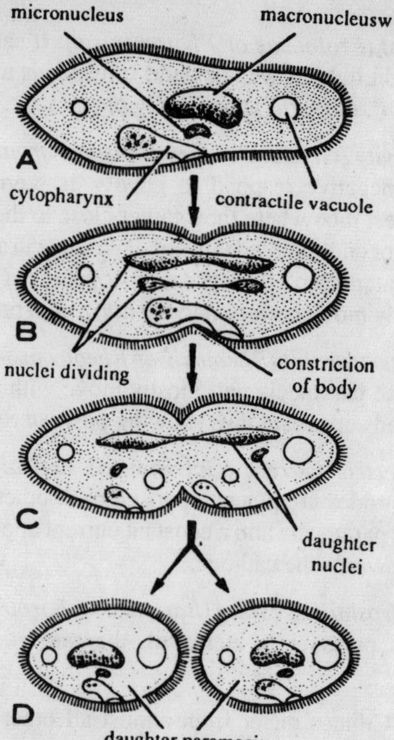

Fig. 14.20 *Paramecium Caudatum* showing binary fission

Before the beginning of fission, *Paramecium* stops feeding and the body elongates to become spindle-like. The oral groove disappear. The micronucleur divides by the complicated process of mitosis and the resulting two daughter micronuclei move towards opposite ends of the cell. All the stages of mitosis i.e. prophase, metaphase, anaphase and telophase take place inside the nuclear membrane. The number of chromosomes is 36-150 depending upon the species. When this is happening, the macronucleus also elongates, divide transversely by amitotic process in which the number of chromosomes may differs. A transverse constriction appears in the middloe. The constriction deepens gradually. In the mean time two oral grooves begin to form one in the anterior half and another in the posterior half. Two original contractile vacuoles of parent remain, one in each half of the dividing parent individual. Two contractile vacuoles are later formed. Two new buccal structures also appear. Now the constriction completely divides the parent into two daughter paramecia of equal size and with a complete set of cell-organelles. These grow to full size before dividing again.

The newly formed daughter paramecia neither resemble each other nor the typical parent, so that for a while they can be distinguished as the anterior daughter (*Proter*) and the posterior daughter (*Opisthe*). Soon they assume the typical form.

The entire process of binary fission is completed within 30-120 minutes depending upon the availability of food and temperature. In *P. caudatum* it may occur 2-3 times a day and in *P. aurelia* about 5 times a day, so that a large number of them can be obtained in a short time. All the individuals which are produced from a single parent, collectively, form a *clone* and all the members of a clone are genetically alike.

The binary fission cannot go on indefinitely. After several generations, the rate of fission declines.

II Sexual reproduction

Paramecium reproduces sexually in which nuclear material of two individuals is interchanged. This method is called *conjugation*, but it is simply a temporary union of two individuals of one and the same species.

Process of Conjugation. The process of conjugation is given by *Herting* and *Maupas* (1889) in *P. aurelia.* The basic pattern is similar with slight variations in different species of *Paramecium.* Here we will discuss the process of conjugation in *P. caudatum.*

At the beginning of conjugation two ciliates come together and become attached first at deciliated areas near the anterior end of their ventral surfaces, and later by attachments formed in the gullet region where membrane fusion leads to the formation of a number of pores providing cytoplasmic bridges between them. Now they are called conjugants. Adhesion probably takes place by the secretion of a sticky substance by the cilia. Attachment lasts for several hours.

The macronucleus of each conjugant starts disintegrating. It simply breaks up into pieces which are digested in cytoplasm. The micronucleus of each conjugant now grows in size and divides twice, the first division being meiotic result in the formation of 4 haploid nuclei.

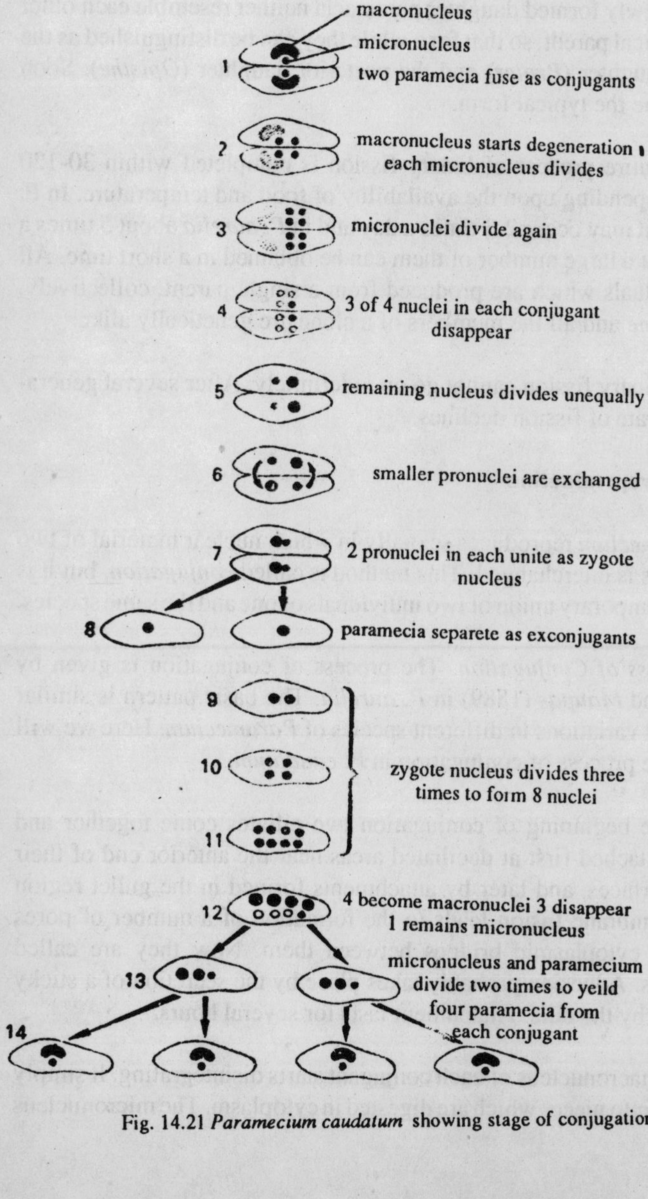

Fig. 14.21 *Paramecium caudatum* showing stage of conjugation

Three micronucli degenerate and disappear from cytoplasm of each conjugant leaving behind one nucleus. The remaning nucleus divides again unequally forming one bigger and other smaller nuclei. The smaller nucleus is active and called *migratory* or *male pronucleus*. The bigger one is inert and called *stationary* or *female pronucleus.*

Now the male pronucleus of one conjugant passes into the other conjugant through the protoplasmic bridge, similarly the male pronucleus of other conjugant passes into the first conjugant. Now the male and female pronuclei fuse results in the formation of *zygote nucleus*. The complete fusion of two nuclei from two different individuals forming a zygote nucleur is called *amphimixis*. The process cannot be called as fertilization because there is no fusion of gametes and their cytoplasm.

After about 12 hours or so the conjugants are now separated and are termed as *exconjugants* The zygote nucleus divides thrice in each exconjugant resulting 8 nuclei. The divisions are mitotic, Out of these 8 nuclei 4 assume large size and become macronuclei. Remaining 4 are micronuclei. Out of these 4 micronuclei 3 disappear leaving behind only 1 micronucleus and 4 macronuclei in the conjugant. The remaining micronucleus divides again forming two and at the same time the exconjugant divides by binary fission into two cells. Each cell having two macronuclei and one micronucleus. The micronucleus again divides into two micronuclei and at the same time the cell divides into two daughter paramecia by binary fission. Each daught having one macronucleus and one micronucleus.

Thus each conjugant produces four daughter individuals in the end of conjugation. The whole process of conjugation takes about 96 hours to complete.

Factors and Conditions of Conjugation

Conjugation is unique type of a sexual process in which following conditions and factors one necessary.

1. Conjugation takes place after a desirable number of asexual reproduction i.e. about 300 generations which have reproduced by binary fission or it occurs after a long continuance of binary fission so that the rejuvination may take place in the exhausted or perishing clone.

2. Conjugation occurs when changes take place in the physiological condition. The individuals are usually smaller in size (210μ in length) than the normal individuals.
3. It occurs during unfavourable conditions such as scarsity of food, light etc.
4. Agglutination favours conjugation. It is a sticky substance secreted by cilia of conjugants.
5. Conjugation starts early in the morning and is continued till the afternoon. It suggests that dark favour conjugation and light supress it.
6. Conjugation never takes place among the individuals of a done. It occurs between the individual of a *variety* or *syngen* belonging to two different *mating types*. Since each syngen has two mating types.

According to *Sonneborn* (1957) in *P. aurelia,* there are 14 syngens and 28 mating types. These are as follows.

Table. Showing the mating types of *P. aurelia* complex-

Syngen number	Mating type	Other mating types with which conjugation occurs (and %)	Group	+ or - type
1	I	II (95%), X (40%)	A	--
	II	I (95%), IX (40%), XIII (10%), V (1%)		+
2	III	IV (95%)	B	
	IV	III (90%)		
3	V	VI (95%), XVI (40%), II (1%)	A	--
	VI	V (95%)		+
4	VII	VIII (95%), XVI (95%)	B	--
	VIII	VII (95%), XV (60%)		+
5	IX	X (95%), II (40%)	A	-
	X	IX (95%), I (40 %		+
6	XI	XII (95%)	B	--
	XII	XI (95%)		--
7	XIII	XIV (95%), II (10%)	B	--
	XIV	XIII (95%)		+
8	XV	XVI (95%), VIII (60%)	B	--
	XVI	XV (95%), VIII (95%), V (40%)	+	

9	XVII	XVIII (95%)	A	.
	XVIII	XVII (95%)		.
10	XIX	XX (95%)	B	--
	XX	XIX (95%)	A	-
11	XXI	XXII (95%)	A	
	XXII	XXI (95%)	.	
12	XXIII	XXIV (95%)	B	--
	XXIV	XXIII (95%)		+
13	XXV	XXVI (60%)	C	.
	XXVI	XXV (60%)		.
14	X VII	XXVIII (95%)	A	.
	XXVIII	XXVII (95%)	.	

In *P. caudatum* there are 16 syngens and 32 meting types.

Significance of Conjugation

The significance of conjugation has been discussed in details, but it still remains uncertain. The following significances are attributed to conjugation.

1. It is the method of *rejuvination* by which the vigour of the species is maintained. If conjugation does not occur for a long time than paramecia become weak and ultimately die.
2. Conjugation is a temporary union in which there is no mixing of cytoplasm of the conjugants but a zygote nucleus is formed by the fusion of their nuclei, therefore, the zygote nucleus of each conjugant possesses the hereditary material of both the conjugatns i.e. new characters come in them.
3. In conjugation, the macronucleus is replaced by the material of zygote nucleus. The number of chromosomes is restored in macronucleus by conjugation due to which the vigour of daughter paramecia is renewed.
4. The function of micronucleus is to maintain balanced chromosomes and gene complex.

Aberrant behaviour in reproduction

Paramecium shows certain variations in its nuclear behaviour during fission and conjugation these deviations are *Autogamy. Endomixis, cytogamy* and *Hemixis.*

1. *Autogamy. Diller* (1934, 36) described a process of self-fertilization or *autogamy* occuring in a single individual of *P. aurelia.* About once a month *P. aurelia* undergoes this process.

During autogamy the macronucleus increases in size to form an irregular structure and later breaks up into small fragments are absorbed

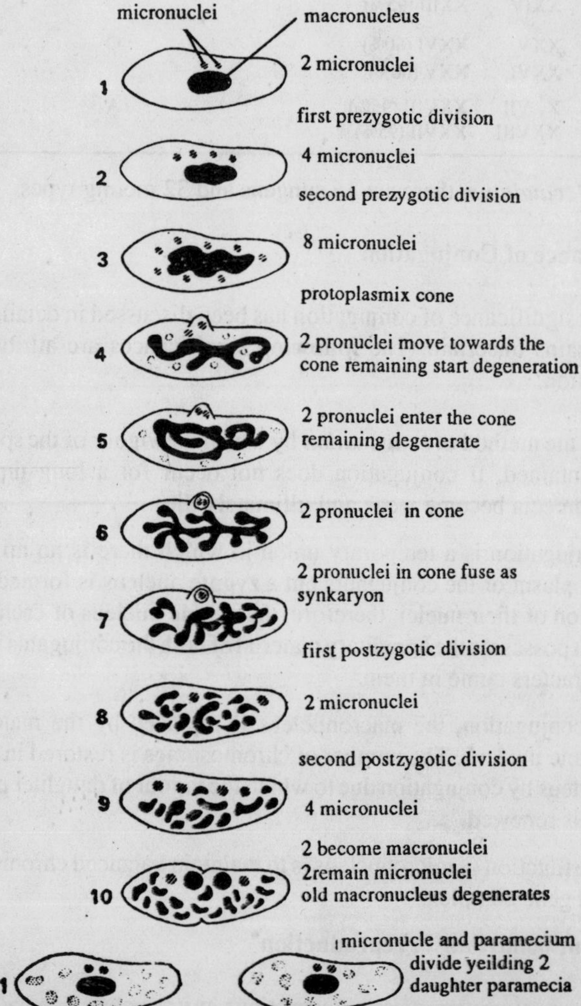

Fig. 14.22 *Paramecium aurelia* showing various stages of autogamy

in the cytoplasm. The two micronucli divide twice forming eight nuclei. As a rule the first division is meiotic thus all the eight nuclei are haploid. Out of these eight nuclei, seven disintigrate and only one nucleus is left. It again divides to form two nuclei. These nuclei are called gametic nuclei. By this time a peculiar cone-like bulge appears on the surface of autogamous animal, slightly to the right and anterior to the cytostome. Because of its position it is called the par *oral cone* or *protoplasmic cone*. The gametic nuclei are at first elongated and spindle like bodies. Often there is slight size variation between them so that they may be termed to as *male* (smaller) and *female* (larger). They approach each other and then move into the cone and fuse to form zygote nucleus. Later it comes out of cone and the cone disappears. The zygote nucleus divides twice forming four nuclei. Two of these nuclei increase in size to become macronuclei and remaining two as micronuclei. The cell body and micro-nuclei -divide by binary fission to form two daughter paramecia eacdh with a macronucleus and two micronuclei.

Significance of Antogamy

1. Rejuvination in the species is accomplished by autogamy.
2. A new macronucleus is formed from the meterial of micronuclei.
3. *Paramecium* regains its lost vigour by this process.

2. *Endomixis*. Endomixis was first observed by *Woodruff* and *Erdmann* (1914) in *P. aurelia*. In endomixis only the nuclear reorganization occurs after which the body of *Paramecium* divides into daughter paramecia. It occurs as follows:

When endomixis starts, the macronucleus disintigrates and disappears. The two micronuclei divide twice resulting in the forrmation of eight nuclei (all diploid). Out of these 8 nuclei six disintigrate and only two are left. Now the *Paramecium* divides by fission into two daughters each with one nucleus. The nucleus of each daughter divides twice forming four nuclei. Two nuclei enlarge to form macronuclei and remaining two micronuclei. The two micronuclei, again divide to form four micronuclei in the mean time the cell divides again by fission into two daughters each containing one macronucleus and two micronuclei. In the end four daughter paramecia are formed from one individual.

According to *Diller* (1936), *Sonneborn* (1947) and *Whichterman* (1953) the endomixis does not occur in *Paramecium.*

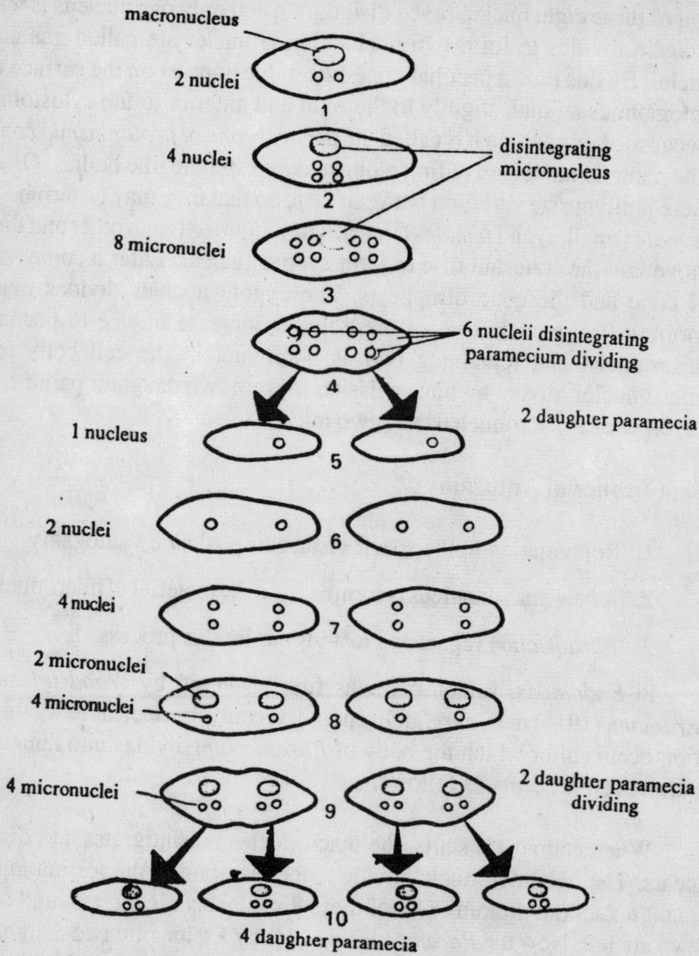

Fig. 14.23 *Paramecium aurelia* showing endomixis.

3. *Cytogamy. Wichterman* (1940) observed the process of cytogamy in *P. caudatum.* The process is something intermediate between conjugation and autogamy. Two individuals stick by their oral surfaces, the micronucleus divides twice (first division in meiotic) of these three disintegrate into the cytoplasm, the remaining one divides

once producing two gametic nuclei. Self-fertilization takes place. The conjugants now separate and called exconjugants. Exconjugants behave and multiply as in conjugation. The process is completed in about 13 hours.

The genetical cousequence of cytogamy is always homozygosity.

4. *Hemixis*. This method of nuclear reorganization was observed by *Hartman* in *P. aurelia* and was supported by *Diller* (1936). During hemixis the micronucli do not participate. The macronucleus throws out two or more equal or unequal fragments to be absorbed in the cytoplasm. The remainder of macronucleus and micronuclei behave normally at the cell division.

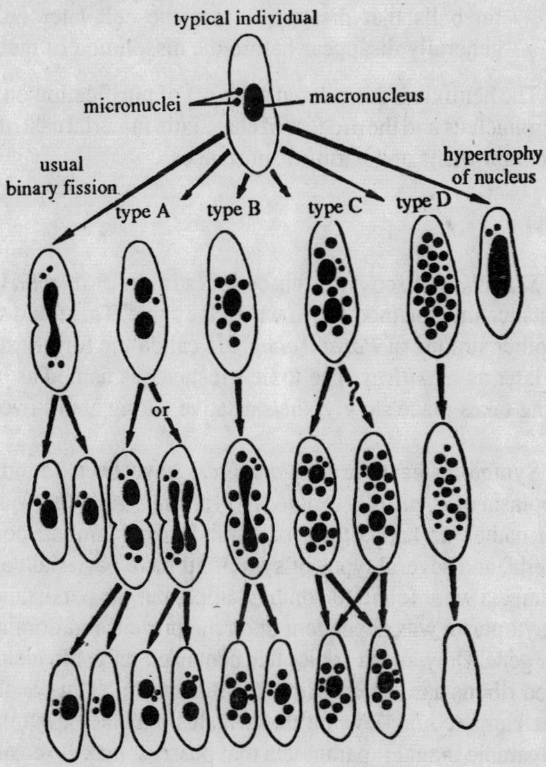

Fig. 14.24 *Paramecium aurelia* showing hexixis.

Diller classified hemixis into four types:

1. *Type A*.. It is the simplest form of hemixis characterized by a division of the macronucleus into two or more parts in the cytoplasm.
2. *Type B*. It is characterized by the extrusion of 1-20 or more chromatin balls from the macronucleus.
3. *Type C*. It is characterized by the simultaneous splitting of macronucleus into 2 or more major portions and extrusion of balls into cytoplasm.
4. *Type D*. It is considered to be the pathogenic conditions in which micronucleus undergoes complete fragmentation into chromatin balls that disappear from the cell later on. Micronuclei generally disappear before the dissolution of macronucleus.

The hemixis is considered as an act of purification on the part of the macronucleus and the process of chromatin material cast off by it are held in contain waste and harmful substances.

Killer

Sonneborn discovered that certain strais of *P. aurelia* liberate a toxic substance into the medium in which they live. This toxic substance kills the other straims of *Paramecium.* He called the former strain as *killers* and later as *sensitives.* The toxic substance is named as *parmecin.* The killing takes place slowly, the sensative taking about two days to die.

Symbiotic bacteria (*Caedibacter*) have been found living in the cytoplasm of *P. aurelia* (*Soldo,* 1974; *Preer,* 1981) and have been given such names as kappa, mu, pi, lambda etc.; lambda possess bacteria flagella, and several types of symbiontt have bacterial cell walls, killer paramecia were found to contain kappa, but the persistance of kappa in the cytoplasm was dependent upon the presence of dominant macronuclear gene. Only kappa bodies that contained refractile inclusions (tightly-coiled ribons associated with virus-like granules) were able to kill, but other kappa could develop these inclusions. Other symbionts act in a comparable manner; paramecia that possess mu kill sensitives at conjugation (they are mate killers), while pi may be inactive kappa, Paramecia that contain lambda produce large lethal granules which are released into the medium. It is assumed that these symbionts derive nutrients from the paramecia. A killer stock has an advantage over a sensitive stock in ecological competition (*Landis*, 1981). Lambda apparently synthesises folic acid, and paramecia with this symbiont have no external requirement for this vitamin.

15

VORTICELLA

Vorticilla has over 200 species described by *Kahl* (1953), most of them are fresh water. Some species are marine, some are epizoic and some are parasitic. The most common species are *V. nebulifera, V. campanula* and *V. microstoma.*

SYSTEMIC POSITION

Phylum	-	Protozoa
Sub-phylum	-	Ciliophora
Class	-	Ciliata
Sub-class	-	Peritrichia
Order	-	Peritrchida
Family	-	Vorticellidae
Genus	-	*Vorticella*

Habits and habitat

It is generally found in lakes, rivers, ponds and streams with aquatic vegetation. It is not free swimming like *Paramecium* but is a sedentary ciliate. Most of them are abundent in stagnant water containing decaying organic substances. *V. campanula* and *V. nebulifera* are found in clean water. It is found attached to the stems and leaves of the equatic plants, mollusc-shells, living oligochaetes and tadpoles.

Morphology

Shape, Size and Colour. Vorticella possesses a bell-shaped or asymmetrical body. It is, therefore, usually called a *bell-animalcule*. To its base is attached a long contractile stalk. Many individual variations

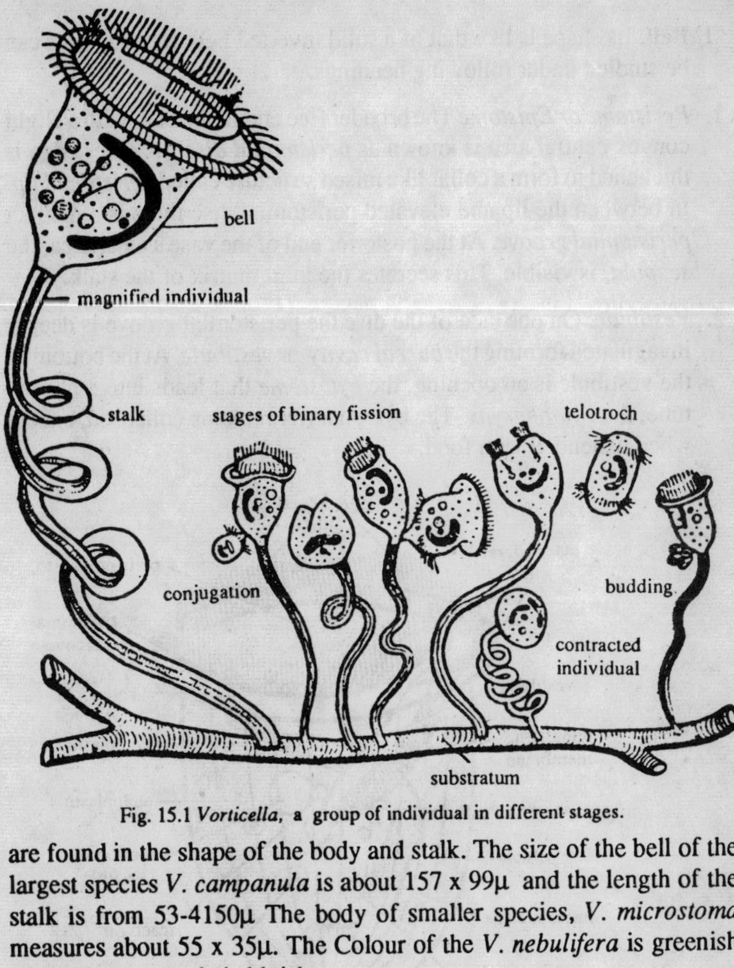

Fig. 15.1 *Vorticella*, a group of individual in different stages.

are found in the shape of the body and stalk. The size of the bell of the largest species *V. campanula* is about 157 x 99μ and the length of the stalk is from 53-4150μ The body of smaller species, *V. microstoma* measures about 55 x 35μ. The Colour of the *V. nebulifera* is greenish while *V. campanula* is bluish.

Structure. The body is divisible into two parts

I. Bell

II Stalk

I **Bell.** Its shape is like that of a solid inverted bell. Its structures can be studied under following headings:

1. *Peristome or Epistome* The broader free end of the bell, with a slight convex central area is known as *peristomial disc.* Its outer edge is thickened to form a collar like raised structure called *peristomal lip.* In between the lip and elevated peristomial disc there is a distinct *peristomial groove.* At the posterior end of the vase a clear area, the *scopula,* is visible. This secretes the inert matrix of the stalk.

2. *Vestibule.* On one side of the disc the peristomial groove is deeply invaginated forming the *buccal cavity* or *vestibule.* At the bottom of the vestibule is an opening, the *cytostome* that leads into a narrow tube, the *cytopharynx.* The cytopharynx remains collapsed, except when distended with food.

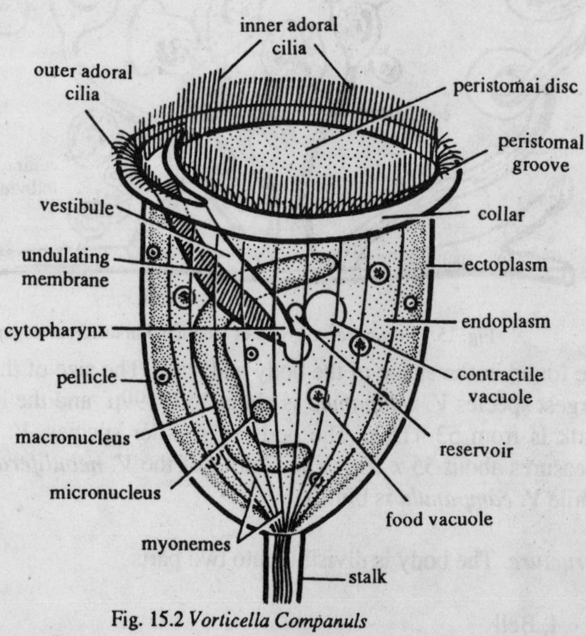

Fig. 15.2 *Vorticella Companuls*

3. *Cilia.* Ciliation in *Vorticella* has been stuided by *Noland* and *Finley* (1931) The cilia are restricted to a certain area that is known as the *adoral zone.* On the body there are no cilia. Three concentric rows of adoral cilia are found in the peristomial groove. Two rows of cilia

are found in the inner circle. These cilia remain straight-and always moving. Only one row of cilia is found in outer circle. The cilia of this row hang horizontally towards the outer side on the lip. They direct the food particles towards the vestibule. The aboral cilia are fused with each other at their base, but are free at their distal ends. Both the spirals of cilia are curved in such a way that they form about one and a half anticlockwise spiral on the peristomial disc. Both the spirals of cilia extend upto the vistibule.

The two inner rows of cilia extend together along the inner wall of the vestibule, where as the outer row of cilia is elongated and its cilia unite together to form an *undulating membrane* along the outer wall of vestibute. The cytopharynx is devoid of cilia.

4. *Pellicle.* The pellicle of the bell is uniformly thin and delicate (thicker than that of *Paramecium*). It is often marked with circular parallel striations and in some species of *Vorticella* it is tuberculate. it covers the body so closely that it becomes difficult to distinguish it. It is devoid of cilia.
5. *Cytoplasm.* The body cytoplasm is differentiated into an outer layer of clear and firm *ectoplasm* and an inner fluid *endoplasm.*

(a) *Ectoplasm.* The ectoplasm is modified to form a *myoneme system.* Innumerable myoneme fibres are found. They are arranged in

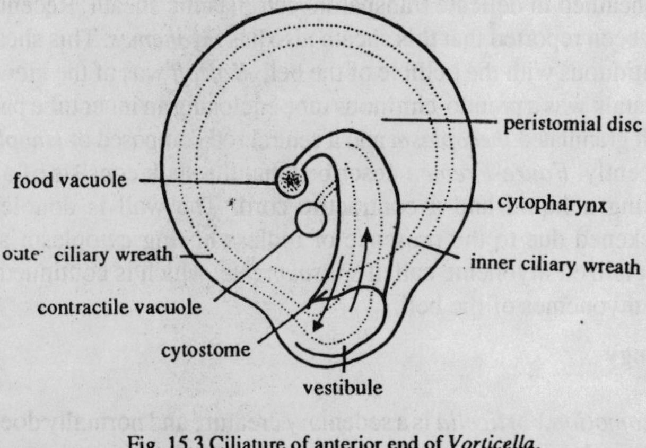

Fig. 15.3 Ciliature of anterior end of *Vorticella*.

both straight and oblique manner and extend across the length and breadth both. All the myonemes of the bell are converged at its base forming a *spasmoneme*. The longitudinally myonemes reduce the length of the body (*Schroder*, 1906). The oblique myonemes pull the disc of bell inwards. The circularly placed myonemes pull the peristomial edge over the cilia and the disc to close the mouth of bell. The parallel myonemes of the sides of the bell are more distinct in its basal part where they enter the stalk.

(b) *Endoplasm.* The endoplasm is granular and fluid like it contains.

(i) *Nuclear apparatus.* There are two nuclei, a *macronucleus* and a *micronucleus*. Macronucleus is large band-like, often horse-shoe-shaped. The micronucleus is small and lying close to the meganucleus.

(ii) *Contracfile vacuole.* A single contractile vacuole with a definit vacuolar membrane of the ectoplasm is placed very close to the gullet and communicates with the gullet through a reservoir.

(iii) *Food vacuoles.* Many food vacuoles are found scattered in the cytoplasm.

(iv) *Cytoproct* It is a temporary or permanent aperture opening into the vestibule. Undigested food is sent out of body through this aperture.

II **Stalk.** The stalk is long, slender and contractile. It is produced by the body itself. It is made up of a central axitlrod known as *spasmoneme* or *stalk muscle,* which consists of a band of spiral myonemes ensheathed in delicate transparent ectoplasmic sheath, Recently, it has been reported that this sheath also has *myonemes*. This sheath is continuous with the pellicle of the bell. *Kolzoff* was of the view that the stalk was a preudo-chitinous tube enclosing an inner tube packed with granulated *thecoplasm* and a central rod composed of *kinoplasm.* Recently, *Faure-Fremeit* described that the stalk consists of a wall having a liquid and a contractile cord. The wall is double and thickened due to the presence of rodless having cytoplasm and a specialized myoneme called *spasmoneme,* which is continued into the myonemes of the bell.

Physiology

Locomotion.Vorticella is a sedentary creature and normally does not move from place to place. However, the cilia remain very active and the movements of the stalk are frequent and rapid. When the animal is feeding, its stalk remains fully extended and the bell sways to and fro. A slight disturbance excites *Vorticella* and immediately its actions are stopped. The stalk contracts to become spiral-shaped. The cilia of disc contract and stop working. The disc is pulled inwards and the peristomial lip contracts to close its upper part. Due to contraction of the stalk, the whole structure is pulled towards the bottom of water where it remains in this state till the danger in over.

Nutrition. The mode of nuttrition is holozoic in *Vorticella.* It feeds on small protozoan, bacteria and organic pieces

Ingestion. A current of water is created with the help of adoral cilia present in the peristome and the disc. The water current brings water along with food materials to the disc from where they are carried to the vestibule and then into cytopharynx. On reaching the bottom of cytopharynx, the food particles along with water take the form of food vacuole in the endoplasm. The food vacuoles do not move in the endoplasm in a regular manner as in *Paramecium*, but show irregular cyclosis.

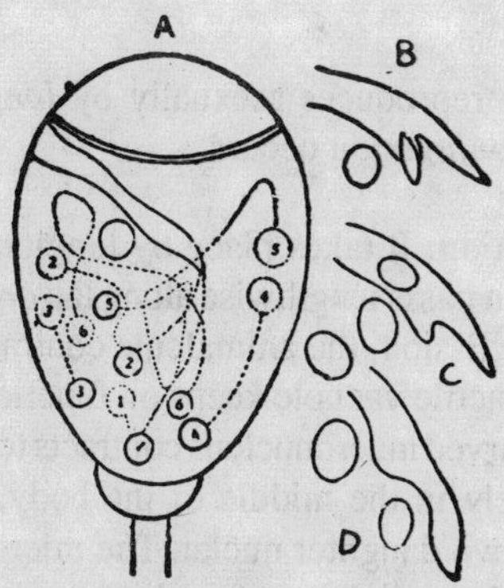

Fig. 15.4 A= Diagram showing movements of food vacuoles in *Vorticella*. B=D stages in removeal of food vacuoles.

Digestion and assismulation. Digestion of the food takes place during movements of food vacuoles. At first the effect is acidic, then neutral and finally alkaline. *Volkovsky* (1931) was of the openion that certain neutral red granules gather on the food. He named them digestive granules. The digested food materials are diffused in endoplasm. The reserve food material is glycogen.

Egestion. Undigested food-materials and old food vacuoles are removed out by the cytoproct into the cytopharynx on its lower wall. From here they pass out.

Respiration. Respiration is aerobic. The oxygen found dissolved diffuses into the body through pellicle and carbon dioxide diffused out in surrounding medium.

Excretion. The nitrogenous waste product formed as a metabolic process are diffused out through general body surface.

Osmoregulation. The excess of water from the endoplasm collects in the contractile vacuole which enlarges in size (diastole). When full, the vacuole contracts quickly and disappears (systole), discharging its contents into the reservoir situated close by. The reservoir then contracts slowly, deriving its contents into the vestibule. From the vestibule the water goes out with the faecal matters. In the mean time the contractile vacuole appears again and the whole process is repeated.

Reproduction.

Vorticella normally reproduces asexually by *longitudinal binary fission* and sometimes *conjugation* occurs.

Asexual Reproduction. It takes place by longitudinal binary fission. This type of division passs lengthwise along the *oral-aboral axis* of the body. During binary fission, the animalcule contracts into a sphere and broadens. The contractile vacuole keeps on functioning during this division. The long and curved macronucleus contracts to become straight and is placed transversely in the middle of the body. Now it divides amitotically and forms two daughter nuclei. The micronucleus divides mitotically into two. With the division of nuclear apparatus, a longitudinal groove appears in the middle of the free end due tto which the peristome is divided into two parts. This groove gradually progresses downward to divide the bell into two parts which are unequal in size the smaller part is devoid of stalk. Whereas the paental remains attached to the larger part. A new spiral of above cilia and a contractile vacuole is formed in the smallar part. Along with this, a spiral of cilia is also formed in the hinder region of the smaller part. This part now becomes cylindrical and breaks off from the parental stalk. It is called *telotroch*. It swims in the water with the help of cilia present on posterior part. After swimming for some times the telotroch attaches itself and undergoes metamorphosis into a normal sessile adult by producing a stalk shortly before developing a stalk, it resembles another ciliate, the *Scyphida*, hence often called *Scyphidia stage*. According to *Finlay* (1939) the binary fission is completed in about 20-30 minutes.

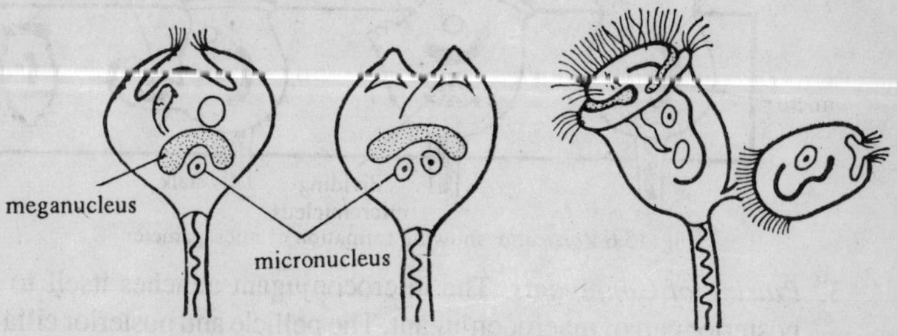

Fig. 15.5. *Vorticella* showing binary fission

During unfavourable condition the normal *Vorticella* develops a posterior ring of cilia to become a telotroch which breaks from its stalk and swims away to some favourable place and then grows a stalk. At times *Vorticella* encysts in a double cyst wall while still fixed to the stalk, then the cyst fall from the stalk, on excystment it swims away as a telotroch.

Sexual Reproduction

Maupas (1888) has described sexual reproduction in *Vorticella nebulifera*. It takes place by conjugation from time to time. The conjugation taking place in *Vorticella* is different to that of conjugation in *Paramecium*. The process of conjugation in *Vorticella* takes place as follows:

1. *Formation of microconjugants*. The *Vorticella* divides twice forming four smaller individuals. Each of them develop posterior circlet of cilia and detached from one another and they swim as independent *microconjugants*. These are similar to telotrochs, but differ from them in being smaller in size and are unable to grow into sessile adults. They die if they fail to reach the macroconjugants within 24 hours.

2. *Formation of macroconjugant.* The vegetative individual becomes a *macroconjugant* after certain nuclear modifications in it. In structure, it is similar to adult individual but has physiological differences from the later. It attracts the microconjugant towards itself.

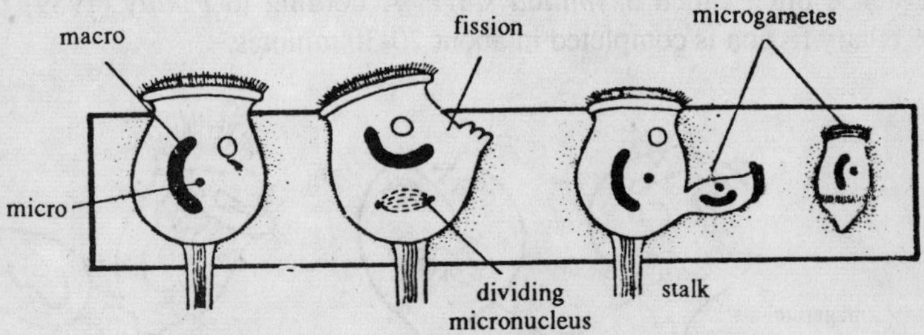

Fig. 15.6 *Vorticella* snowing tormation of microgameter

3. *Pairing of Conjugants.* The microconjugant attaches itself to the posterior part of macroconjugant. The pellicle and posterior cilia are lost. After attachement the macronucleus of each conjugant disintegrates and is absorbed in the cytoplasm. The micronucleus of microconjugant divides thrice so as to form 8 nuclei while the micronucleus of macroconjugant divides twice to form 4 nucler. 7 nucei in microconjugant and 3 nuclei of macro conjugant disintigrate. Thus only one nucleus is left in each conjugant. Both the nuclei more toward the wall separating them. In the mean time the wall separating them dissolved and two conjugants become one. The nuclei of both the conjugants now divide meiotically to form four nuclei. One nucleus of each conjugant remains while other disintegrates. The remaining nucleus of the microconjugant is called *male pronucleus* or *migratory pronucleus,* while that of macroconjugant is called *female pronucleus* or *stationary pronucleus.* Now both male and female pronuclei unite to form *zygote nucleus* or *synkaryon.*

The macroconjugant is now called the *zygote* The zygote nucleus divides thrice to form 8 nuclei. All the divisions are mitotic. Out of these 8 nuclei, 7 increse in size and become macronuclei. The remaining one is the mircronucleus. The micronucleus and the body of zygote divide in such a manner that one of the daughter cell possesses 4 macronuclei and one micronucleus and other daughter possesses 3 macronuclei and one micronucleus. Now the daughter cell which has 4 macronuclei and one micronucleus divides twice to form four daughter individuals while the

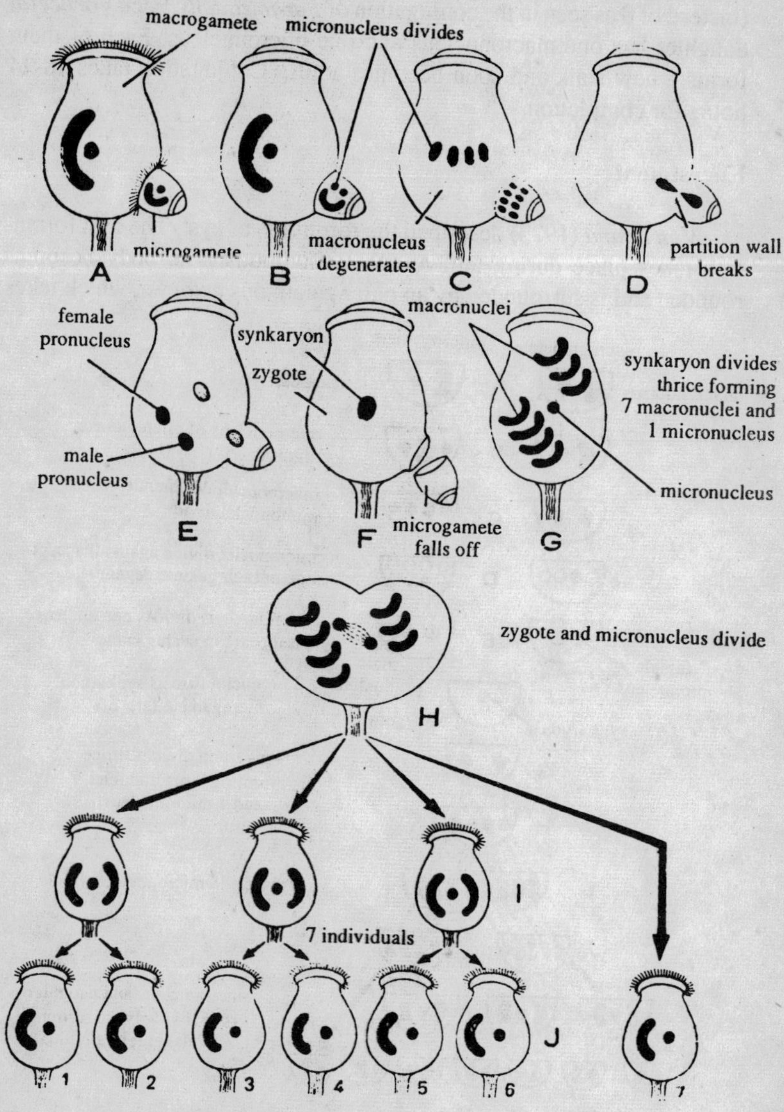

Fig. 15.7 *Vorticella*, stages of conjugation

daughter cell that has 3 macronuclei and one micronucleus forms only three daughter individuals.

Thus as a result of conjugation 7 daughter *Vorticellae* are formed (instead of 8 as seen in the conjugation of *Paramecium*). Each *Vorticella* daughter has one macronucleus and one micronucleus. Each of them forms a new stalk and soon becomes adult. Conjugation takes 18-24 hours for completion.

Encystment

Von Brand (1923) described the formation of cyst. The cyst formation takes place during unfavourable conditions. The body becomes rounded and is surrounded by an outer gelatinous *ectocyst*, which takes

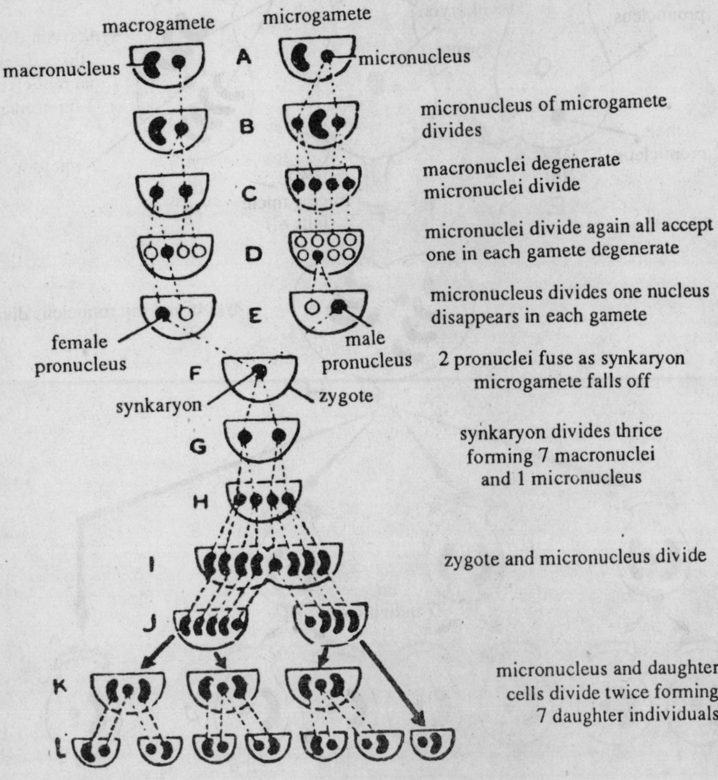

Fig. 15.8 *Vorticella*. Diagrammatic representation of nuclear changes and divisions during conjugation

on a wrinkled appearence. The *endocyst* is a double membrane, the outer is much wrinkled. Myonemes and pellicular striations become indistinguishable and contractile vacuoles pulsate more rapidly. The complete cyst is marked by two scars, one shows the previous point of attachment to the stalk and opposite one indicates the point of escape for excreted water.

During favourable conditions the vacuole enlarges and as a result of pressure, organism comes out of cyst with an explosion, forms a posterior circlet of cilia and swarms off as a telotroch.

16

SUCTORIA

The suctoria are common in fresh and salt water attacked to various objects. Some attach to a living or life-less object, while others attach to a particular animal and frequently to a particular organ of it. Some are parasitic living as parasite in the endoplasm of *Paramecium* like *Sphaerophrya* while *Endosphaera* lives in Vorticellids.

Morphology. The form of the body varies greatly, it may be spherial, conical, club-shaped, vermiform or iregularly shaped. Typically it is said to be vase - like, with or without a stalk. In the sessile form the body is attached by its base to the substratum. In peduneclate form the body is raised on a straight non-contractile stalk of secreted substance. The stalk consists of a sheath, a cortical layer and a medually substance stratified longitudinally. The base of stalk rest on a cushion of secreted substance.

The body is protected by a secreted theca continuous with the sheath of stalk in peduncelate forms. The house may be in the form of cup open distally as in *Thecacineta* or may fairly heavy wall enclosing the body as in *Squlophrya*. In *Astrophrya*, the house is build of foreign particles.

As in other cliates, colonies may be formed of considerable size, extent and various sizes as in *Dendrosoma*. The non-peducelate form like *Dendrocometes* produces spreading colonies.

The most characteristic feature of globe is the presence of tentacles and the absence of cilia in the adult state. The tentacles are lacking very rarely as in *Endosphaera* which is an endoparsite of cliates. The tentacle is single in *Rhyncheta* and one or two in *Urnula*. In other they are many. The tentacles may be distributed over the entire surface or they may arise

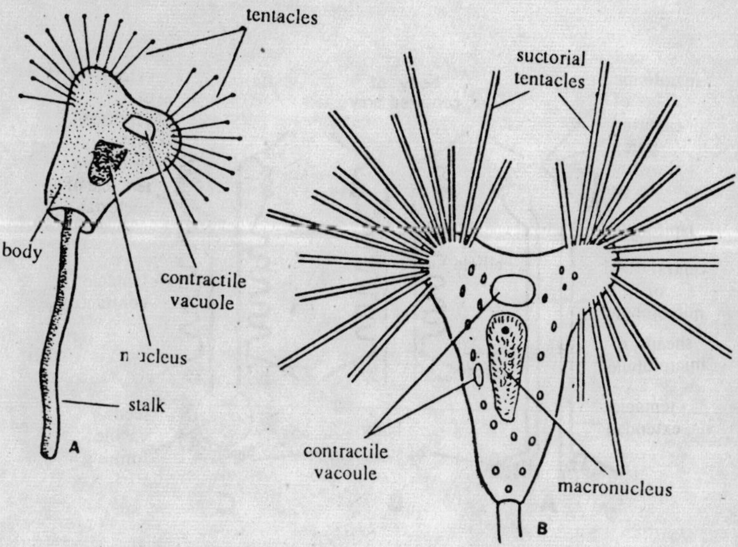

Fig. 16.1. A = *Tokophrya bengalensis* and B = *T.quadripartita*

in clusters or from lobes on extensile arms. Two varieties of them are known namely *capitate* which ends distally in a flattened or rounded expansion. The other type is *lanceolate* which tapers more or less to a point. The exterior of tentacle is covered by delicate pellicle. Its axial region is farmed of semifluid protoplasm while the outer is firm contractile and probable contains myonemes. In attest some species, like *Tokophrya,* the tentacle contains an inner tube which passes into the endoplasm of body.

In correlation with the sedentary habit the organization of the suctorian is very much simplified. The ectoplasm is freely developed and pellicle is as in other ciliates. Contractile vacuole one or more in number. The macronucleus takes various shape. In *Dendrosoma* it is branched to the same extent as the colony.

Nutrition. Ciliates like (*Paramecium* and *Didinum*) and other organisms are caught by the tentacles. The tentacles adheres to the pray and are powerful enough to hold it, even it is large then himself. The prey is paralysed and its cuticle is pierced or dissolved where the tentacle touchs it. The semifluid protoplasm of the prey-flows down the tentacle into the body of capture. The suggestion that some sort of suction is applied by

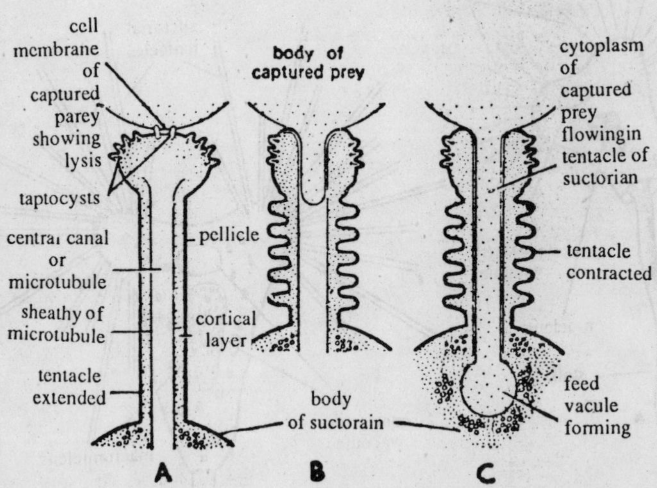

Fig. 16.2 Feeding in Suctoria by hollow tentacles.

the sunctorian in this process his not been confirmed. *Dragesco* and *Guilcher* (1950) have noted that the wall of inner canal may undergo contractions suggesting a sort of peristaltic activity. Whether such activity plays a major part is ingestion is not yet certain.

Encystment is thus common occurring when food is scare.

Reproduction The methods of reproduction varied. Simple binary fission is rare. The reproduction takes the form of *budding* which may be *exogenous* or *endogenous*. In most primitive case as shown by *Podophrya* there is no fission. The entire suctorian withdraws its tentacles develops cilia and leaves the stalk as a holotroch (free-swimming bud) with longitudinal rows of cilia. In case of exogenous budding the distal half may absorb tentacles and swims - off or a number of elevations may come out from the distal surface which develop cilia and swim off as embroys like in *Ephelota*.

In endogenous budding a space or brood pouch appears in the enterior by evagination or by split. In this space adjacent cytoplasm protrudes and develops into a single or a number of embryo like in *Acineta*. Although it may appear unusual that both exogenous and endogenous budding, occur in some species as in *Anarma*. In all the budding the nuclei behaves as in ordinary fission. The macronucleus

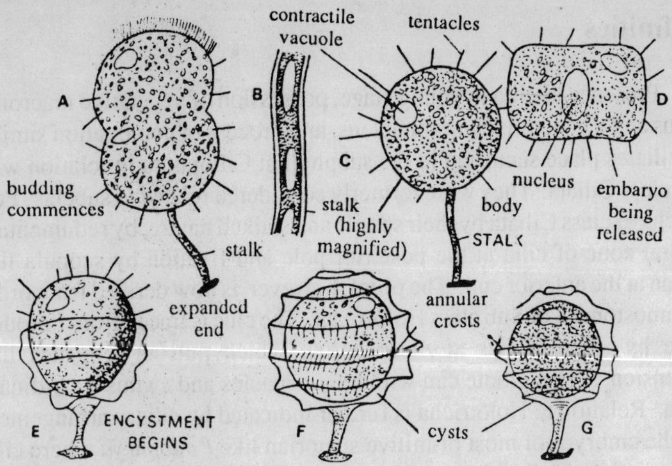

Fig. 16.3 Budding and encystment in *Podophryafixa*.

extends into each bud and divides directly while the micronucleus undergoes some sort of mitosis. The embryos bear cilia. After swimming for some times the embryo undergoes metamorphosis. They loose cilia and develop tentacles. Some suctorian like *Ophryodendron* give off vermiform buds in addition to ciliated embryos by exogenous budding.

The disappearance of cilia during metamorphosis of the bud may not includes the basal granules. The granules persist in adult stage of *Podophrya* and in reproduction each bud receives its share of granules which multiplies and gives rise to larval cilia.

In the life-cycle of some forms like *Podophrya* an intermediate stage between the larva and atult exist. In *Podophrya* the result of metamorphosis is a sphaerophrya stage which floats until it makes a contact with a ciliated host.

Conjugation comparable to typical ciliate is known from several forms like *Acineta* and *Dendrocometes.* In contrast to conjugation in *Tokophrya* in which the two conjugants eventually separate, the complete fusion and the production of a single synkaryon has been reported in *Lernaeophrya. Encystment* is known in a number off species. I *Tokophrya (Noble, 1932)*, encystment involes the deposition of a transparent secretions, at first basally, and finally over the apical end of the body.

Affinities

Presence of cilia in young stage, possession of vegetative macronucleus and a generative micronucleus, and process of conjugation similar to ciliates place suctorian in the subphylum Ciliophora in relation with the class Ciliata. They were formerly considered related to subclass Peritricha of class Ciliata by their sessile and stalked nature, by redumentary adoral zone of cilia at the posterior pole and fixation by scopula like organ at the anterior end. The proof, however, is now derived from order Zymnostometa of Sub class Holotricha. The clue is traced from Porodon type by such ciliates at *Actinobolina* which putsout tentacles like extensions. This ciliate can withdraw tentacles and swims by ordinary cilia. Relation to Holotricha is further indicated by ciliary arrangement on the embryos of most primitive suctorian like *Podophrya* where cilia occur in longitudinal rows and not in girdle.

17

SKELETON IN PROTOZOA

In some protozoa, the body is naked and not covered or enclosed in any covering, in others, the body is covered by a firm layer produced as differentiation or secretion from the most superficial part of the body.

A beginning of such a differentiation is seen in *Amoeba verrucosa* in which a firm pellicle covers the body sarface but is elastic to allow amoeboid movement. Advances is seen in some flagellates where the thin cuticle of the body allows euglenoid movements. Still in others a firm covering gives shape to the body which varies from a thin flexible cuticle to the rigid inflaxible *lorica*. The lorica takes the form of rigid, cellulose covering in Phytomastigina. In some of the Dinoflagellates it consists of two pieces, the upper epitheca and lower hypotheca. In other

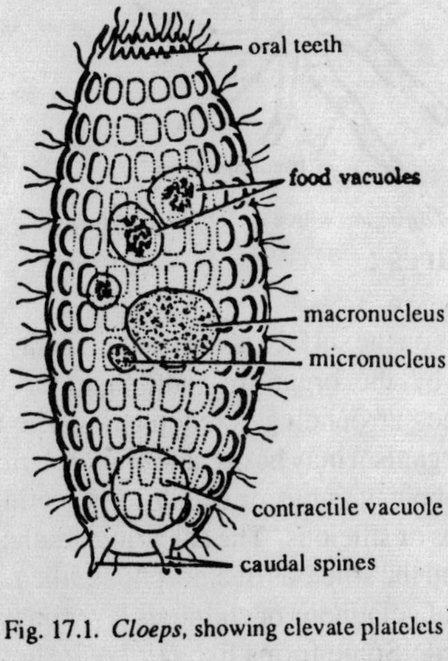

Fig. 17.1. *Cloeps*, showing elevate platelets of pellicle.

Dinoflagellates the cover may be more complex, formed of a number of plates : cingular, pracingular, post-cingular, apical and end-apical plates. In Ciliaphora such as Peritricha and Suctoria, the cover takes the form of continuous lorica of various forms.

A close fitting cuticle or lorica which is a part of the body is different from the structure built externally or internally as shell, test etc. Such structures exhibit infinite variety but may be classified under two headings :

1. External structures
2. Internal structures.

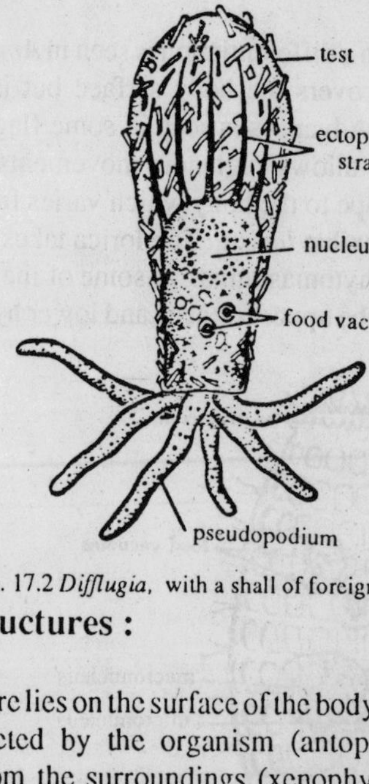

Fig. 17.2 *Difflugia*, with a shall of foreign bodies.

1. External structures :

Such a structure lies on the surface of the body and may be composed of materials secreted by the organism (antophya) or from foreign particles taken from the surroundings (xenophya). The skeleton elements secreted by the organism may be organic or inorganic in nature. In the former case it may be gelatinous or cellulose or tectin. In the later case it may be calcarious or silicious. The calcarious skeleton as formed of calcium carbonate and the silicious of amorphous silica. In xenophya the skeleton has a base of gelatinous or chitinous layer on which flowing particles are commented. Some forms like *Difflugia* shows no special projections for any particular materials. Many species, however, forms test of particular materials. In *Haliphysma* the test is formed of sponge spicules and in *Technitella* of calarious plates from echinoderms. The simplest Architecturae type of shell is a single chambered or *monothalamous* chinous shell of *Sqammulina*. In majority of foraminifera the simplest

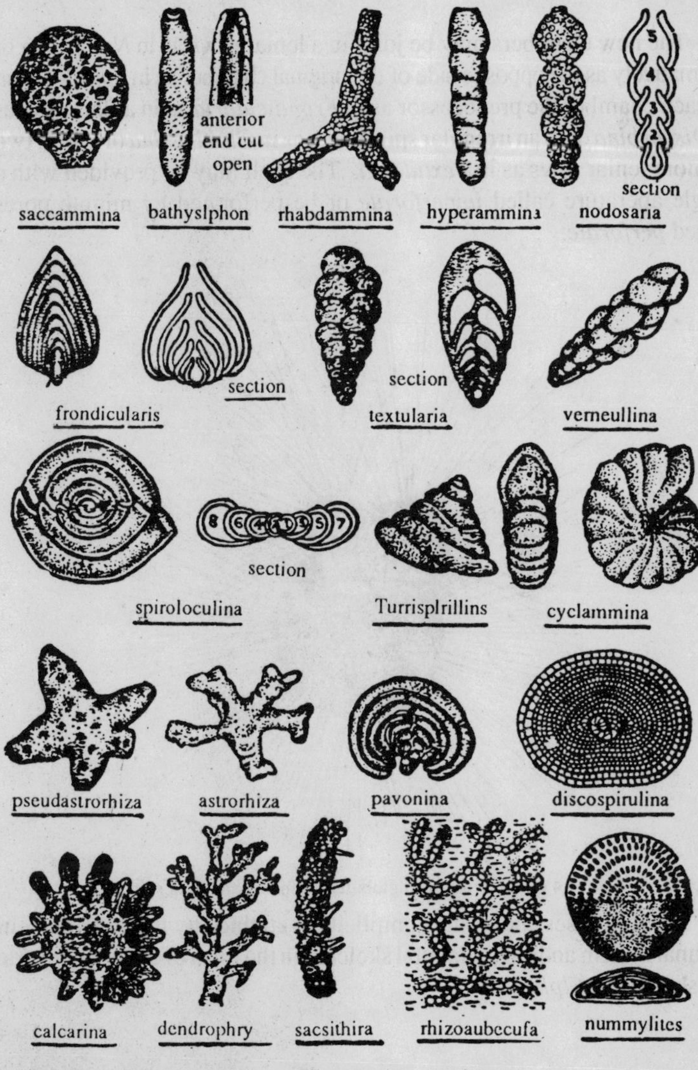

Fig. 17.3 Various types of shells of Foraminifera

form gets complicated. The protoplasm of the original chamber overflows through its aperature and the extruded protoplasm surrounding itself secretes a shell in connection with the original chamber and communicating with it through its aperture. In this way a two chambered cell in produced and by repetition of this process many chambered or *polythalamous* shell is formed.

The new chambers may be joint in a leniar rows as in *Nodosaria* or alternatively as the opposite side of the original chamber as in *Spiroloculina* or each chamber the predecessor as in *Frondicularia* or in a flat spiral as in *Discorbina* or in an irregular spiral such as in *Globigerina* or laid in two or more leniar rows as in *Textularia*. The shell may be provided with a single aperature called *imperforate* or be perforated by minute pores called *perforate*.

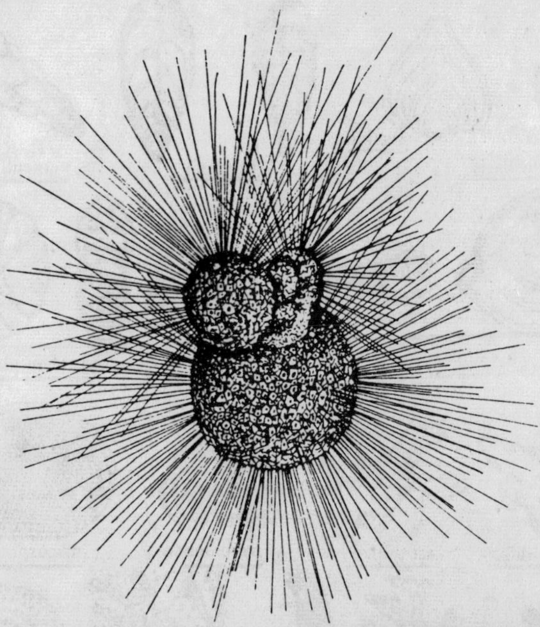

Fig. 17.4 *Globigerina* with globular chambers and long spines.

In many cases additional complicity is attained by the development of canal system and supplemental skeleton in the form of deposits outside the shell as in *Elphidium*.

Some polythalamous shells exhibit dimorphism are distinguished into the *microspheric* and shells. The microspheric shall has a small proloculum white the megalospheric shell has a large initial chamber. In some forms like *Discorbia*, in addition to the size of proloculum the microspheric shell is coiled clock-wise and the megalospheric shell is anti-clock-wise.

In some Heliozoa like *Nuclearia* the skeleton is in the form of gelatinous investment. In other Heliozoa like *Rapidiophrya* the body is surrounded by a capsule of loosely woven fibres.

2. Internal structure

In many cases the body form is maintained by the internal supporting structure. In some forms like *Allogromia* the protoplasm of the body flows round it, so that the shall becomes an internal structure. Such a internal structure is, however, different from the one which arises inside the body.

The true internal structure in many Heliozoa are of temporary in nature like the axial rods in pseudoipodia. These rods stiffens the pseudopodia and are absorbed when pseudopodia are retracted. In other cases the structures are of permanent nature. From such structure lacks the axostyle and axial rods of pseudopodia, it is easy to derive in which the organic bases in indurated with mineral substances or spicules composed entirely of mineral substances. In either case the spicule grows by deposition of fresh inorganic substances upon that already laid. If such a deposition takes place at one end of the rod - shaped spicules the opposite extremity pushed out wards and ultimately projects beyond the limits of the body.

As regards the material, the spicules are calarcous, silicious or of strontium sulphate. In the simplest case the spicuules are rod-shaped or needle like disposed radially or tangentially. A simple type spicular skeleton is seen in *Acanthocystis*, in which the silicious spicules are arranged like rods and project some distance beyond the surface. In others like *Sphaerozoeum* the spicules are arranged tangentially.

From this simple type more complicated types evolved by fusion of the spicules to form a frame work. The commonest type is a net-work

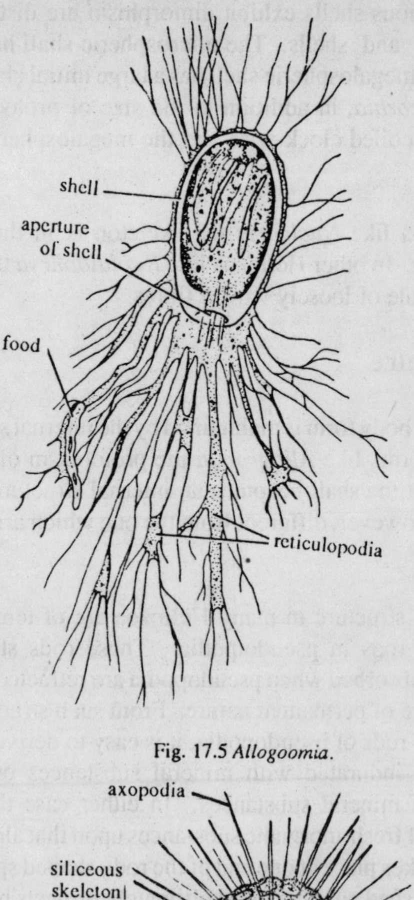

Fig. 17.5 *Allogoomia.*

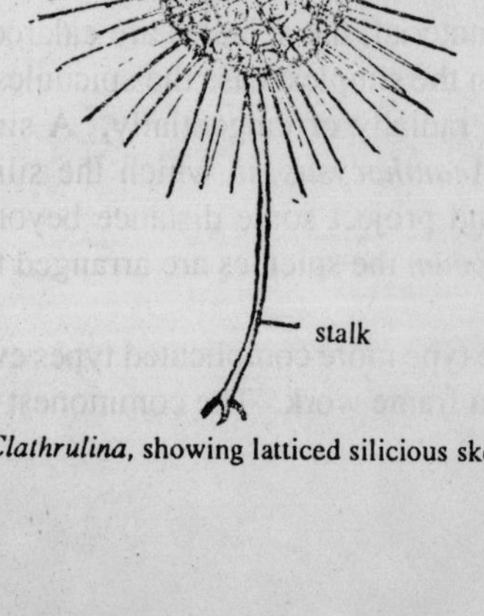

Fig. 17.6 *Clathrulina*, showing latticed silicious skeleton.

formed by the fusion of tangentially disposed spicules as in *Clathrulina.* The lattice-like skeleton may be strengthened by radially disposed branches, as the animal grows the first frame - work forms concentric frame - work out-side the original. A beautiful form of such a skeleton is exhibited by *Actinocoma*, in which there are three concentric spheres connected by radiating spicules.

In some radiolaria like *Aulocanthus* foreign bodies are used in building of the skeleton. In *Acantharia* it is formed of strontium sulphate laid in the form of 20 rods radiating from the centre. These rods emerge out of the animal in fine circles comparable to the equatorial tropical and circumpolar circle of the globe. This arrangement is known Mullar's law.

The central capsule of Radiolaria may also be included in the internal skeleton. It is composed of chitinous, pseudochitinous or tectinuous materials and has perorations in it one or groups which permit cytoplasmic continuity between the extra - capsular and intera - capsular part of the cytosome.

Functions of Skeleton

The skeleton performs several functions. These are :

1. It gives a definite shape to the organism.
2. It provides surface for the attachment to myonemes.
3. It protects the animal from its enemies.

18

LOCOMOTION IN PROTOZOA

Locomotion means movement or progression through the medium. It includes a study of the locomotory organelles and the mechanism by which movement is achieved.

Locomotory Organelles

Protozoa possess a variety of locomotory organelles, and usually each type is characteristic of a class. They include :

I. Pseudopodia

II. Flagella

III. Cilia

IV. Myonemas.

I. Pseudopodia

These are temporary cytoplasmic extensions of the body surface of naked protozoa which do not possess a definite pellicle. They can be with drawn quickly and formed afresh rapidly depending on the activity of the animal. Four main types of pseudopodia are distinguished as follows :

1. ***Axopodia***. These are also called *actinopodia*. The axopodia are long, stiff, semitransparent extension of cytoplasm with pointed ends. They radiate out from the circular body in all directions. Each axopodium is strengenthened by an internal axial filament consisting of a number of fibrils. The axial filament usually originate from a central granule in the middle of the protozoa body, sometimes within the nucleus.

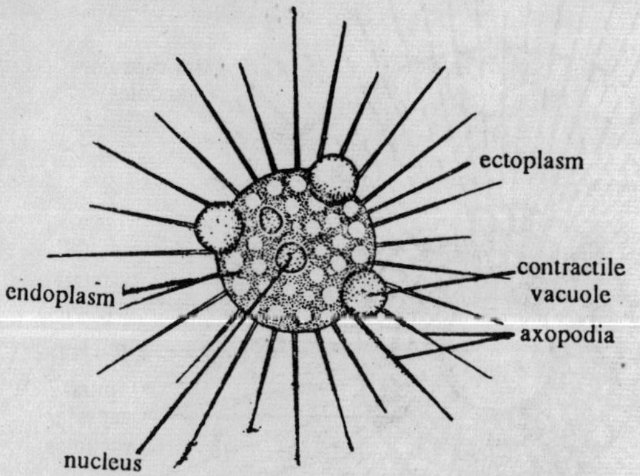

Fig. 18.1 *Actinophrys*, showing axopodia.

2. ***Lobopodia.*** These are characteristic of amoeboid protozoa. These are broad, finger-like or lobe-like projections of both the ectoplasm and endoplasm. The advancing pseudopodium is tipped by a hyaline cap. The lobopodia are also seen in some flagellates and testaceans (*Arcella*).

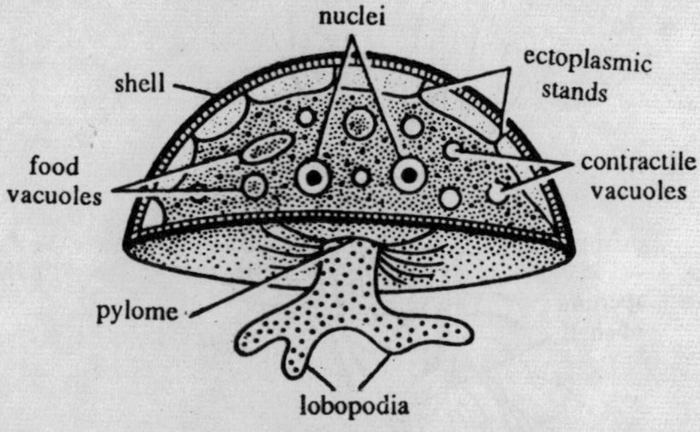

Fig. 18.2 *Arcella*, showing lobopodia.

3. *Filopodia.* These are slender, thread-like or filamentous hyaline projections, formed exclusively of ectoplasm. These have pointed tips and a tendency to branch and radiate in all directions. Filopodia are commonly met with in *Euglypha* and many testaceans.

4. *Reticulopodia.* These are also known a rhizopodia or myxopodia. These are also filamentous structures of ectoplasm which branch and anastomose to form a network. Retiurlopodia are characteristic of Foraminifera, such as *Elphidium, Chlamydophrys* etc.

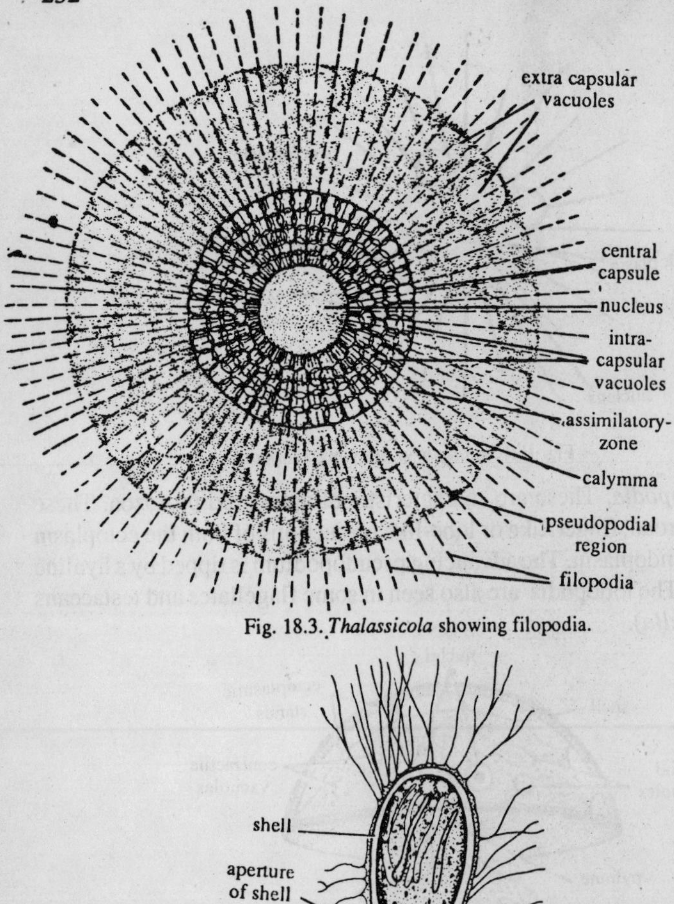

Fig. 18.3. *Thalassicola* showing filopodia.

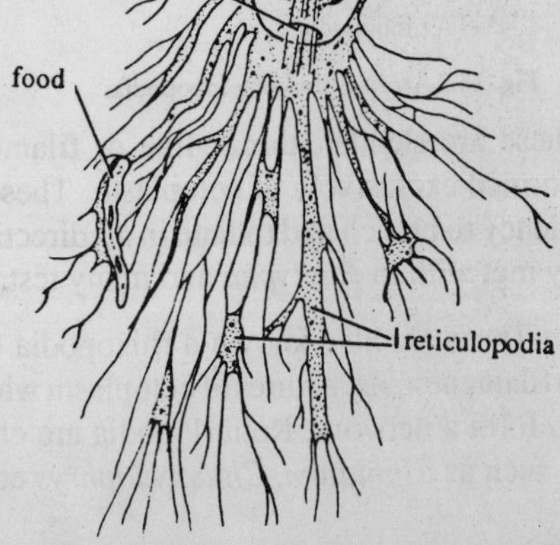

Fig. 18.4 *Allogromia* showing reticulopodia

II. Flagella

Flagella are extremely fine, delicate, highly vibratile extensions of the protoplasm characteristic of all Mastigophora and the flagellate stages of some sarcodine protozoa.

Pitelka (1949) studied the ultra structure of flagellum of *Euglena gracilis.* A flagellum consists of a central stiff but elastic shaft, the axoneme surrounded by a protoplasmic sheath. The axoneme surrounded by a protoplasmic sheath. The axoneme is formed of nine peripheral and two central fibres. The central fibres are enclosed in a membraneous inner sheath. Each peripheral fibre is made up of two fibrils and double

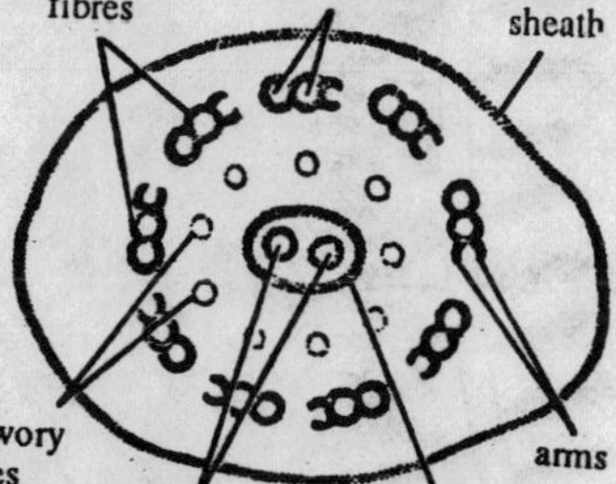

Fig. 18.5. T.S. flagellum

row of short arms. In many flagella lateral filaments, arising out of one, two or more of the peripheral fibrils, may be found. These are called *mastigonemes.* On the basis of arrangement of mastigonemes the following five types of flagella are distinguished :

1. ***Stichonematic.*** It bears a single row of lateral mastigonemes, as in *Euglena.*

2. ***Pantonematic.*** The mastigonemes are arranged in two or more rows as found in *Monas* etc.

3. ***Acronematic.*** The lateral mastigonemes are absent but the distal end bears a terminal frilament and hence assumes a whip-like appearance that is why the flagellum is known as *lash* or *whip flagellum.* It occurs in *Polytoma, Chlamydomonas, Cercomonas* etc.

4. ***Pentachronematic.*** It bears a terminal filament or mastigoneme as well as one or two rows of mastigonemes. e.g. *Paranema.*

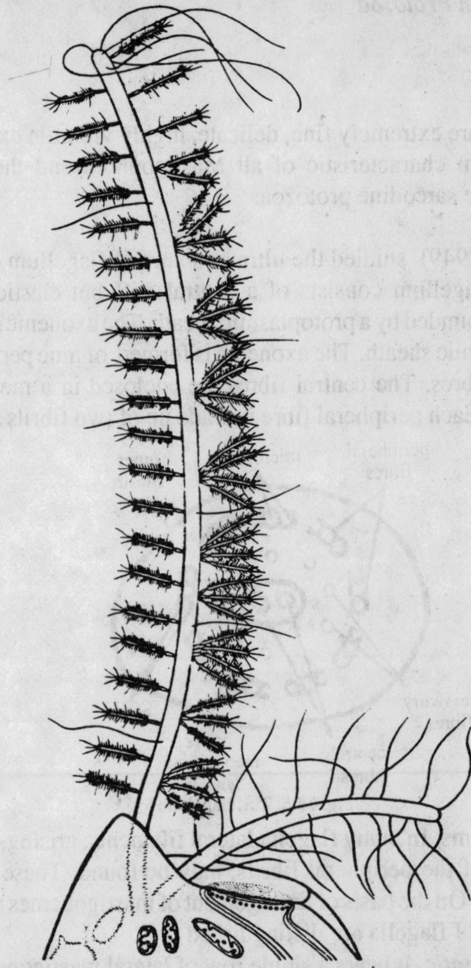

Fig. 18.6. *Ochromoras*, anterior end of the call with flagellum showing mastigonemas.

5. ***Simple or Animatic.*** Flagellum without mastigonemes and terminal filament e.g. *Noctiluca.*

Number of Flagella. The number of flagella varies from species to species. In most cases only one or two flagella are present, but some parasitic forms have more e.g. *Giardia limblia* has 8 flagella and *Trichonympha* has numerous. When 2 or more flagella are present they may be equal in length and similar in function as in *Volvox* or one dragging the body and is directed forward and known as *tractellum* while

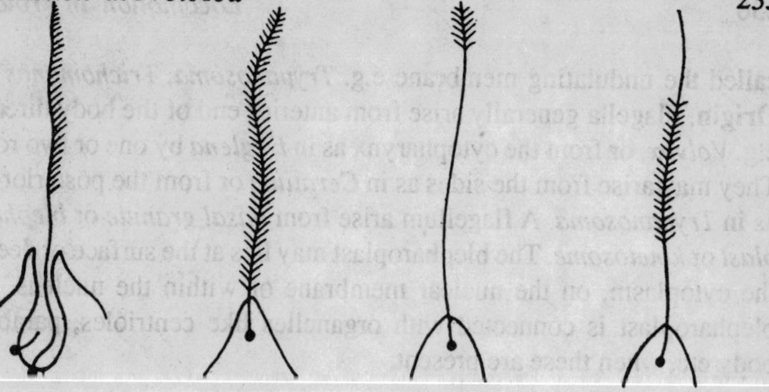

Fig. 18.7 Arrangement of mastigonema on flagella.

the other is directed backward along the side of the body and acts as an organ of propulsion. It is known as *trailing flagellum* (e.g. *Bodo, Anesonema etc.*). When there are numerous flagella often one of them is the *main flagellum* and rest are *accessory flagella.*

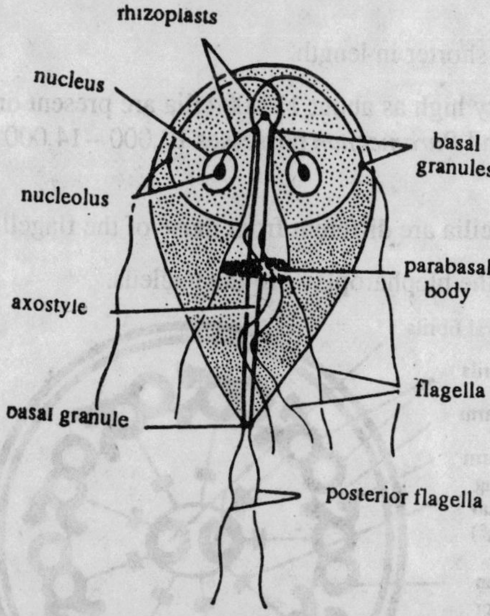

Fig. 18.8. Flagella in *Giardia*.

In few forms (Choanoflagellata) a flagellum is found at the posterior end of the body for pushing it forward. It is known as a *pulsellum.* The terms pulsellum, tractellum and trailing are coined by *Lankester*. In some parasites the flagellum borders a thin wavy protoplasmic extension

called the undulating membrane e.g. *Trypanosoma, Trichomonas* etc.
Origin. Flagella generally arise from anterior end of the body directly, e.g. *Volvox,* or from the cytopharynx as in *Euglena* by one or two roots. They may arise from the sides as in *Ceratium* or from the posterior end as in *Trypanosoma.* A flagellum arise from *basal granule* or *blepharoplast* or *kinetosome*. The blepharoplast may lies at the surface or deep in the cytoplasm, on the nuclear membrane or within the nucleus. The blepharoplast is connected with organelles like centrioles, parabasal body etc, when these are present.

III. Cilia

The ciliary locomotion is characteristic of ciliata. The cilia are extremely fine thin extensions of the ectoplasm. These are fine vibratile processes, which resembles the flagella in structure but differ from later in following respects :

1. The cilia are much shorter in length.
2. Their number is very high as about 11,600 cilia are present on the body of *Protodon* and *Paramecium* possesses 10,000 --14,000 cilia are different.
3. The movements of cilia are different from those of the flagella.
4. Lack of relation to the blepharoplasts to the nucleus.

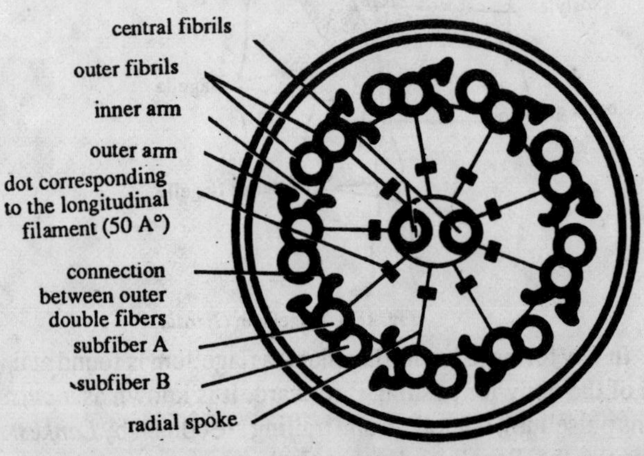

Fig. 18.9. T.S. of a cilium,

The cilia cover the entire body as in Holotricha; restricted to within a few circles as in Peritricha; or confined to the ventral surface as in Hypotricha.

Ordinarily the cilia are arranged in longitudinal, digonal or spiral rows, being inserted to individual kinetosome disposed on a ridge or furrow. The manner of arrangement is rather constant for a given species. Cilia of a particular row have their kinetosomes inter connected by fine fibrils called kinetodesma. The cilia, kinetosomes and *kinetodesma* of a single row constitute a functional locomotor unit, the *kinety*.

The ultra-structure of a cilium is similar to that of a flagellum. Its ultrastructure is given by *Satir* (1968). The axoneme consists of 9 paired peripheral fibres and 2 central fibres. The central fibres are enclosed in the protoplasmic sheath. The peripheral fibres made up of two fibrils each (doublet). One of the fibril give double row of short arms (dynein arms).

In some cases the cilia becomes fused variously forming compound organelles such as *undulating membranes, membranelles* and *cirri*.

(a) ***Undaulating Membrane.*** It is a delicate membrane formed by the union of long row of cilia. Their wave-like rippling movements drive food particles down to cytopharynx e.g. *Cyclidium, Pleuronema* etc.

(b) ***Membranelles.*** These are rectangular plates formed by the fusion of several short rows of cilia in the region of the peristome. They typically occur in series and serve to direct the food into cytopharynx. In *Stentor* and *Vorticella*, several membranelles are present forming an *adoral zone*, which encircle the peristome.

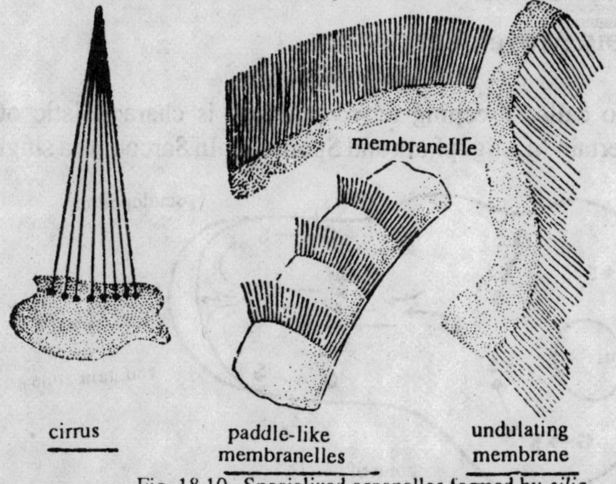

Fig. 18.10. Specialized organelles formed by *cilia*..

(c) Cirri. These are present in Hypotricha. They are present on the ventral surface of the body and may replace the cilia completely although both occur in same species. A cirrus is a long, stiff, bristle - like organ composed of a tuft of cilia arranged in 2 or 3 rows, fused completely. The cirri are independently movable in all directions and are used in crawling.

IV. *Myonemes*

The myonemes are extremely fine and highly contractile structures running in all the directions in the pellicle or ectoplasm of flagellates, ciliates, sporzoa and infusorians. The myonemes are best developed in ciliate. In certain stalked ciliates (*Vorticella*) the longitudinal myonemes of the body proper converge basally into the stalks forming the *spasmonemes*.

MODES OF LOCOMOTION

There are four modes of locomotion as there are 4 lomotory organelles. These are :

1. Pseudopodial or Amoeboid movement.
2. Flagellar movement
3. Ciliary movement
4. Metaboly or gliding.

1. Amoeboid Movement

It is also called creeping movement and is characteristic of all Sarcodina, certain Mastigophora and Sporozoa. In Sarcodina a single or

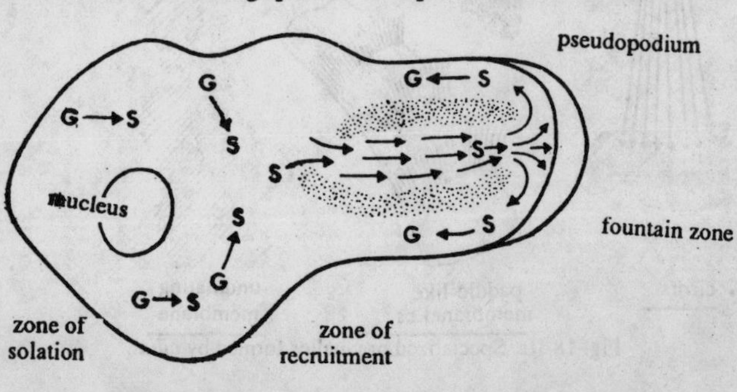

Fig. 18.11 Schematic diagram of psedopodial formation

few pseudopodia are formed by the streaming flow of the cytoplasm in the direction of movement. A number of theories have been proposed to explain the formation of the pseudopodia. These are :

(i) Surface tension Theory of *Rhumbler (1898).*

(ii) Contraction Theory of *Dellinger* (1906).

(iii) Change of viscocity or sol get Theory of *Hyman (*1917) supported by *Mast* (1923-25) and *Pentin* (1923)

(iv) Front contraction Theory or Fountain zone Theory of *Allen (1961)* and *Goldacre* (1961-64)

(v) Rolling movement theory of *Jennings* (1904).

Of the various theories advanced about the formation of pseudopodia, the sol-gel. Theory of *Hyman* (1917) enjoys a wide acceptance. All the theories are described in the chapter of *Amoeba.*

2. Flagellar movement

The flagella are so extremely fine structures so that their movement is difficult to observe under light microscope. The mechanism of flagellar movement is not properly understood. A number of theories have been proposed some of them are as follows.

(i) ***Screw propeller Theory.*** According to *Butschlii* a flagellum undergoes a series of lateral movement during which pressure is exerted at right angles to its surface. This can be resolved into (a) a force parallel to the body and (b) a force at night angles to the main axis. The first drives the animal forward and the second rotates the animal on its axis. The animal is propelled like a screw.

(ii) ***Circular beat Theory.*** *Metzner* (1920) suggested that flagellum beats in a circle tracing a cone producing sufficient suction to drive the animal forward.

(iii) ***Sidewise lashing Movement.*** According to *Ulhela* and *Krijsman* (1925) a flagellum ordinarily moves in a sidewise lash during rapid locomotion consisting of an *effective stroke* and a *recovery stroke.* In the effective stroke the flagellum is held out rigidly with slight concavity in the direction of stroke, while in recovery stroke the

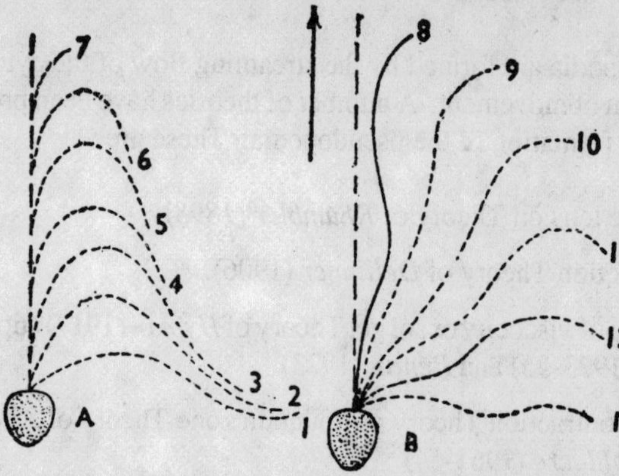

Fig. 18. 12 Action of flagellium. A = Recovery stroke, B = Effective stroke

flagellum is relaxed, strongly curved and is brought forward again. This draws the body forwards. Usually the flagellum beats obliquely so that during forward movement, the animal also rotates on its longitudinal axis.

(iv) *Undulating Movement.* This is the latest theory. *Verworn* observed it in *Paranema*. According to this theory the flagellum undergoes spiral undulations with the wave on it usually passing from base to

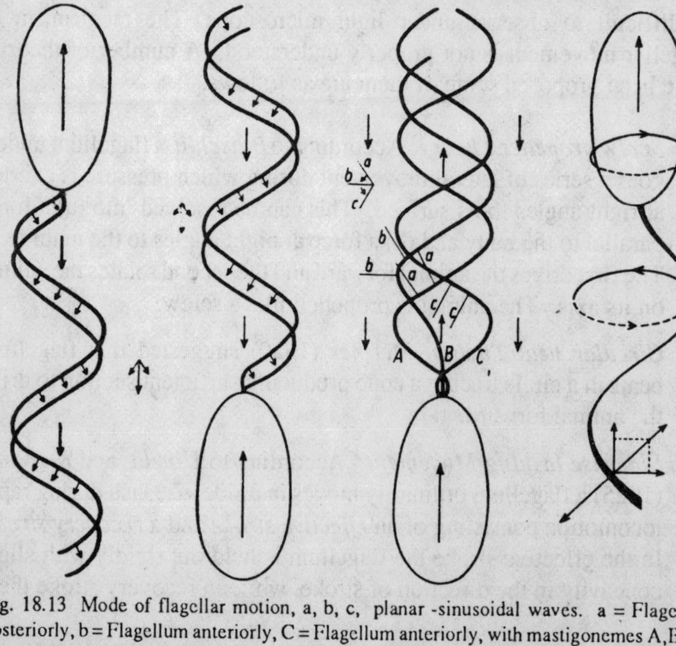

Fig. 18.13 Mode of flagellar motion, a, b, c, planar -sinusoidal waves. a = Flagellum posteriorly, b = Flagellum anteriorly, C = Flagellum anteriorly, with mastigonemes A,B,C = Three different phases of a sine wave progressive proximal -distal. d=Flagellum posteriorly, helicoidal wave.

tip. The flagellum is usually held backward of the body during swimming. The lateral undulations are *sinusoidal* and the spiral undulations are *helicoidal.* Helicoidal motion adds a gyrational (rotating on its axis) component to movement and the animal is propelled forward in water like a screw (*Grell,* 1973). The locomotive force produced by a given flagellar wave form depends on the amplitude, wavelength and frequency of the movement of waves along the flagellar length. Mastigonemes increase the surface area of a flagellum and consequently its effectiveness also. A backward movement can be caused if the undulations progress from base to tip.

The flagellates can cover distance of 15-300 μ per second.

3. Ciliary Movement

Ciliary movement is brough about by the cilia. The cilia work like oars. Each beating of cilia consists of an effective stroke and a recovery stroke. In the effective stroke the cilia beat back as straight rigid rods. In the recovery stroke the cilia return to their original position in a limp flexed state. The effective stroke push the water back and propel the body forwards. During the effective stroke water strikes the cilia at right angle to its length, the animal moving in the direction opposite to it.

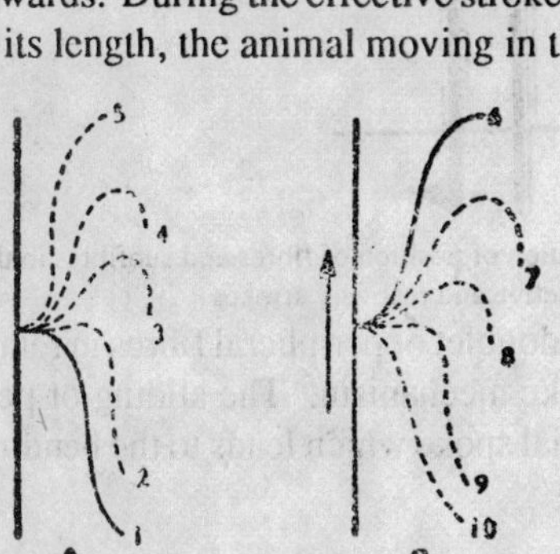

Fig. 18.14 *Cilia.* A= Effective stroke, B = Recovery stroke.

Cilia are usually disposed in linear series. A cilium in a particular row beats slightly in advance of the cilium behind it and slightly later than the cilium just ahead of it. Thus the cilia of a longitudinal row beat metachronously, while those of a transverse row beat synchronously.

The bending of cilia is caused by contraction of their peripheral fibrils. According to *Giese* (1979) ciliary movement is caused by the

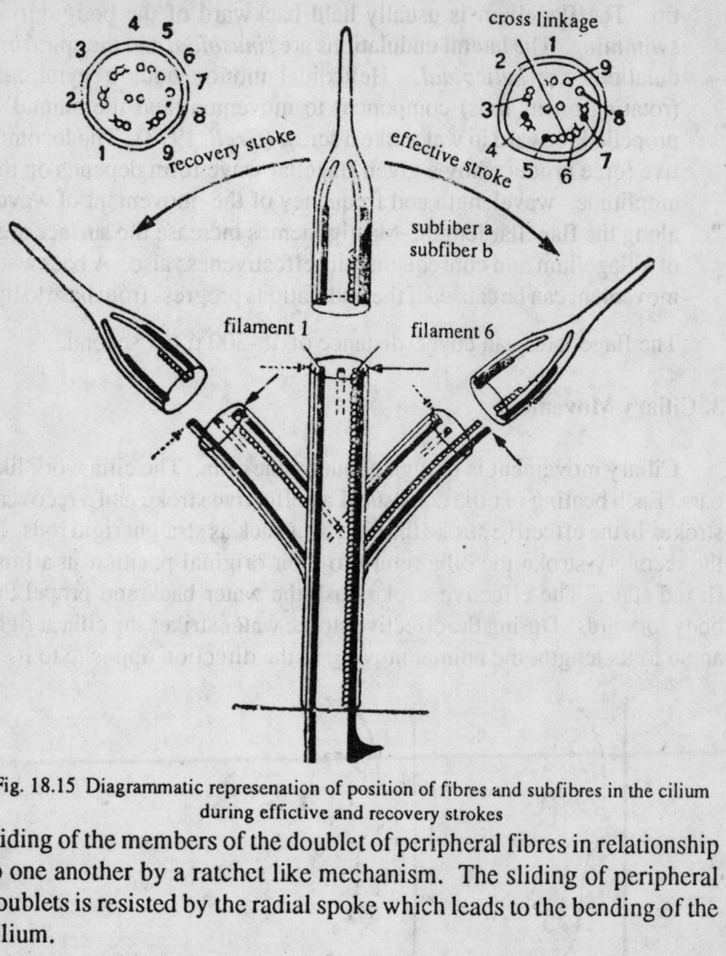

Fig. 18.15 Diagrammatic represenation of position of fibres and subfibres in the cilium during effictive and recovery strokes

sliding of the members of the doublet of peripheral fibres in relationship to one another by a ratchet like mechanism. The sliding of peripheral doublets is resisted by the radial spoke which leads to the bending of the cilium.

4. Metabolic Movements

This type of locomotion is met within sporozoa such as gregarians and flagellates like *Euglena*. These animals are capable of slow wriggling movements brought about by rhthymic contractions of myonemes. The contraction of myonemes bring about a change in shape of animal which moves a little from its place. By contraction and expansion of the body wriggling movements arise in *Euglena* as a result of which it moves forward. This type of locomotion is called *gregrine* or *Euglenoid* or *metabolic* movements.

19

OSMOREGULATION IN PROTOZOA

Osmoregulation means the control of water content of animal body. It is brought about by contractile vacuole in Protozoa. It is usually found in fresh water protozoa. The fresh water has lower osmotic pressure than the cytoplasm, with the result, water continually enters their body through general body surface by endosmosis. Some water enters the body with the food vacuoles during ingestion. Therefore elimination of excess water in necessary. For his purpose contractile vacuoles are formed by filling with water and suddenly contract expelling out their contents to the exterior.

In marine protozoa the situation is different owing to considerable salt content the osmotic pressure of see water is quite high as compared to fresh water. Marine protozoa thus do not face the osmotic stress. The constant fluxes of ions and water keep the osmotic pressure of marine protozoa is approximation to that of sea water. Although some marine protozoa also possess contractile vacuole. In such cases it probably removes the water taken with ingested food. Certain parasitic protozoan like *Balantidum, Nyctotherus* etc. also possess contractile vacuoles but there is no satisfactory explanation about their presence.

The contractile vacuole in protozoa is variously distributed. In most rhizopods the contractile vacuole is single and simple and is located more towards the posterior end which is called the uriod. In Testacea and Heliozoa the vacuole is located in the ectoplasm. In Mastigophora it appear to be situated near anterior end of the body. Normally it is situated near the cytopharynx. In *Englena virids.* it is surrounded by a number of secondary vesicles. The secondary vesicles accumulate fluid and pass it into the contractile vacuole. The vacuole then empties into the reservoir. In ciliates one to few contractile vacuoles are disposed in the body in a constant location. A pore in the pellicle is situated directly above the discharge canal leading out from contractile vacuole. In *Balantidium* and *Nyctotherus* the contractile vacuole is situated close to the permanent cytopyge located at the posterior end through which it empties its contents.

In *Paramecium* there are two contractile vacuoles but in *P. multimicornucleatum* their number varies from 2-7. Each vacuole is surrounded by 5-7 radially disposed feeder canals. In *Paramecium trichium* the contractile vacuole is without radial canal. The contactile vacuole in *Stentor* is associated with two feeder canals one lies longitudinally and the second surrounding the peristomial rim. In *Spirostomum* the vacuole is large in size and its single feeder canal is very large and extends to the anterior end of the animal.

The activity of contractile vacuole is carried in two steps: The *diastole* or enlargement of the vacuole to a maximum size and the *systole* or discharge of the contents. The formation of the contractile vacuole in some amoeboid protozoa like *Naegleria gruberi* begins with a number of small viscles which ultimately fuse to form a large single contractile vacuole (diastole). Under electron microscope *Pappas* (1959) studied the memebrane of these vacuoles in *Amoeba verrucosa.* The membrane is structurally similar to a cell membrane. The disappearance of the contractile vacuole (systole) preceded not only by the contraction of the membrane but also its break down. Isolated contractile vacuole of *Amoeba proteus* keep their rhythmic contraction if provided with ATP and Mg^{++}. The contractility of contractile vacuole is therefore energy dependent and the external surface of the contractile vacuole.

The rate of contraction of the contractile vacuole is not only influenced by the salt and solute content of the medium in which protozoa live but also by temperature and presence of other substances like metabolities and toxicants. The pulsations are slower at low temperature and fast at high. This also shows that energy in required in vacuolar function.

Many dinoflagellates possess small vacuoles called *psule.* Pusule open by canaliculi near the bases of the transverse flagellum. Their pulsations are less frequent and irregular as compared to the contractile vacuole. According to *Dodge* (1922) they are a unique osmoregulatory organelles of the dinoflagellates.

The rate of contraction varies from a few second to several minutes in different protozoa. In *A. proteus* a contractile vacuole measures 50µm in diameter and it pulsates every 5 minutes. In *Paramecium* it measures 10µm and it pulsate every 5-10 seconds. The contractile vacuole is very large in *Spirostomum* and it pulsates every 30-40 minutes.

The exact mechanism of diastole is not known, however, four mechanisms have been proposed :

1. Hydrostatic pressure cause filling of the contractile vacuole.
2. Water is secreted by the cytoplasm into the vacuole.
3. Active transport of water fills the vacuole.
4. Solute is first secreted into vacuole and water diffuses subsequently.

Besides osmoregulation, contractile vacuole also brings about some excretion. Some nitrogenous waste products are reported in contractile vacuole *Sleigh* (1973) reported that although urea has been found in the vacuolar fluid of *Spirotomum* in most other protozoa of small size wastes diffuse out of the general body surface.

20

NUCLEUS IN PROTOZOA

Due to a great variety and form of the body and organization in protozoa the nuclei are of various size, forms and structures. Majority of protozoa contain a single nucleus though many organisms possess two or more nuclei through out their life - cycle. *Zelleriella sp; Protoopalina sp.* etc. contain two similar nuclei while in Suctoria and Euciliata the two nuclei are different one is larger than other and known as macronucleus or meganucleus while later is small and known as micronucleus. The macronucleus control the vegetative activities of cell body while the micronucleus performs the reproductive functions. Many protozoa contains a large number of nuclei such as *Opalina, Pelomyxa, Actinospherium* etc.

The main morphological components of nucleus are : the *nuclear memebrane, chromatin, nuclear sap* and *plastin.*

Two main types of nuclei are found in protozoa :

1. Vesicular nucleus
2. Compact nucleus.

1. ***Vesicular nucleus*** . The nucleus consists of a nuclear memebrane which is sometimes delicate and distinct, nucleoplasm, achromatic and chromatin. Besides there is a distinct intranuclear body which is more or less spherical. It may be composed of chromatin or plastin (an acidophilic proteinous part) or of the mixture of both. The first type when it contains chromatin it is called *Karyosome* and second type when it contains plastin is called *Plasmosome.*

However, it is possible to recognize two general type, one is endosome and second is without endosome. In endosome type the

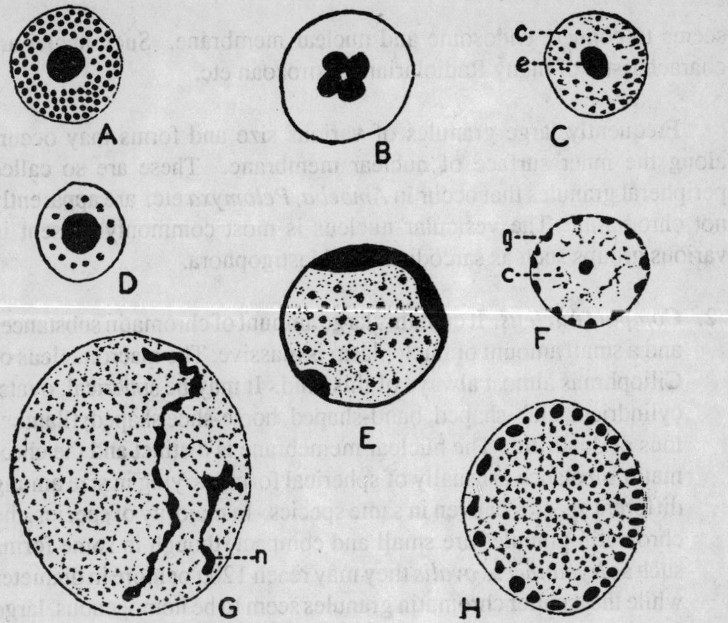

Fig. 20.1. Nuclei. A= *Heteronema*, B= multiple endosomes in Heteronema, C = *Haematococums*, D = *Iodamoeba*, E = *Chilomastix*, F = *Entamoeba*, G = *zelleriella*, H = (c = chromatin, c= endosome *pelomyxa*, n = nucleolus g= peripheral endosome.

chromatin lies between the nuclear memebrane and more or less centrally located endosome. It does not apparently contribute directly in the formation of chromosomes. A negative Feulgen reaction shows that the deoxyribouncleic acid (DNA) is absent in the endosome of Englemnida, Phytomondina, trypanosomes, *Entamoeba coli, E. histolytica* and *Endolimax nama*. The nucleus in *Entamoeba* contains a small endosome and relatively little chromatin. The well defined peripheral granules, united to the nuclear membrane and commonly considered chromatin granules. The nucleus of Englinuda contains abundant Feulgen positive chromatin and a rather large endosome which is sometimes fragmented. The endosome disappears in early mitosis in Endamoebidae but persists and undergoes division in Euglenida.

Nucleus without endosomes may contains several nucleoli which often disappear in mitosis although they present in *Zelleriella*.

When viewed in life, the nucleoplasm is ordinarilly homogenous and structureless. But upon fixation there appears a network which seems to connect endosome and nuclear membrane. Such nuclei are characteristic of many Radiolarian, Sporozoan etc.

Frequently large granules of various size and forms may occurs along the inner surface of nuclear membrane. These are so called peripheral granules that occur in *Amoeba, Pelomyxa* etc. are apparently not chromatin. The vesicular nucleus is most commonly present in various groups such as sarcodina and Mastigophora.

2. ***Compact nucleus.*** It contains large amount of chromatin substances and a small amount of nucleplasm is massive. The macronucleus of Ciliophra is almost always of this kind. It may be spherical, ovate, cylindrical, club-shaped, band-shaped, horse-shoe shaped, filamentous or dendritic. The nuclear memebrane is distinct and the chromatin substance is usually of spherical form, varying in size among different species or even in same species. In majority of species, the chromatin granules are small and compact, though in some forms such as *Nyctotherus ovalis* they may reach 120μ or more in diameter while the smaller chromatin granules seem to be homogenous, large forms contains alveoli of different sizes in which small chromatin granules are suspended (*Kudo,* 1936).

There is no sharp democration between the vesicular and compact nuclei, since there are numerous nuclei, the structure of which are intermediate between two. Several experiments show that the number, size and structure of the endosome in the vesicular nucleus, and the arrangement of chromatin in the compact nucleus vary according to the physiological state of the entire organism. The macronucleus may divide into two or more or even 200 or more than 200 in *Dileptus anser.* Each contains a plastin core and chromatin cortex. Such nucleus is called *dispersed nucleus.*

In compact nucleus the chromatin granules fill the intra-nuclear space completely. There may be one or more endosomes. In many nuclei the chromatin granules appear freely suspended while in others it may be reticulate. The chromatin of compact nucleus gives Feulgen's positive reaction.

In *Paramecium caudatum and P. aurelia* the meganucleus stains deeply then the micronucleus while it is reverse in the case of *Colopoda,*

Urostyle and others. This gives the support that the nuclei of Euciliate and Suctoria are made up of different chromatin substances idiochromatin in micronucleus and trophochromatin in mega nucleus. In other classes of Protozoa both idiochromatin and trophochromatin are found in the same nucleus. *Moss (1950)* found in *Paramecium caudatum* the micronucleus and macronucleus possess similar concentration of total protein, non-histone protein and deoxyribonucleic acid and ribonucleic acid. The nucleic acid RNA is present in large amount than DNA in both nuclei.

Chromidia

By certain nuclear stains the chromidia are detected. Several investigators found that in many protozoa extra-nuclear granules are present. *Hetwig* (1902) coined the term chromidia to these granules and found them in the cytoplasm of *Arcella vulgaris, Actinosphaerium* and others. He mentioned that under unfavourable conditions, such as lack of food material, the nucleus gets disappears and these granules are spread in the cytoplasm. In *Arcella vulgaris* both the nuclei breakdown and produces a network which later on transform into smaller secondary nuclei. *Belar* found in *Actinosphaerium* the encystment occurs rather than the chromidia formation. In *Difflugia* the chromidial net is vacuolated or alveolar. In each alveolus appears glycogen granules. These act as reserve food material in the time of reproduction. The chromidia occurring in *Actinosphaerium* are made up of carbohydrate and protein (*Rumjantzew* and *Wermel*, 1925). In *Sphaerita* (a cystozoic parasite) volutin is present. *Calkins* (1933) found that the chromodia of *Arcella vulgaris* gives negative taste to the nuclear reaction, but by omitting acid - hydrolysis and treating with fuchsin sulpurous acid for 8-14 hours the chromidia and secondary nuclei give positive taste and believed that chromidia are chromatin. The real nature of chromodia is known but many protozoologists think that the substance is not chromatic but, in some way, is connected with the metabolism in protozoa.

Nuclear division

Ways and modes of nuclear division are also various. There are three principle types of division (1) Amitosis (2) Mitosis and (3) Meiosis

(1) *Amitosis.* Occurs normally and regularly in many forms but it is not so common as was previously believed to be. It occurs in the meganucleus of ciliates.

(2) *Mitosis* It occurs in the micronucleus of Ciliophora and nuclei of other protozoa. Mitosis, though essentially the same as in Eumatazoa, is somewhat of primitive type. Protozoa mitosis are of three types :

(i) *Eumitosis.* In this mitosis the chromosomes are distinct.

(ii) *Paramitosis. Westphal (1976) has reported in Aggregate eberthi* (a coccidian sporozoa) mitosis occur without formation of a metaphasic equatorial plate and without shortening of chromosomes.

(iii) *Cryptomitosis.* It was described by *Belry* (1926) and *Grasse* (1952) in many cases. In cryptomitosis the chromosomes are not distinct and chromatin concentrates as a mass upon the equator of the spindle which is often not properly developed, the mass then divides into halves that are drawn apart. Most protozoologists agree that appearance of chromosomes may actually be masked by inert chromatin material.

The first indication of mitosis is the multiplication of centriole or its equailvalent. In many protozoa like *Pelomyxa, Cryptomonas* etc. no centrioles are seen, the spindle - fibres are starched between the opposite ends on an enlongated nuclear memebrane. In many protozoa like *Euglena* the nuclear membrane persist throughout mitosis and surrounds an internal spindle of microtubules. In certain spore - producing sporozoa the chromosomes are divided repeatedly within the unbroken nuclear memebrane, each resulting mass posses numerous sets of chromosomes which rare later separated into as many daughter nuclei. Nuclei having several sets of chromosomes are known as *polyenergid nuclei.*

The endosome in many protozoa like *Englena* and dinoflagellates persists throughout mitosis during which they also divide into two (*Sleigh, 1973*). The endosome in most other protozoa disappears during the early phase of nuclear division and reappears in the daughter nuclei as in *Euglypha* and *Monocystis.*

3. ***Meiosis.*** It is reported in those protozoa who reproduce sexually. Meiosis of protozoa is more or less similar to that of Eumetzaoa. In some protozoa chromosomal halving is carried out by a single division in a so-called *one step meiosis* (*Grell*, 1973). Usually it takes place the time of gamete formation but in Telosporidea the first division of zygote is meiotic.

21

NUTRITION IN PROTOZOA

The process by which an organism takes in food, digest it and assimilate it is known as *nutrition.*

There are different methods by which protozoa take their food, because food provides energy to organism to make-up for the wear and tear of body components lost during its manifold activities. Following types of nutrition takes place in protozoa :

(1) Holozoic

(2) Holophytic

(3) Saprophytic or Saprozoic.

(4) Parasitic

(5) Mixotrophic

(6) Coprozoic

1. *Holozoic nutrition*

Holozoic (Gr. *holos* = whole; *zoone* = animal), as the name indicates animal like. All the higher animals obtain their nourishment by this method. They depend upon ready-made food. This ready-made food is either small organism or plant. The nutrition involves into the capturing of food, its ingestion, digestion and assimilation, and rejection of waste undigested products.

Food capturing and ingestion. The methods of food capturing is different in different species of different classes. The locomotory organelles play an important role for capturing the food. But in addition to it there are few other structures which also help in this process.

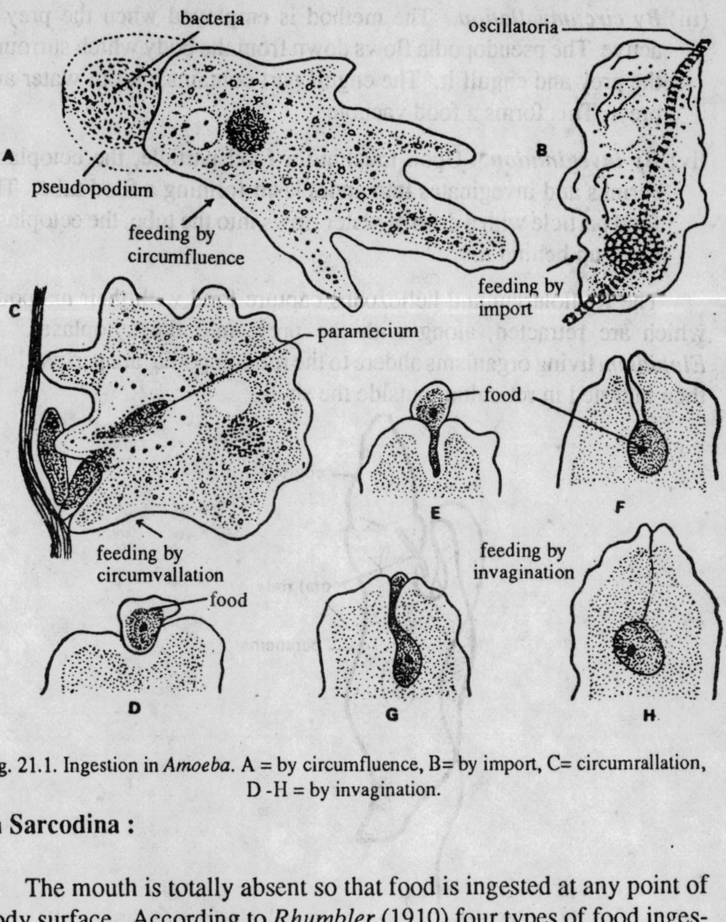

Fig. 21.1. Ingestion in *Amoeba*. A = by circumfluence, B= by import, C= circumrallation, D -H = by invagination.

In Sarcodina :

The mouth is totally absent so that food is ingested at any point of body surface. According to *Rhumbler* (1910) four types of food ingestion occur in *Amoeba* namely :

(i) ***By import.*** In this method the food is taken into the body upon contact with very little movement on the part of animal. For example, the algal filament is taken inside the body by *Amoeba verrucosa.*

(ii) ***By circumfluence***. The cytoplasm of body flows around the prey and engulf it. the prey is usually immobile like bacteria.

(iii) ***By circumvallation.*** The method is employed when the prey is active The pseudopodia flows down from the body which surround the prey and engulf it. The engulf part contains a lot of water and prey. This forms a food vacuole.

(iv) ***By invegination.*** Upon touching a food particle, the ectoplasm shrinks and inveginates into endoplasm forming a food tube. The food particle with a drop of water flows into the tube, the ectoplasm heals up behind it.

The Radiolarian and heliozoans capture food with their axopodia which are retracted, along with the prey, into the cytoplasm. In *Elphidium* living organisms ahdere to the reticulopodia, are parlysed and then ingested in reticulum outside the shell.

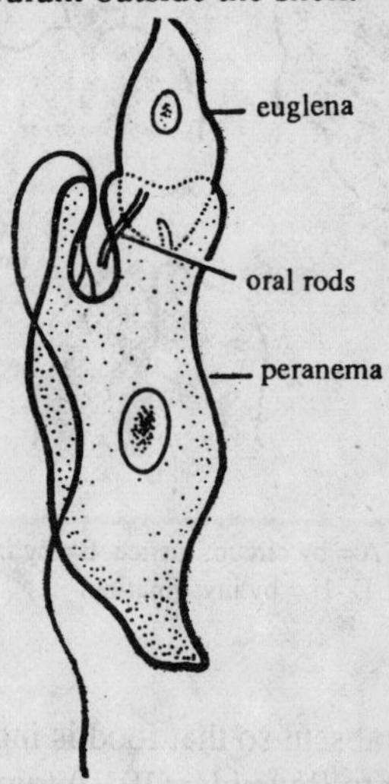

Fig. 21.2 *Paranema* feeding on *Englena*

In Matigophora :

They generally bear a cytostome which opens into the endoplasm by means of a tubular cytopharynx as in *Englena.* By the lashing movement the food-particles are directed into the cytostome. The method of ingestion in *Paranema* is observed by *Ch*en (1950). *Paranema* traps *Englena* by first harpooning the victime with its oral rods. The flagellum beats actively which contacts the body, followed by elongation. This process goes till the *Englena* comes wholly in the body of *Paranema*

In Ciliata :

There are many types of cytostomes and associated structures, but the food capturing seems to be two types. In some ciliates the cytostome remains permanently open e.g. *Paramecium.* It is a continuous feeder. The food particles are small in size that pass through the cytostome and goes into endoplasm. In another kind the cytostome remains closed e.g. *Coleps* and *Didinium.* The cytostome opens only when a prey is to be ingested (*Kidder and Dewey,* 1940). In reptorial ciliates like *Porodon* a circlet of trichites surrounds the cytostome. The cytostome leads into cytopharynx. The cytopharynx is ciliated which creates water current and allows to come food particles with water. Trichocysts of *Paramecium* and myoneme of *Vorticella* also help in ingestion of food.

In Suctoria:

In some suctorian the tubular tentacles help in ingestion of food. The tips of these tentacles are devoid of pellicle and possess haptocysts which pierce into the body of the prey. They suck liquid and solid materials as in *Choanphrya* . The tentacles also secrete toxic substance which paralyses the prey.

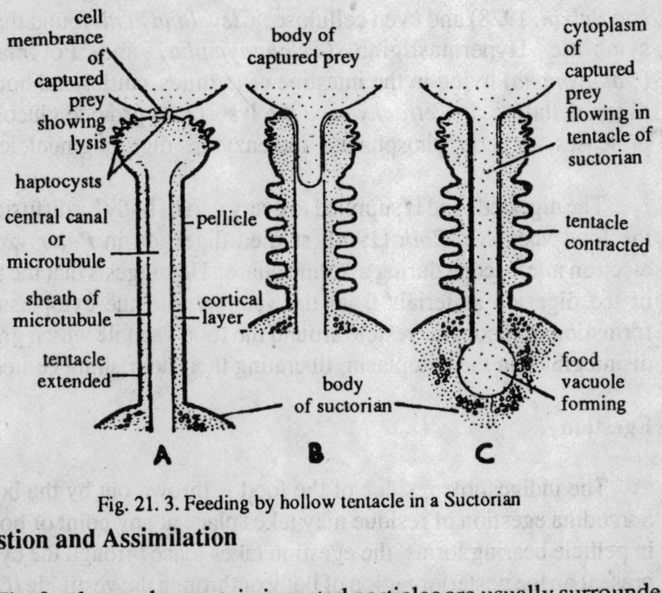

Fig. 21. 3. Feeding by hollow tentacle in a Suctorian

Digestion and Assimilation

The food vacuoles contain ingested particles are usually surrounded by water. The quantity of water vary in different vacuoles. The food vacuoles finally reach in the endoplasm. In Amoebina the vacuoles are carried about by the moving endoplasm. In ciliates the fluid endoplasm shows streaming movement called cyclosis. In *Vorticella* species food vacuoles are formed one by one at the end of cytopharynx migrate throughout the cytoplasm. The old vacuoles disintegrate at the base of cytopharynx and thence to out side (*Hall*, 1931).

The motile or active prey which is taken in by the protozoa losses its movement and finally dies. It is believed that acid, alkalies and digestive enzymes are poured in food vacuoles (*Engelmann*, 1878,*Brandt*, 1881). According to *Nirenstein* (1925) the food vacuoles of *Paramecium* undergo in changes in reaction first acidic and then alkaline. On the other hand *Khainsky* (1910) observed that the food vacuoles of *Paramecium* is acidic during the entire period of protein digestion; becomes neutral and finally becomes alkaline when the solution of food substance is ended. *Shapiro* (1927) studied the reaction changes in food vacuoles of *Paramecium caudatum* by using phenol red, Congo red and neutral red. Various types of enzymes are produced by different protozoa. viz. in *Amoeba proteus*, peptidase (observed by *Holter* and *Kopac*, 1937) protienase (observed by *Anderson* and *Holter*, 1949) and amylase (by *Hotler* and *Doyle*, 1938) In *Paramecium caudatum* peptidase and amylase enzymes observed by *Holter* and *Doyle* (1938). They can also digest fat (*Dawson* and *Belkin*, 1928) and even cellulose. *Cleveland et al.* found that many symbiltic Hypermastigina (*Trichonympha)* and Polymastigina (*Colonympha)* living in the intestine of termites, cattles and horses can digest cellulose. In *Tetrachymena* the lysosomes provide glucosidases, proteases, amylase, phosphatase and enzymes digesting nucleic acids.

The digested food is supplied to every part of body by diffusion from the food vacuoles. *Roth* (1960) studied digestion in *Pelomyxa* under electron microscope during alkaline phase. He suggests that the transfer of the digested materials from the vacuole into the cytoplasm is by formation of pinocytic vesicle around the food vacuole which gradually disintegrates in the cytoplasm, liberating their nourishing content.

Egestion

The indigestible residue of the food is thrown out by the body. In Sarcodina egestion of residue may takes place at any point of body; but in pellicle bearing forms, the egestion takes place through the cytopyge present on the posterior region of body or through the vestibule (*Carche-*

sium). Permanent cytopyge is absent in some forms. he permanent cytopyge is absent in some forms. The permanent cytopyge is absent in some forms. The permanent cytopyge bearing ciliates are *Balantidium* and *Nyclotherus* .

2. Holophytic nutrition

(Gr., *Holes* = whole; *phyton* = plant) as the name indicates plant-like nutrition. In this process the animal, like plants, decompose carbon dioxide by means of chlorophyll Contained in chromatophores in the presence of light, so as to form carbohydrate with the liberation of oxygen. The animal forms a special type of starch called paraglycogen. In a number of other cases the organism itself is without chromatophoses but is apparently not holozoic, because of the presence chlorophyll bearing organisms within it. For example in the testacean, *Paulinella,* absence of food vacuole but chromatophores of peculiar size *Paullinella* appears to be a species of algae which holds a symbiotic relationship with the testacean. In *Paramecium burseria* and *Stentor polymorphus* the chromeatophores are absent but they are also not holozoic. They are able to obtain nourishment with the aid of symbiotic algae, Zoochlorellae, in the cytoplasm.

3. Saprozoic nutrition

(Gr., *Sapros* = rotten; *zoon* = animal). This type of nutrition is seen in many parasitic protozoa, and colourless, free-living flagellates. The organic food materials are dissolved in water. The organism does not possess any special organ for the ingestion. The ingestion takes place through the body surface by the process of diffusion. *Pringshein* (1937) observed in *Polytomella* that sodium acetate is needed from which the starch (among others) is produced and carbohydrates have no direct bearings upon the nutrition.

The complicated system of food vacuoles, called *pusule,* found in dinoflagellates (*Ceratium)* is the example of special organelles employed for saprophytic nutrition, since they are filled with decomposed organic matters.

4. Parasitic nutrition

The protozoa which live within the body of another oganisms or hosts are able to nourish themselves by absorbing the digested or decomposed food of the host is termed as parasitic nutrition.

According to the type of food substances, the parasitic protozoan are of two types;

(i) ***Food robbers.*** The protozoa of this category feed on the digested food material found in the alimentary canal of the host by holozoic manner e.g. *Balantidium* and *Nyctotherus* or by saprozoic method e.g. *Opalina*. They are usually commensels.

(ii) ***Pathogenic.*** The protozoa of this category feed on the living tissues of the host and cause different types of diseases. *Entamoeba histolyttica* obtain food by holozoic manner, and *Plasmodium* and *Trypanosoma* employ saprozoic method.

5. Mixotrophic nutrition

Some protozoa utilize more than one method of obtaining nutrients. The common example is *Euglena gracilis*. According to *Zumstein* (1900) and *Lwoff* (1932) *E. gracilis* losses green pigments in dark. This indicates that it can survive without chlorophyll by saprophytic nutrition. *Entamoeba histolytica* engulfs RBC and tissue fragments holozoically and dissolves host cells by enzymes it secretes. The dissolved material is than absorbed saprozoically.

6. Coprozoic nutrition

Many free-living protozoa feed upon the faecal matter of other organisms and are teamed coprozoic. e. g. *Rhymchomonas nasuta, Chlamydophrys stercorea* etc.

Reserve food materials

- By the anabolic activity of organism result in growth and increase in volume, and also in the formation and storage of reserve food material.

The reserve food material is glycogen, paraglycogen, certain crystals etc. These are used during scarcity of food material or growth or reproduction.

The anabolic product of holophytic nutrition are starch, paramylon, oil and fats. The paramylon bodies of various sizes among different species but in a given species they are similar. These are insoluble in boiling water but soluble in concentrated sulphuric acid, potassium hydroxide etc. It does not stain with iodine or zinc chloride.

Fat occurs widely in i protozoa as globules. *Zingher* (1934) found that in Sarcodina and Ciliata each species shows morphological characteristic in having the fat. In *Amoeba* and *Pelomyxa* abundant fat bodies occur which can be seen with staining sudan III.

Volutin occurs in every protozoa except Ciliophora. *Reichenow* and others hold that volutin appears to be a free nucleic acid and is special reserve material for the nuclear substance. *Sassuchin* (1935) studied the volutin in *Spirillum volutans* and found that during strong period of growth, nourishment and multiplication they increase and disappear in unfavourable conditions of nourishment and gives a series of carbohyrate 22reactions.

22

REPRODUCTION IN PROTOZOA

In many protozoa, reproduction occurs at frequent intervals, with relatively short periods of growth intervening under favourable conditions. In some cases growth may extend over a period of several to many days so that reproduction occurs at long intervals.

In the less complex protozoa many modes of reproduction are seen. In some they are very simple while in others they are much complex. In either case they are essentially similar to the cell-divisions as seen in higher animals and always begin with the division of nucleus.

Basically reproduction in Protozoa takes place by two distinct methods;

I. Asexual Reproduction

II. Sexual Reproduction

I. Asexual Reproduction

In asexual reproduction, no specialized sex cell are involved. The stage of asexual reproduction is found in all the groups of this phylum. Even those protozoa which reproduce sexually also multiply by asexual processes. Generally the asexual reproduction is of the following types:

1. Binary Fission
2. Multiple Fission
3. Plasmotomy
4. Budding.

1. ***Binary Fission***. This is the most common type of reproduction found in protozoa. Here division of parental cell into two or more daughter cells occur. The nuclear division is mitotic (*Karyokinesis)* followed by the division of cytoplasm (*cytokinesis)* by a constriction.

The plane of the division of the cytoplasm of the body varies in different groups of protozoa. On the basis of this, binary fission is of the following types:

(i) *Simple binary fission*. Here the plane of division can-not be definitely ascertained because no special event takes place at the time of division. Simple binary fission occurs in Sarcodina (*Amoeba)*.

(ii) *Longitudinal binary fission*. In flagellates and peritrichous ciliates (*Vorticella)* the plane of division is parallel to the longitudinal axis of the body.

(iii) *Transverse binary fission*. In Cilophora (*Paramecium*) the body of animal is transversely divided into two daughters. In a few flagellates (*Oxyrthis)* transverse binary fission is seen.

(iv) *Oblique binary fission*. Here the division of body is somewhat oblique as in *Ceratium* and *Cochiodinium*.

(v) *Repeated binary fission*. In this method the daughter individuals produced by first binary fission fail to separate and undergo repeated binary fission giving rise to four or more individuals e.g. Colonial protozoa include *Dinobryon, Polykrikos* etc.

(vi) *Encysted binary fission*. During unfavourable condition this mode of reproduction occurs. The animal secretes a cyst around itself and divides by binary fission as in *Amoeba, Colpoda, Tillina* etc.

2. ***Multiple Fission***. In multiple fission the nucleus divides many times either by mitosis or by fragmentation into many nuclei without the division of cytoplasm. The daughter nuclei arrange themselves along the periphery of the animal and are surrounded by a bit of cytoplasm forming daughter individuals. Their number varies in different animals and may reach thousands in some. They become free by the rupture of parental body and lead their independent lives. Multiple fission is very common in Radiolaria, Foraminefera, Sporzóa and Mastigophora.

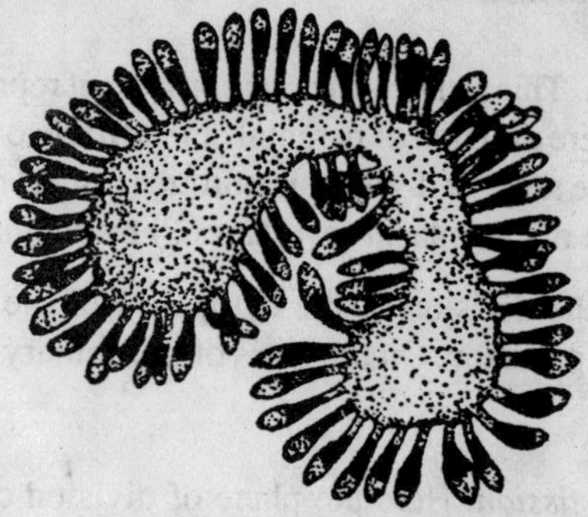

Fig. 22.1 Schizogony in *Ovivorra*.

The multiple fission occurs at different phases in life-cycle of different individuals and are named accordingly. The multiple fission can be divided into following three groups:

(i) *Agamogony or Schizogony*. Asexual process not related to the sexual phenomenon. Product of schizogony grows into trophozoite. e.g. *Plasmodium.*

(ii) *Gamogony*. Multiple fission yields gametes which take part in sexual reproduction e.g. *Monocysts.*

(iii) *Sporogony*. The multiple fission in the zygote is known as sporogony. The offsprings are known as spores which are infective stage.

On the basis of presence of characteristic organelles, the individuals formed are variously named:

(a) ***Pseudopodiospores***. The spores which posses pseudopodia are called pseudopodiospores.

(b) ***Flagellospores***. Those protozoa in which flagella are formed as a result of sporogony are called flagellospores.

(c) ***Swarm spores***. When the individuals, formed from the multiple fission, are motile, they are called swarm spores.

3. ***Plasmotomy***. It is common in multinucleate protozoa such as *Opalina, Pelomyza* and *Actinosphaerium*. Here the cytoplasm is divided into several pieces, each containing several nuclei. Here the nuclear division does not occurs. Each of the daughter regains the normal number of nuclei by the division of parental nuclei.

Fig. 22.2 Plasmotomy in *Pelomyxa*.

4. ***Budding***. Here the nucleus divides and cytoplasm emerges out in the form of one or more projections each have a daughter nucleus, these projection detache and develop into individuals. The projections are called *buds*. The process of budding is not common in protozoa. It occurs regularly in Suctoria. According to the number of buds there are two types of budding

(i) ***Single gemmation or Monotomic budding***. When only one bud is formed at a time e.g. *Vorticella*.

(ii) ***Multiple gemmation***. When many buds are formed simultaneously e.g. Suctoria.

According to the formation and position of buds, the budding is of two types:

(i) ***Exogenous buds***. When the buds are formed as outgrowths on the parental body which later pinch off as in many Suctoria, *Noctiluca* etc.

(ii) ***Endogenous buds***. Here the buds are produced inside the body of parent in special spaces called *brood-chamber* as in some Suctoria, Testacea etc. The apical surface of the parent is invaginated to form the brood pouch which is lined ectoplasm.

II. Sexual Reproduction

Sexual reproduction takes place among Sporozoa, Ciliata and Suctoria. It has been reported from many Foraminifera, Heliozoa, Hyper

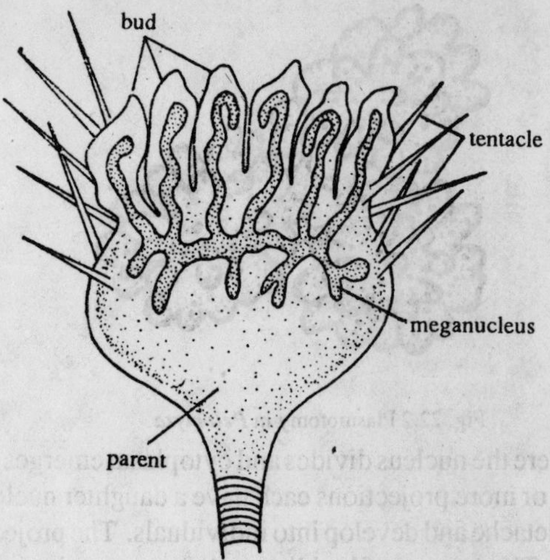

Fig. 22.3 Exgenous budding in*Ephaleta*

mastigina and Phytomastigina. Two types of sexual reproduction have been distinguished besides some aberrant forms of conjugation are found in the species of *Paramecium.*

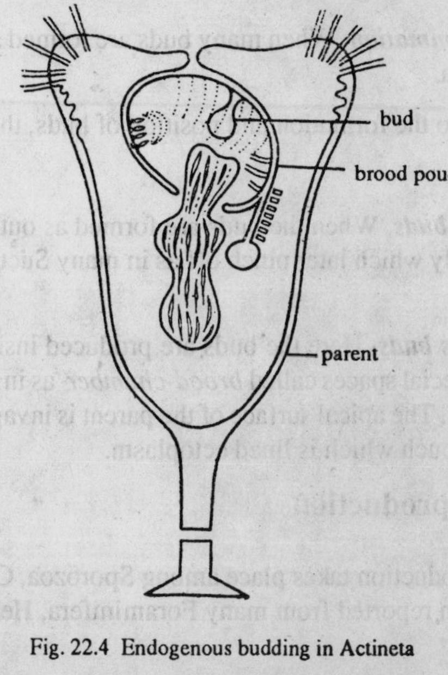

Fig. 22.4 Endogenous budding in Actineta

1. Syngamy.
2. Conjugation

1. ***Syngamy***. It involves the complete fusion of two gametes by which the zygote is formed. Depending upon the degree of differentiation of gametes syngamy is distinguished into following types;

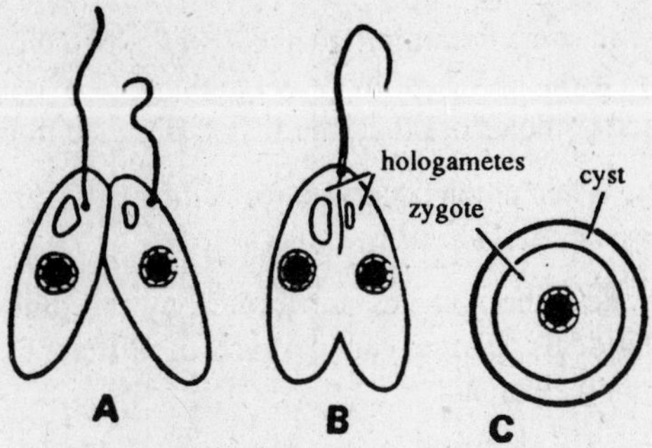

Fig.22.5 Hologomy in *Copromonas*.

(i) ***Hologamy***. It is the fusion of two mature organisms which do not produce gametes, but behave as gametes as in *Trichonympha* and in certain rhizopods.

(ii) ***Paedogamy***. If the fusion of gametes takes place in young individuals it is called paedogamy.

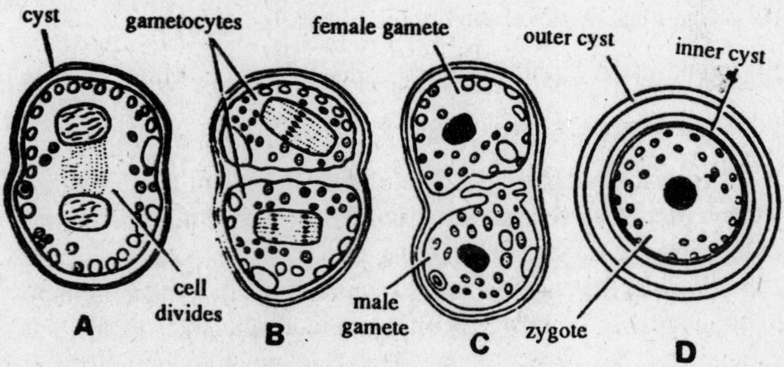

Fig. 22.6 Paedogamy in *Actinophrys*.

(iii) ***Isogamy***. When the gametes are similar in size and form. They are called isogametes and their union is isogamy. It is common in Forminifera, Gregarinida and Phytomonadina.

(iv) ***Anisogamy.*** The gametes fusing in this type differ from each other in shape, size and behaviour. The gametes are called anisogametes and their fusion is known as isogamy. The smaller, motile and active gametes are the *microgamtes* or male *gametes* and the bigger, non-motile and passive gametes are use *female gametes* or *macrogametes.* Anisogamy is more common in *Plasmodium Eimeria, Volvex, Protopalina* etc.

(v) ***Microgamy.*** In some Foraminifera and *Arcella,* two types of micro -gametes are formed. When the microgametes of one animal fuse with the microgametes of other, the fusion is called microgamy.

(vi) ***Macrogamy.*** When union occurs between the macrogametes of the same species e.g. *Actinophrys.*

(vii) ***Merogamy.*** Here the gametes are formed by the division of the normal individuals. These gametes are of different shape and structure from the parents.

(viii) ***Exogamy.*** It takes place between the offsprings of different parents.

(ix) ***Autogamy.*** Fusion of gametes derived from the same parent cells. e.g. *Actinosphaerium.*

Significance of Syngamy. In syngamy the fusion of the two nuclei takes place results in synkaryon. Syngamy has following effects:

(i) Two different types of hereditary characters come together.

(ii) Variations are developed in the individuals.

(iii) It renews the vigour which is lost due to repeated binary fission.

(iv) The fusion of nuclei initiates the development of egg.

2. *Conjgation:* Conjugation is a temporary pairing of two individuals, the *conjugants,* that separate after mutual exchange of male pronuclei. Conjugation is Characteristic of Ciliates e.g. *Paramecium, Vorticella* etc. In *Paramecium* the conjugants are separated but in *Vorticella* there are two conjugants one large macroconjugants and other microconjugant which is smaller. The microconjugant dies after conjugation. (For detail see chapter *Paramecium* and *Vorticella)* .

Significance of Conjugation. The significance of conjugation has been:-

(i) It is a process of rejuvination and reorganization by which the vitality of race is restored.

(ii) New nuclear combination and new hereditary combinations.

(iii) The metabolic activities of the animal are increased by the formation of a new nucleus in place of the old one.

Aberrant Behaviour in Reproduction

1. ***Automixis*** Here the fusing nuclei are derived from the same micronuclei of the same individual. This can be:

   (a) *Autogamy.* In this the fusion of two pronucli of the same individuals. It corresponding to self-fertilization in higher animals. e.g. *Paramecium aurelia.*

   (b) *Paedogamy.* It is the fusion of two young daughters produced by the same parent. e.g. *Actinophrys.*

2. *Endomixis.* It is a type of nuclear organization with a single individual of *P. aurelia* without nuclear division and nuclear fusion.

3. *Hemixis.* It is a process of recoganization of macronucleus without any change in the micronucleus. In certain species of *Paramecium* small fragments of macronucleus fall apart and are absorbed in the cytoplasm. The remaining part of macronucleus behaves normally.

4. ***Regeneration.*** The body protoplasm possess the power of regeneration of the lost part. Any small a fargment of nucleus can regenerate into a new individuals.

5. ***Parathenogenesis.*** In certain animals like *Actinophrys, Chlamydomonas* etc. The gametes fail to units so as to form synkaryon so that the gametes develop into adult without fertilization. Such a development in called parthengensis.

6. ***Plasmogamy.*** In some rhizopods it has seen observed that two individuals unit with each other and their cytoplasm become one but their nuclei remain separate. After they spearate from each other without any change. This process is called plasmogamy.

23

ECONOMIC IMPORTANCE OF PROTOZOA

The protozoa are acellular, microscopic organisms found everywhere, in water, soil, on the body of plants and animals. Although they are minute and apparently insignificant, but they are of considerable economic value to the mankind. The activities of these organisms can be divided the following two heads:

1.Beneficial

2.Harmful.

BENEFICIAL PROTOZOA

The protozoa are useful in the flowing ways:

1. ***Food.*** Protozoa provide food for insect lavvae, crustaceans and worms, which are taken by large animals like fishes, lobsters, clams, and crabs, which are eaten by man. Thus they form sources of food supply to man both directly and indirectly.
2. ***Symbiotic Protozoa.*** Certain protozoa like *Trichonympha* and *Colonymphya* etc. live in the gut of termite which help in the digestion of cellulose. The digested cellulose is uitalized by the host.
3. ***Insect Control.*** Several protozoa control harmful insects by persisting their bodies.
4. ***Helpful in Sanitation.*** A large number of protozoa living in polluted water feed upon waste organic matters and thus purify it. Many protozoa feed upon bacteria and play important role in the sanitary betterment and keeping water safe for drinking (*kudo*, 1947).

5. ***Industry.*** The skeletal deposits of marine protozoa (Foraminifera and Radiolaria) from oceanic ooze at the sea-bottom. About 30% of oceanic bed is covered with the *Globigerina* ooze. These skeletal deposits are put to many uses. Some are employed as filtering agents, others are made into chalk and still others are used for abrasives.

6. ***Building material.*** The skeletal deposits in due course of time change into the limestone rock. Limestone is provided by the *Globigerina* ooze, such as that of cliffs of Dover, which have played an important role in the defence of England. *Camerina* (Formally known as *Nummulites*, are the largest protozoa, which form lime-stone. The pyramids of Egypt are made largely of this Nummulitic lime-stone. Thus, limestone-beds are used as building material.

7. ***Oil Exploration.*** Petroleum is organic origin. The skeletal deposit of Forminifera and Radiolaria are often found in association with oil deposits. In this manner they help in the exact location of oil.

8. ***Scientific Study.*** Many protozoa are used in biological and medical researches. A Holotricha, *Tetrachymena geleii* is used in nutritional research. The effects of various foods and poisons have been investigated on this protozoan.

HARMFUL PROTOZOA

The protozoa are harmful in following ways:

1. *Pollution of water.* Drinking water in natural condition is made unpalatable by the reproduction of some free-living protozoa in it. For example, *Uroglenopsis* brings fleshy odour like that of cod-liver oil. *Peridinium* emits smell resembling that of clam-shells.

2. *Destruction of animals of food value* Biolumeniscent dinoflagellates like *Noctiluca* and *Gonyaulax,* when become abundant, are responsible for turning the ocean red. The water becomes foul and cause toxic reaction to molluscs like clam oysters and mussels and they become unfit for eating by human beings.

3. ***Destruction of Wooden articles.*** Some flagellater like *Trichonympha* and *Colonympha* live in the gut of termites and help in the process of cellulose digestion. In the absence of these flagellates the termites will die or change their diet. Thus such protozoa indirectly help in distruction of wooden articles and books.

4. ***Reduction in fertility of soil.*** It has been observed that about 200 - 300 species of protozoa are present in soil. These protozoa feed on nitrogen-fixing bacteria thus reduce the fertility of soil.

5. ***Parasitic Protozoa.*** Parasitism is an association of two organisms in which one organism lives at the expense of the other giving nothing in return. The first animal is called parasite and other the host. Parasitism is a specialized mode of life and various level of it are found in protozoa. There are two type of parasites on the basis of their survival *Facultative parasites* are those that can live for various lengths of time without the host. The *obligatory parasites* are those which can-not survive when separated from their host.

Depending upon the nature and degree of intimacy and the placement of the parasites on their hosts. These are two types of parasities.

Ectoparasites. When the parasite lives on the external surface of the host, it is called ectoparasite. A few species like *Hydramoeba hydroxena* (found on epidesmis of *Hydra)* ; *Costa necatrix* found in the epidermal cells of fishes, and *Ichthyophthirius multifilis,* inhibiting in the skin of fishes. They feed on the liquified host cells causing various degrees of cellular necrosis. These parasites are transferred directly from host to host.

Endoparasites. When the parasite lives inside the body of the host such parasite is called endoparasite. Following types of endoparasites are there depending upon their location:

1. ***Coelozoic.*** When the parasite lives in the alimentary canal or cavities of host body. For example *Entamoeba histolytica, Giardia lamblia, Balantidium coli* etc.

2. ***Histozoic.*** Which occupies space between the host cells. For example young trophozoie of *Monocystis* lives in sperm morula of earthworm. *Trypanosoma gambiense* lives in the blood plasma of verebrates.

3. ***Cytozoic.*** These live inside the host's cell. For examples *Plasmodium* and *Babesia* in the blood cells of hosts. *Sarcocystis* lives in the striped muscles of host.

Effects of Parasitism on the parasite.

Parasitic protozoa are usually simplified in their structure. They do not possess any elaborate organelles, except the ciliate parasite. The endoparasites show maximum structural simplicity. They lack contractile vacuoles, locomotory and ingestory organelles. However, coelozoic forms like *Trichomonas, Giardia* and *Balantidium* possess locomotory and ingestory organelles. Surplus food is stored as glycogen, volutin granules and chromatoid bodies. Many parasites are aerobic as well as anaerobic depending on the availability of oxygen. In some. protozoa like *Gregarina* develops the organelles of attachment. Occurrence of complicated life-cycle often involving cyst-formation and repid multipletion by merogony and schizogony.

Effects on hosts.

The parasites are of two types non-pathogenic and pathogenic. The non-pathogenic parasites have no noticeable ill-effects on the host. For example *Entamoèba coli, Trichomonas hominis, Endolimax nana* etc. The pathogenic parasites are those which cause ill-effects or diseases to the hosts.

Transmission of parasite to other host and its infection. The infection of new host takes place in two steps first the exist of parasite from the first host, and second the transmission in the second host. Following ways are there in which the parasitic protozoa reach new hosts:

1. ***Active migration.*** This occurs in epizoic parasites of aquaitic hosts e.g. *Ichthyophthirius* which actively comes out of the host i.e. fish and infects the other host.
2. ***Direct transfer.*** This type of infection takes place on coming in contact of the other host e.g. *Trichomonas vaginalis* etc.
3. ***Accidental infection.*** in this method the new host in infected accidently swallowing of the parasite in the encysted condition along with food and weter e.g. *Conccidia, Entamoeba histolytica* etc.
4. ***Infection by intermediate host.*** This takes place in blood-parasites like *Plasmodium* and *Trypanosoma* in which a vector transmits the parasite through blood sucking (like *Anopheles* female mosquito,

tse tse fly). The developmental stages may or may not be found in the vector.

5. ***Congential transfer.*** The germinative infective, taking place at or before birth, may occur in several ways as placental transfer takes place in case of *Plasmodium* in man, ovarian transfer as takes place in *Nosema* in silkworm etc.

6. Pathogenic Protozoa

The disease causing protozoa are called pathogenic protozoa. Such protozoa are met within all the classes of this phylum. In man alone, more than 25 types of parasitic protozoa are found. Depending upon the location inside the host body, these are divided into various groups. These are:

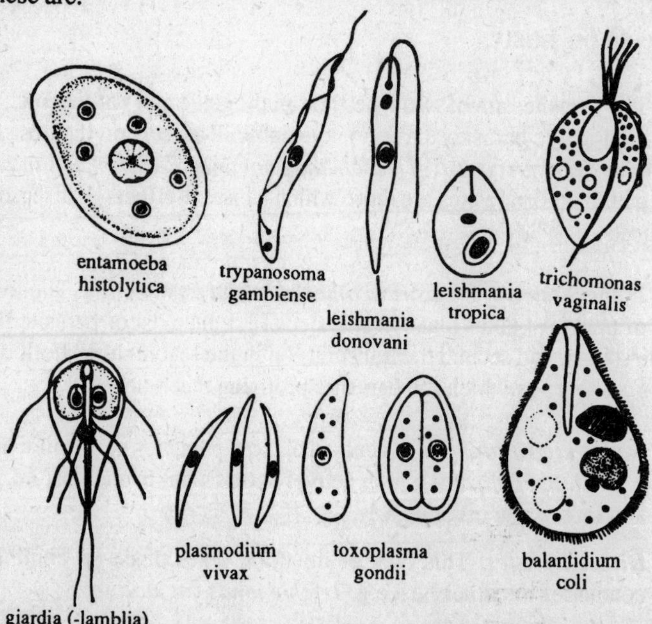

Fig. 23.1 Various Pathogen Protozoans.

1. Parasites of mouth.

(i) ***Entamoeba gingivalis.*** A Sarcodina lives in the tartar of teeth and pockets of gums. They probably aggravate pyorrhoea in man. They are also found in dogs and cats. The human infection is caused by kissing.

(ii) ***Trichomonas tenax.*** A flagellate also lives in pus pockets formed between teeth and gums. It causes *pyorrhoea.* It is transmitted during kissing.

(iii) ***Leishmania braziliensis.*** A flagellate lives in the ro-nasal mucous membrane. It causes *espundia.* The infection is transmitted to man by the bite of sand-fly.

2. Parasites of Digestive Tract.

(i) ***Entamoeba histolytica.*** It lives in the colon of man, sometimes in dogs and cats. It causes *amoebic dysentery.* It may reach liver, spleen, lungs and liver and causes *amoebic abscesses.* Infection takes place by swallowing of cysts.

(ii) ***Trichomonas hominis.*** It lives in large intestine of man, monkeys, cats and dogs. It causes *diarrhoea* and *dysentery.* It is transmitted through infective cysts.

(iii) ***Chilomonas.*** It lives in large intestine of man and causes *diarrhoea* and allied troubles. Transmission is by means of infective cysts.

(iv) ***Giardia intestinalis.*** It lives in the large intestine of man causing *diarrhoea.* Transmission takes place by tetranucleated cyst with contaminated food and water.

(v) ***Isopora hominis.*** It lives in the large intestine. Some other species occur in dogs and cats. They cause *diarrhoea* and abdominal discomforts. Infection occurs by occysts through contamination.

(vi) **Sarocystis.** It occurs in the muscles of oesophagus of sheep causing weakness of muscles. Transmission by cysts.

(vii) ***Nosema apis.*** It occurs in the intestine and Malpighian tibules of honey bee causing *Nosemia.*

(viii) ***Balantidium coli.*** It lives in the colon of man causing chronic ciliary dysentery. Transmission is by infective cysts.

(ix) ***Eimeria.*** It is found in the digestive tract of sheep, birds and cattle. It causes *diarrhoea* to haemorrhage. Transmission by *oocysts.*

(x) ***Trichomonas gallinae.*** It is found in the epithelial lining of oesophagus and crops of fowl, pigeons and turkeys. It causes necrotic nodules. Transmission through droppings.

(xi) ***Histomonas meleagridis.*** It lives in the calcium and liver of turkeys and fowls causing black-head disease. Transmission by ingesting contaminated eggs of caecal worm.

3. Parasites of Blood and Lymph.

(i) ***Trypanosoma sps.*** A number of species of *Trypanosoma are* parasitic in the blood of man and other verebrates. These are:

(a) ***T. gambiense.*** It occurs in the blood of man causing African sleeping sickness. It is transmitted through the bite of tsetse fly, *Glossina palpalis.*

(b) ***T. rhodesiense.*** In man it causes Rhodesian sleeping sickness. It is transmitted through the bite of tsetse fly, *Glossina morsitans.*

(c) ***T. cruzi.*** In early stage it is found in the muscles, heart, brain, special cord and in later stages in the blood. It causes *Chaga's disease* in South and Central Africa. Infection transmitted by the bite of bug, *Tritoma magista.*

(d) ***T. brucei.*** It causes *nagana disease* in domesticated cattles and it is transmitted by *Glossina morsitans.*

(e) ***T.evensi.*** It occurs in the blood of horse, drogs, camels etc. It causes *surra disease.* It is transmitted by *Tabanus fly* .

(f) ***T. equiperdum.*** It is found in the blood and germinal epithelium of horses, dogs and donkeys. It causes *dourine* disease in horse. The disease is transmitted during copulation.

(ii) ***Leishmania sps.*** It is a parasite of man and other animals. A number of species are there:

(a) ***Leishmania donovani.*** It is found in the endothelial cells of blood and lymph capillaries and leucocytes in the spleen, liver, bone marrow and lymph gland causing *visceral leishmaniasis* or *kala-azar*. Infection occurs by the bite of sand-fly, *Phlebotomus.*

(b) ***L. tropica.*** It occurs in the endothelial cells of dermal tissues causing *Oriental sore* or *cutaneous leishmaniasis* in man. Infection occurs by the sand-fly.

(c) ***L. Infentum.*** It occurs in the spleen of children. It causes enlargement of spleen. Transmission by the bite of sand-fly.

(iii) ***Plasmodium sps.*** A number of species of *Plasmodium* occur as intracellular parasites in red blood corpuscles of man and birds. The *Anopheles* female is the victor. These are.

(a) ***Plasmodium vivex.*** It causes *benign terrain malaria.*

(b) ***P. ovale.*** It also causes benign terrain malaria.

(c) ***P. falciparum.*** It causes *malignant tertian.*

(d) ***P. malariae.*** It causes quartan fever.

(e) ***P. praecox.*** It causes malaria in birds.

(f) ***P. gallinaceum.*** It causes malaria in chickens and pheasants.

(iv) ***Babesia sps.*** Three species of *Babesia* found in cattles, dogs and horses, These are:

(a) ***Babesia bigemina.*** It occurs in R.B.C. of cattles causing *Texas* or *Red water fever.* It is transmitted by the bite of tick *Boophilus.*

(b) ***B. canis.*** It occurs in the R.B.Cs. of horses causing *jaundice* and sometimes paralysis of hind-limbs. Transmission through the ticks.

(v) ***Haemoproteus.*** It is found in the R.B.Cs. and endothelial cells of blood vessels in birds

(4) Parasites of Urogenital tract.

(i) ***Trichomonas vaginalis.*** It is found in the vagina of female causing *vaginitis* (annoying itch and abnormal discharge) and in man it is found in urethra. Transmission is direct through inter-course.

(ii) ***Trichomonas foetus.*** It occurs in the urogenital system of sheep, horse and cattles. It causes uterine and penial disorders. Transmission is direct through intercourse.

(iii) ***Eimeria truncata.*** It occurs in the kidneys of goose and often causes death of the host.

INDEX

A

Accessory Flagellum, 235
Actinopodia, 230
Adoral Zone, 208
Alar process, 126
Allen, 1, 105
Amoebula, 114
Amphitrophic, 77
Anaplasma, 2
Anopheles, 148
Antheridia, 24
Archegonia, 24
Autotrophic, 77
Axonema, 71
Axopodia, 134
Axostyle, 31

B

Bacterium coli, 188
Bano, 155
Barnes, 3
Berthold, 102
Binary fission, 89, 119, 136
Biurets, 109
Blepharoplast, 71, 85
Brandit, 256
Brokaw, 76
Butschli, 1, 75

C

Caedibacter, 204
Calkin, 249
Capitate, 219
Carbonyl diurea, 109
Cephalonts, 169
Chadefaud, 69, 78
Charl Laveran, 147
Christian Huygens, 175
Chromidia, 36
Cladothrix, 35
Cleveland and Sanders, 116, 121
Congo Red, 190
Conoid, 145
Coccidiosis, 165
Condensation membrane, 100, 189
Craig, 117
Cryptomerozoites, 150
Cryptomitosis, 112
Cunninghum, 94
Cyclosis, 100, 89
Cytolysin, 55, 118
Cytopharynx, 70
Cytostome, 70
Cytopyge, 107, 186

D

Dellinger, 103, 239
Deutomerite, 166, 169
Diastole, 78, 109
Difflugia, 224
Diller, 200, 202, 203
Dobell, 116, 121
Dujardin, 1

E

Ectocyte, 138
Effective stroke, 76, 186, 239
Elmassion, 116
Endocyst, 216
Endosome, 86, 118
Endospheres, 171
Englmann, 256
Epicuticle, 69

Epimerite, 169
Eruptive type, 101
Euglenomorpha, 69
Exconjugants, 197
Exogamous, 141

F

Feltcher *et al.*, 116
Finlay, 213
Flagella, 2
Flagellula, 130
Fritz Schaudinn, 160

G

Gambusia, 159
Gamogony, 149
Gamont, 130, 153
Gelei, 178
Germplasm, 2
Globigerina ooze, 39
Glossina, 84, 92
Glycogen Vacuole, 119
Goldacre, 104, 237
Goldfuss, 1
Golgi, 149
Grassi, 1
Gray, 75
Gregarine movement, 140, 242
Grell, 129, 241, 251

H

Haemozoin, 148
Hahn, 142
Hall, 1
Halo, 118, 130
Hegner, 123
Herting and Maupas, 195
Hoarse, 121
Hologamete, 162
Holophytic, 77
Holozoic, 105, 118, 129
Honiberg *et al* 3
Hyman, 101, 103, 239

I

Idiochromatin, 171
Incubation period, 152
Intra capsular, 138
Isogametes, 142

J

Jennings, 102, 239
Jepps, 129
John Hill, 175

K

Kahl, 296
Kala azar, 96
Karyosome, 246
Kathrobic, 68
Kazloff, 3
Kinetic appartus, 86
Kinetodesma, 182
Kinetoplast, 85
Kleine, 84
Koch, 116
Kofoid, 116
Kolzoff, 210
Kortulis, 116
Krijisman, 76
Kudo, 1, 136, 248, 248

L

La Calvez, 129
Lamble, 116
Lanceolates, 219
Landis, 205
Laveran and Mesnis, 94
Leedale, 69
Leeuwenhoek, 1
Leishman and Donovan, 94
Leishmaniasis, 96
Linnaeus, 1
Lister, 129
Lobopodia, 101, 118

Lorica, 223
Losch, 116
Lowends, 75
Luciferase, 25
Luciferin, 25
Lumbricus, 137
Lumnisencence, 25

M

Macroconjugants, 214
Manton, 72
Marshall, and Williams, 141, 149
Mast, 103, 239
Mastigonemes, 72
Megalospheric, 126
Merozoites, 162
Mesosaprobic, 68
Metacycleic, 90
Metcalf, 173
Microconjugants, 213
Micronemes, 145
Microspheric, 126
Mieschner's tube, 48
Miller *et. al.*, 116
Minuta form, 119
Muciferous bodies, 69
Mucron, 138
Muller, 134
Multiple Fission, 2, 136
Myocyte, 138
Myonemes, 69
Myres, 129

N

Neuromotorium, 182
Neuromotor System, 74
Nevicella, 143
Ni Colle, 94
Noirot - Timothee, 171
Noland and Finlay, 208
Noller, 116
Nosema, 49
Nummulites, 2

O

Oligosaprobic, 68
Omnivorous, 106
Oocyst, 154, 164
Ookinete, 154
Opisthe, 195
Osmoregulation, 100

P

Paedogamy, 46
Pantin, 103, 239
Paraflagellar body, 71,
Parthenogonidia, 24
Pasteur, 1
Pateka, 1, 119, 176,233
Pebrine, 49
Pellicle, 2
Peramecin, 204
Periplaneta americana, 166
Pfifer, 144
Phlebotomus, 94
Pheretima posthuma, 137
Plasmosome, 246
Plasmosome, 246
Polyenergid nucleus, 114, 250
Polysaprobic, 68
Polythalamous, 126, 226
Precystic form, 119
Primite, 167
Pringsheim, 69
Profluent type, 101
Proter, 195
Protomerite, 166
Pseudonevicella, 168
Pseuopodia, 2
Pseudopodiospores, 114, 130
Pulsellum, 235
Pusule, 257
Pylome, 36
Pyrenoid, 74
Pyrenophores, 73
Pyrenosomes, 74

R

Rainey's corpuscles, 48
Recovery stroke, 76, 186, 239
Rejuvination, 199
Reservoir, 87

Residium, 164
Reticulopodia, 41
Retral proceses, 126
Rhumbler, 106, 239, 253
Rhumbler and Butschli, 103
Rinaldi and Jahn, 105
Ronald Ross, 1

S

Sacllie, 147
Sarcophytic, 77
Sarcocystin, 49
Sarcocyte, 138
Sarcosporidiosis, 49
Satellite, 167
Schaudinn, 116, 130
Schizonts, 150, 161, 162
Schuffner's dots, 152
Scyphida, 213
Sleigh, 250
Soldo, 204
Somatoplasm, 2
Sonneborn, 198, 202
Spasmoneme, 209
Spike, 31
Spirostomum, 2
Sporogony, 149
Sporozoites, 143, 161
Stentorin, 59
Stichonematic type, 72
Symmetrogenic, 81, 89, 173
Syngen, 198
Synkaryon, 154, 215
Systole, 78, 110

T

Tail piece, 69
Tchakotine, 80
Telotroch, 212
Tetrasporous, 164
Thecoplasm, 210
Tractellum, 71
Trichocysts, 175
Tricystosome, 181
Triuretes, 109
Trophochromatin, 171
Trophozoite, 138, 148, 161, 166
Trypanosome Chancre, 92
Trypanosomiasis, 93

U

Undulating membrane, 85, 209
Utricularia, 159

V

Verworn, 240
Vestibule, 185, 208
Vickerman, 87
Volkovsky, 185, 211
Volvox Volutin, 1, 259
Von Rossel Rosenhof, 98
Vorticella, 3

W

Walker and Sellards, 116
Wessenberg, 171
Westphal, 250
Wich terman, 202
Woodruft and Erdmann, 201
Wright, 94

X

Xanthosomes, 126

Z

Zelleriella, 247
Zingher, 259
Zoochlorallae, 175
Zoospores, 130
Zootrophic, 108
Zumstein, 258

Recent Trends in

INVERTEBRATES

VOL. 2

H. S. Bhamrah
Kavita Juneja

ANMOL PUBLICATIONS

ANMOL PUBLICATIONS
4378/4B, Ansari Road,
Darya Ganj, New Delhi-110002.

Recent Trends in Invertebrates

ISBN 81-7041-473-3 (Set)
ISBN 81-7041-474-1 (Vol. I)
ISBN 81-7041-475-X (Vol. II)
ISBN 81-7041-476-8 (Vol. III)
ISBN 81-7041-477-6 (Vol. IV)
ISBN 81-7041-478-4 (Vol. V)
ISBN 81-7041-479-2 (Vol. VI)
ISBN 81-7041-480-6 (Vol. VII)
ISBN 81-7041-481-4 (Vol. VIII)

PRINTED IN INDIA

Published by J. L. Kumar for Anmol Publications, New Delhi and Printed at Sangita Printers, Maujpur, Delhi-53.

1

GENERAL CHARACTERS & CLASSIFICATION

Sponges which constitute phylum Porifera, are the most primitive of the multicellular animals. Neither true tissues nor organs are present, and the cells display a considerable degree of independence. All the members of the phylum are sessile and exhibit little detectable movement. This combination of characteristic convinced *Aristotle, Pliny,* and other ancient naturalists that sponges were plants. In fact it was not until 1765, when internal water currents were first observed by *Ellis,* that the animal nature of sponges became clearly established. As a result, *Linnaeus, Lamarck* and *Cuvier* placed the sponges under Zoophytes or Polypes in their systems, regarding them as allied to anthozoan coelenterates, and many investigators sought to find in sponges the polyps they thoughts must be there. *de Blaiville* (1816) recognized the lack of affinities of sponges with coelenterates and separated them into a group spongiaria allied to Protozoa, but this idea gained little attention, and, through much of the 19th century, sponges were placed with coelenterates, usually under the name Coelenterata or Coelentera or Radiata. Morphology and physiology of sponges were first understood by *Robert E. Grant* (1825), who coined the term Porifera (*L. porous* = pore; *ferro* = to bear) *Huxley* (1875) and *sollas* (1884) proposed the complete separation of sponges from other Metazoa on the ground of many peculiarities, but this view did not find its present general accept until about the beginning of 20th century. Sponges are now recognized as constituting a separated isolated branch of the Metazoa, named Parazoa, after *Sollas.*

General Characters

1. All sponges are aquatic, mostly marine but a few are fresh-water.
2. They are multicellular organisms with cellular grade of body organisation without forming tissues or organs.

3. They are sessile, solitary or colonial.
4. They have asymmetrical or radially symmetrical bodies.
5. The body is cylindrical, tubular or vase-like.
6. The body wall is *diploblastic* with outer dermal epithelium and inner gastral epithelium with a gelatinous non-cellular *mesenchyme* in between.
7. The body surface bears numerous minute pores i.e., *ostia* for the ingress of water.
8. They possess peculiar *canal system* through which water current flows drawing food and oxygen inside the body and carrying away excretory and reproductive products.
9. Internal skeleton is present in the mesenchyme in the form of calcareous or siliceous *spicules* or of proteinous *spongin fibres.*
10. Digestion is entirely intracellular.
11. Nervous and sensory cells are lacking.
12. All sponges possess great power of regeneration.
13. They are mostly bisexual but asexual forms also exist.
14. Reproduction takes place both by asexual and sexual methods.
15. Asexual reproduction takes place by budding, fission or *gemmule* formation and sexual reproduction by *sperms* and *ova.*
16. Fertilization is internal and at the same time it is cross fertilization.
17. Cleavage is holoblastic and development is indirect followed by a free swimming ciliated larva, the *amphiblastula* or *parenchymula.*

CLASSIFICATION

The classification of sponges presented great difficulty, and no one scheme has been unanimously accepted by specialists on sponges. The classification is based on the type of skeleton, and this serves to mark off rather sharply two groups of sponges, the calcarious and the glass sponges. The remaining silicons and horny sponges are, however, less clearly delimited, and they are variously subdivided by different workers. The classification here adopted is that given by *Topsent, Wilson* and *de Laubenfels.*

Phylum porifera is divided into three classes :

Class-1. Calcarea

The members of this class possess their skeleton composed of separate calcareous spicules made of calcium carbonate (*L., Caleis* = lime; Gr., *sponges* = sponge).

1. Members of this class are smaller in size, extending up to 15 cm in length; may be solitary or colonial. All are marine.
2. Shape of body usually cylindrical or vase-like.
3. Osculum is narrow, terminal and fringed with spicules.
4. The colour of the body is white or dull-brown.
5. Skeleton consists of spicules which may be *monaxon, triradiate* or *tetraradiate.*
6. Collar cells or *choanocytes* are comparatively large.
7. The canal system is ascon, sycon or leucon type.

Order - (i) Homocoela

1. Body is small, cylindrical and thin walled without any foldings.
2. Canal system ascon type.
3. Spangocoel is lined throughout by choanocytes.
   Examples : *Leucosolenia, Clathrina.*

Order - (ii) Heterocoela

1. Body is comparatively bigger, vase-like, thick walled with internal foldings.
2. Canal system sycon or leucon type.
3. Choanocytes are restricted to the flegellated chambers.
4. Spongocoel is lined by dermal epithelium.
   Examples : *Scypha* or *Sycon, Grantia.*

Class 2. Hexactinellida or Triaxonida

The members of this class are called glass-sponges having *triaxon* or *hexactinal* spicules made up of silica.

1. Members of this class are of moderate size, extending up to 1 metre in length and are usually solitary.
2. The body shape is usually cylindrical, cup-shaped or funnel-shaped.
3. The osuclum is wide and usually covered over by a siliceous sieve plate.
4. The skeleton consists of *triaxon* or *hexactinal* siliceous spicules, often fused resembling spun glass.
5. Dermal epithelium is wanting.
6. The canal system is simple rhagon type.
7. The choanocyte cells are restricted to finger-like flagellated chambers.
8. They are exclusively marine and occur chiefly in tropical deep sea water.

Class Hexactinellida is divided into two orders :

Order (i) - Hexasterophora

1. The beautiful glass like body is cylindrical, very light and attached to the substratum directly.
2. The spicules are *hexasters* having star-like appearance.
3. The flagellated chambers are radially and regularly arranged.
Example : *Euplectella* (Venus's flower basket).

Order (ii) - Amphidiscophora

1. The body is usually rounded or oval and attached to the substratum by root tufts.
2. The spicules are *amphidiscs* having a convex disc, bearing a crown of backwardly directed marginal teeth at both ends.
3. The flagellated chambers are irregularly arranged.
Example : *Hyalonema* (Glass rope sponge).

Class 3. Demospongiae

This class comprises diverse and complicated sponges with skeleton of siliceous spicules or spongin fibres or a combination of both or totally absent. (Gr., *demos* = frame; *spongos* = sponge.)

1. Members of this class are highly organised, varying from small to large size and may be solitary or colonial.
2. The body shape is variable being rounded, oval, cup-like, funnel-like.
3. The skeleton is made up of silicious spicules or spongin fibres or both. The spicules are monaxon or tetraxon.
4. Canal system is complicated and leucon type.
5. Choanocytes are restricted to small flegellated chambers.
6. They are mostly marine but few sponges are fresh water.

Class Demospongiae is divided into three sub-classes :

Subclass-I. Tetractinellida

1. Body is rounded or flatenned.
2. Silicious spicules, if present are tetraxon, or absent.
3. Spongin fibres are absent.

Order (i) Myxospongia

1. Spicules are absent

2. Body surface is velvet-like
   Example : *Oscarella.*

Order (ii) Carnosa

1. Homogenous leuconoid structure.
2. Spicules are of equal-size.
   Examples : *Chondrilla, Plakina, Halina.*

Order (iii) Choristida

1. Body form is usually elaborate
2. Both small (microscleres) and large (megascleres) spicules are present
   Example : *Geodia, Ancoria.*

Subclass-II. Monaxonida

1. Body form variable, occurring in shallow and deep water.
2. The skeleton consists of monaxon spicules distinguished into megascleres only.
3. Spongin fibres may or may not be present.

Sub-class Monaxonida is divided into four orders :

Order (i) Hadromerina

(i) megascleres are mostly *tylostyles* i.e., the broad end is knobbed.
(ii) Microscleres are usually absent and if present, are star-like.
(iii) Spongin fibres are absent.
Examples : *Cliona* (Boring sponge), *Suberites, Poterion.*

Order (ii) Halichondrina

(i) Megascleres are always of more than one kind.
(ii) Microscleres are usually absent.
(iii) Spongin fibres are present but they are very scanty.
Example : *Halichondria.*

Order (iii) - Poecilosclerina

1. Megascleres are usually of two or more kinds and united by spongin fibres into a regular network.
2. Microscleres may be curved, bow-shaped or C-shaped.
   Examples : *Myxilla, Microciona.*

Order (iv)- Haploscerina

(i) Megascleres are always diactinal and pointed at both ends.
(ii) Microsclers may or may not be present.
(iii) Spongin fibres are usually present.
Examples : *Chalina, Spongilla* (fresh-water).

Sub-Class 3. Keratosa

(i) The body form is usually rounded and massive with a leathery surface and dark colour, occurring generally in warm shallow water.
(ii) The skeleton is exclusively composed of spongin fibres.
(iii) The spicules are totally absent.
Examples : *Euspongia* (Bath sponge), *Hippospongia* (Horse sponge).

CLATHRINA

Phylum	-	**Porifera**
Class	-	**Calcarea.**
Order	-	**Homocoela**
Genus	-	***Clathrina***

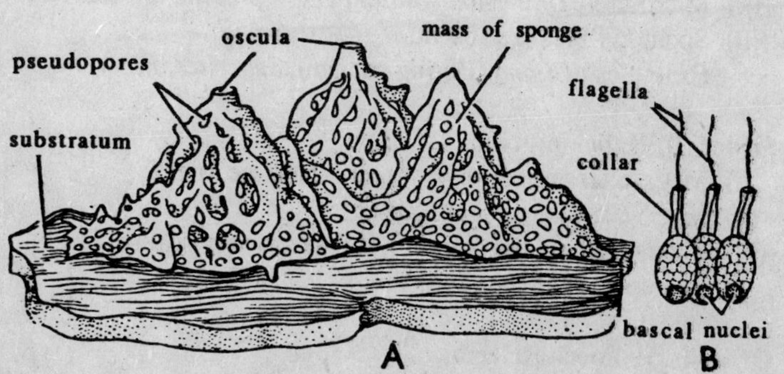

Fig. 1.1. Clathrina
A-Colony of *Clathrina clathrus*.
B-Choanocytes of *Clathrina coriacea*.

1. It is a simple, marine sponge attached to the hard substratum.
2. It is sedentary sponge with a free end containing *osculum*.

3. The surface is perforated by minute apertures called *ostia*, through which the water enters the body. Canal system asconoid type.
4. Skeleton consists of caleareous spicules which are triradiate with equal rays.
5. Choanocytes lines the spongocoel.
6. Dermal membrane or cortex is lacking.
7. Development includes a paranchymula larva and the so called olynthus stage.

GRANTIA

Phylum	-	**Porifera**
Class	-	**Calcarea**
Order	-	**Heterocoela**
Genus	-	***Grantia***

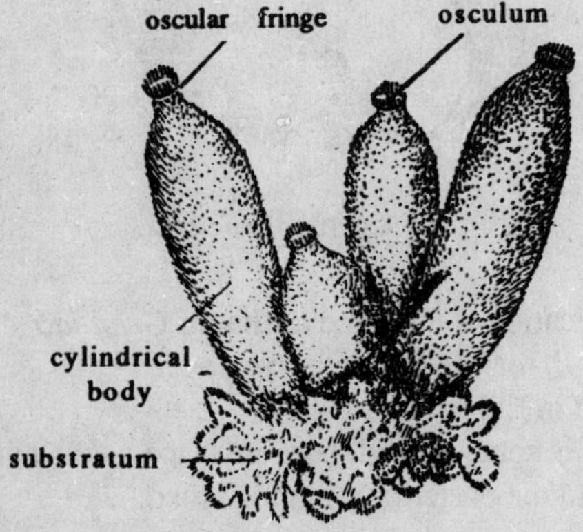

Fig. 1.2. Granitia

1. It is a small, solitary or colonial, marine sponge.
2. It is found in the Gulf of St. Lawrence.
3. The body is complex vase-like.
4. Each cylinder bulges in the centre and opens to the exterior by an osculum.
5. The osculum is encircled by a fringe of large giant monaxon spicules forming oscular fringe.
6. The body surface is covered by the dermal membrane forming a cortex of varying thickness having special cortical spicules. The spicules are tangential radiate.

7. Body wall is composed of outer dermal epidermal, middle mesenchyme and inner flatenned epithelium lining the spongocoel.
8. Canal system is sycnoid.
9. Choanocytes are restricted to the radial canal, and the flegella arise directly from nuclei.
10. Bisexual. Reproduction by asexual and sexual methods. Asexual reproduction by budding and regeneration. Sexual by gametes. Larva is an amphiblstula.

LEUCILA

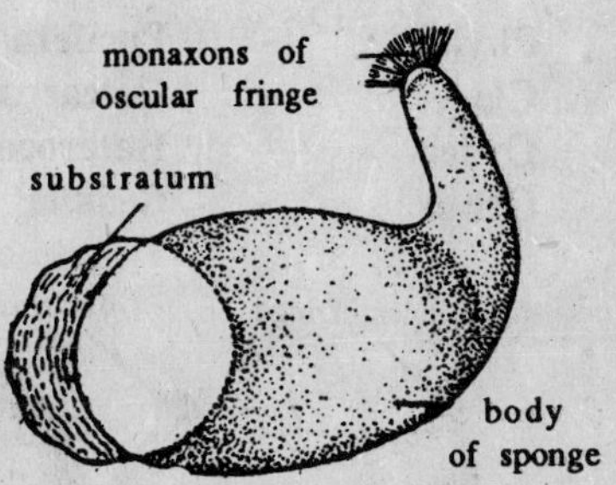

Fig. 1.3. Leucila

The classification is the same as that of *Grantia*.

1. It is a small, marine and solitary sponge.
2. It is found in Europe and U.S.A.
3. The body is somewhat globose in shape as the anterior region is narrow and the posterior region is broad.
4. The osculum is situated at the anterior most end and is surrounded by oscular fringe of monaxon spicules.
5. The spongocoel is narrow while the body wall is quite thick.
6. Canal system is primitive leuconoid type. Each radial canal is subdivided into elongated flagellated chambers grouped around a common excurrent canal.
7. The skeleton is of inarticulate type.

LEUCANDRA

The systematic position is the same as that of *Grantia*.

1. It is a small, marine and solitary sponge.

2. Body is globose in shape.
3. Body wall is thick. Osculum is situated at the anterior end and is surrounded by oscular fringe of monaxon spicules.
4. Dermal skeleton is formed of tangentially arranged triradiate spicules which may sometimes develop an apical ray.
5. The flagellated chambers are small and spheroidal.

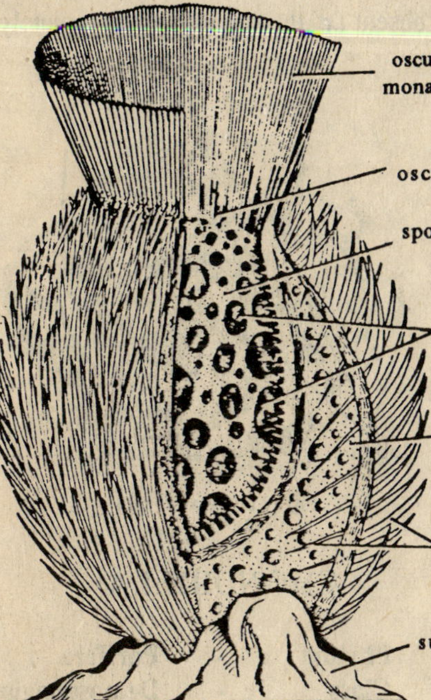

Fig. 1.4. Leucendra

OSCARELLA

Phylum	-	Porifera
Class	-	Demospongia
Subclass	-	Tetractinellida
Order	-	Myxospongia
Genus	-	*Oscarell*

1. It is found in shallow water forming encrustation over the rocks and submerged objects.

2. It occurs in British water.
3. It is a bright coloured small sponge whose upper surface is lobulated.
4. The canal system is leuconoid type.
5. The canal system is leuconoid type.
6. The lobules are covered with thin dermal layer. Each lobule contains as excurrent canal surrounded by flagellated chambers and the depression between the lobules house the incurrent canal.
7. Skeleton is absent *i.e.* the spicules are completely absent.

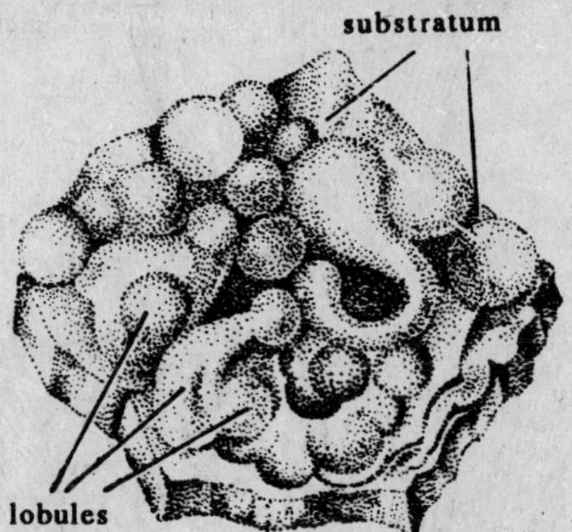

Fig. 1.5. Oscarella

CHONDRILLA

Phylum	-	**Porifera**
Class	-	**Demospongia**
Subclass	-	**Tetractinellida**
Order	-	**Carnosa**
Genus	-	***Chondrilla***

1. It is a marine sponge found in British water and U.S.A.
2. The entire body looks like a rounded mass hence commonly called *chicken-liver* sponge.
3. The body remains attached to rocks etc.
4. Free surface contains few oscula.
5. Body surface is smooth, but mesogloea is thick and stony due to spherasters.
6. Canal system is leuconoid type.

7. The spherasters found in cortex between the cones and are well developed while other spicules are absent.

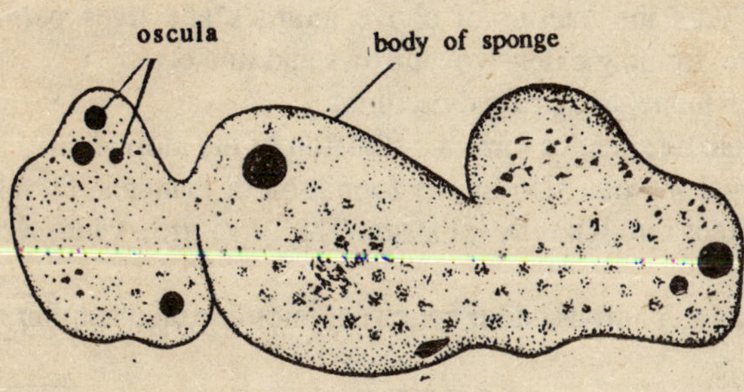

Fig. 1.6. Chondrilla

CLIONA

Phylum	-	Porifera
Class	-	Demospongia
Sub class	-	Monaxonida
Order	-	Hadromeria
Genus	-	*Cliona*

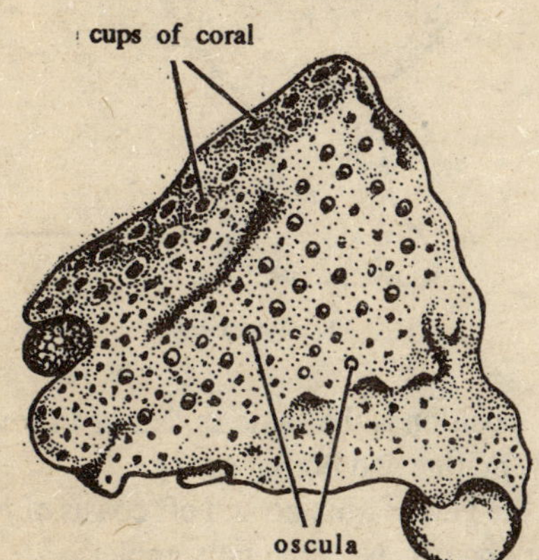

Fig. 1.7. Cliona

1. It forms low encrustation on rocks, molluscs shells, coral skeleton and other calcareous objects.
2. It is cosmopolitan in distribution.

3. It is commonly called 'Boring sponge' or 'Sulphur sponge'.
4. It is light yellow in colour but may be green or purple, and forms a mass upto 20 cm in diameter.
5. It enters the interior of above animals and lives permanently there, forming extensive burrows and tunnels.
6. Each tunnel has a raised osculum.
7. Monaxon spicules form the skeleton of the individual.
8. Sexual reproduction by the formation of gametes. Development indirect through a larval stage. The larva grows into a compact mass.
9. It has sulphurous odour hence commonly called sulphur sponge.

POTERION

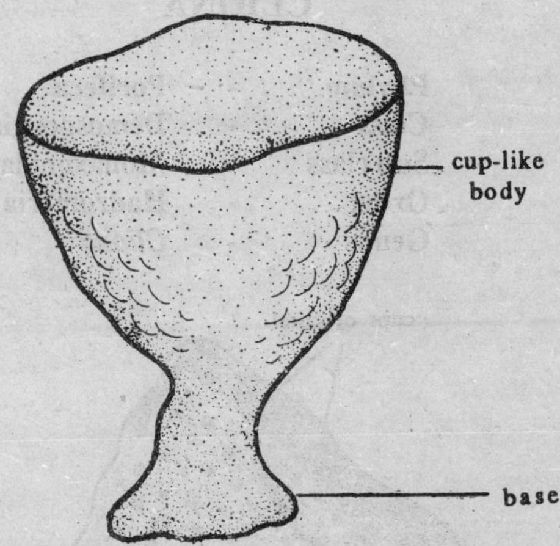

Fig. 1.8. Poterion

The systemetic position is the same as that of *Cliona*.

1. It is a solitary, marine and boring sponge.
2. It is found in the gulf of Mexico and off coasts of North Carolina.
3. It is commonly known as 'Naptune's goblet'.
4. The shape of body is cup like with a base. The body remains fixed to the substratum by base.
5. The incurrent apertures are present on the outer surface while excurrent apertures on the inner surface.
6. The skeleton is formed of tylostyle (monaxon) spicules.

7. Osculum at the anterior end is very large.
8. It is supposed to be one of the largest sponges, measuring 1-2 metres in hight.

TETHYA

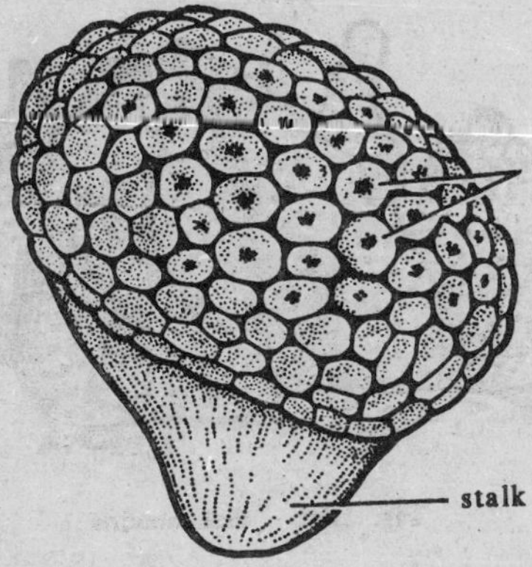

Fig. 1.9. Tethya

The systematic position is the same as that of *Cliona*.

1. It is a marine sponge inhibiting deep seas.
2. It is a rounded sponge.
3. Body surface tuberculate with radiating bundles of monaxon spicules and several exasters.
4. The styles have indistinct tylote ends.
5. Spongin fibres are absent.

HALICHONDRIA

Phylum	-	**Porifera**
Class	-	**Demospongia**
Subclass	-	**Monaxonida**
Order	-	**Halichondrina**
Genus	-	***Halichondria***

1. It is a colonial sponge. It forms low encrustations on rocks in shallow waters.
2. It is cosmopolitan in distribution.
3. It is commonly called the crumb of 'breed sponge'.

4. It is soft and yellowish or green or brown in colour.
5. The surface presents one or more large rounded openings called oscula.
6. Skeleton consists of oxeas lying without any arrangement.

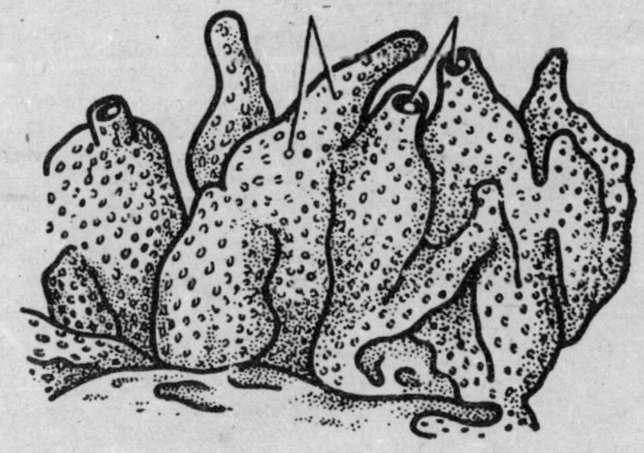

Fig. 1.10. Helichondria

MICROCIONA

Phylum	-	Porifera
Class	-	Demospongia
Subclass	-	Monaxonida
Order	-	Poecilosclerina
Genus	-	*Microciona*

1. It is a marine, dimorphic, colonial form. Found in shallow water as a soft, thin encrusting layer on rocks, shells and calcareous objects with ascending lobes.
2. It is found in long Island Sound and also present in South Carolina to Cape Cod.
3. In deep sea water the colony becomes massive.
4. It is bright red in colour. The animal comprises of several finger-like projections.
5. The spicules are of various forms and united by spongin into a regularly network.
6. The animal is used in regeneration and reassociation experiments. This animal has tremendous capacity to reassociate and regenerate.
7. Sexual reproduction by gametes. Development through a free swimming larval stage in early August.

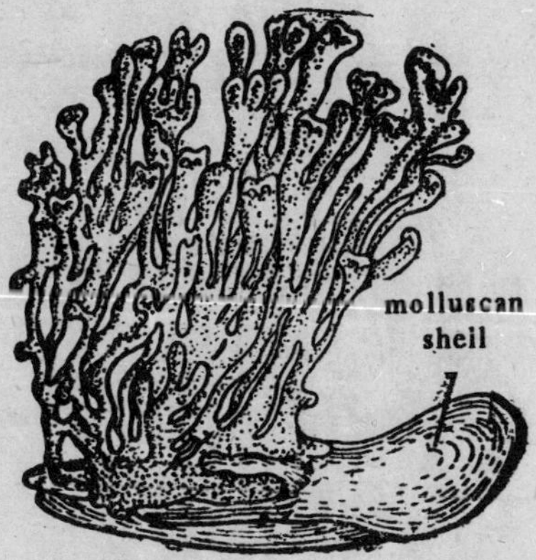

Fig. 1.11. Microciona

CHALINA

Phylum	-	**Porifera**
Class	-	**Demospongia**
Subclass	-	**Moxaxonida**
Order	-	**Haploscerina**
Genus	-	***Chalina***

1. *Chalina* is commonly called *Dead man's finger* found in deep water, orange-brown or red in colour.
2. It is found in Europe, common from Rhode Island to Labrador. North Carolina to Cape Cod.
3. It is much branched and each branch looks like a finger, so also called *mermaid's glove.*
4. Each branch is perforated by numerous openings or oscula.
5. Canal system is leucon type.
6. Monaxon siliceous spicules embedded is spongin fibres constitute the skeleton of this animal.
7. Asexual reproduction by budding and sexual reproduction includes a free-swimming larva.

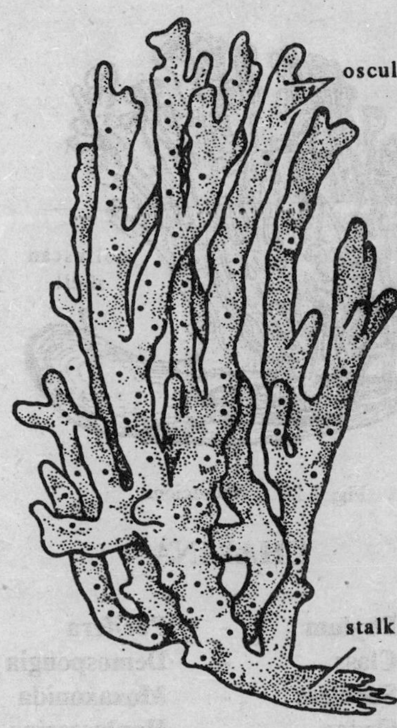

Fig. 1.12. Chalina

SPONGILLA

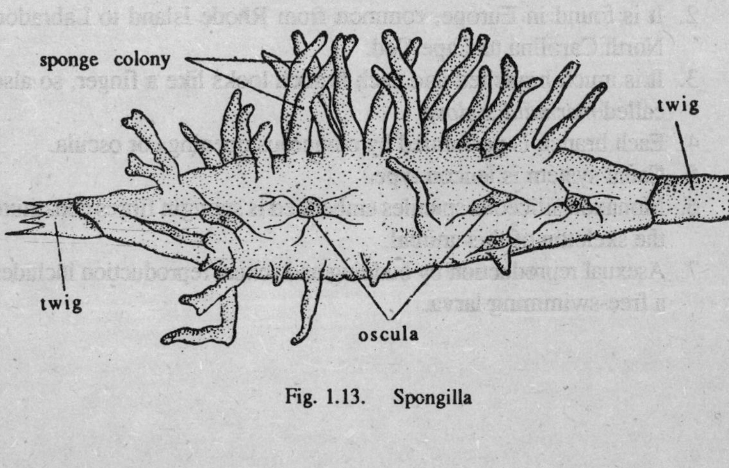

Fig. 1.13. Spongilla

The classification is the same as that of *Chalina*.

1. *Spongilla* is a colonial fresh-water sponge found attached to submerged sticks and plants of ponds, lakes and streams.
2. It is cosmopolitan in its distribution. It is found mainly in Atlantic, European and American waters.
3. Siliceous spicules form the skeleton. The spicules are oxeas, embedded in spongin.
4. Canal system rhagon type.
5. Highly branched colony with numerous oscula.
6. Asexual reproduction by budding and gemmule formation; sexual reproduction also occurs.
7. Gemmules are protected by the amphidisc spicules.

EPHYDATIA

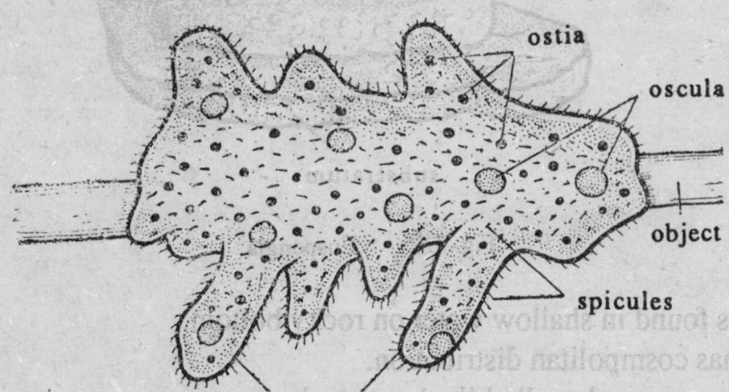

Fig. 1.14. Ephydatia

The classification is the same as that of *Cliona*.

1. It is a greyish or greenish fresh water sponge found in clear water in rivers, ponds and streams.
2. It is found in Vancouver Island, Eastern and Central States.
3. The body consists of delicate tubes each bears an osculum.
4. Surface is perforated by several large oscula and minute pores the osita.
5. The green colour of animal is due to the symbiont zoochlorellae.
6. Asexual reproduction by gemmules.

EUSPONGIA

Phylum	-	**Porifera**
Class	-	**Demospongia**
Subclass	-	**Keratosa**
Genus	-	***Euspongia***

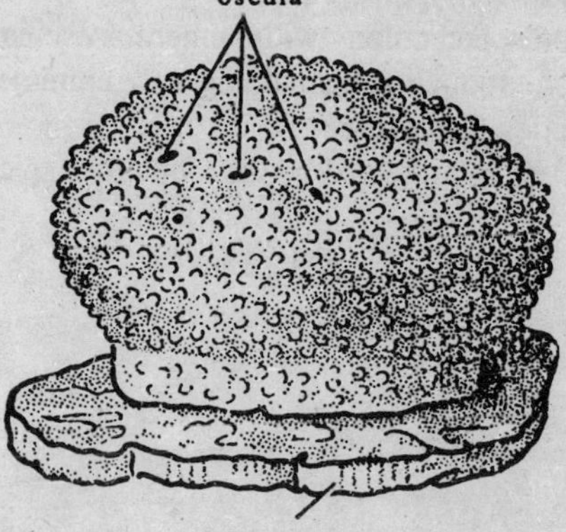

Fig. 1.15. Euspongia

1. It is found in shallow water on rocky bottom.
2. It has cosmpolitan distribution.
3. It is commonly called 'bath sponge'.
4. It forms a large brownish more or less spherical mass covered by a dark, non-living lathery membrane of spongin.
5. The surface bears small projections, the *conules,* supported by stiff spongin streams.
6. The body surface contains large openings called as oscula, and small openings called inhalant canals.
7. Internally the sponge resembles a liver in both consistency and colour.
8. It is unisexual and males are rare.
9. Skeleton is made of only spongin fibres. Spicules are absent. Many small worms and crustaceams live inside it as commensals.
10. It can live upto 50 years.
11. It is commonly used bathing, washing automobiles, furnitures, applying face creams etc.

HIPPOSPONGIA

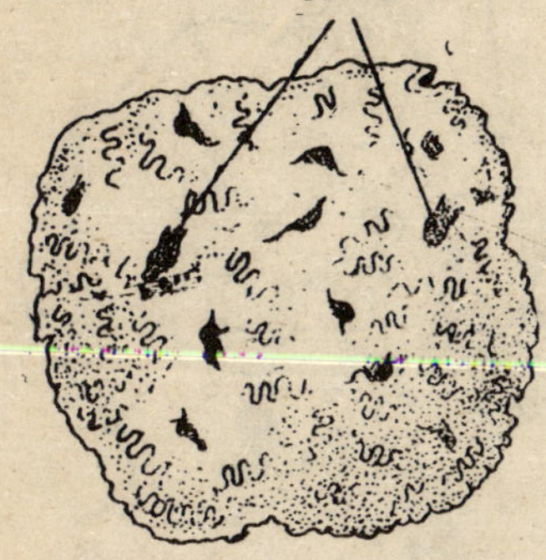

Fig. 1.16. Hippospongia

The classification is the same as that of *Euspongia.*

1. It is a horny sponge found 9-16 meters deep.
2. It is found in Mediterranean west Indies and common in Florida.
3. Body is massive and permeated by large, often cavernous canals.
4. Body is covered by a dark and dead membrane provided with numerous flagellated chambers.
5. Body surface is raised at several places bearing oscula. A single sponge is made up of few to many individuals, provided with few to many oscula.
6. Skeleton is made up of spongin only, spicules being absent.
7. It grows very large and may live for 50 years.
8. It harbours several small commensal crustanceans and worms.

HIRCINIA

The classification is the same as that of *Euspongia.*

1. It is a marine and sedentary sponge.
2. It has world-wide distribution.
3. It is commonly called horny sponge.
4. It attains a considerable size and rounded massive form. Body surface contains conules.
5. The body colour is due to presence of symbiotic blue-green algae.
6. The canal system is diplodal leucon type.
7. Skeleton consists of spongin fibres only.

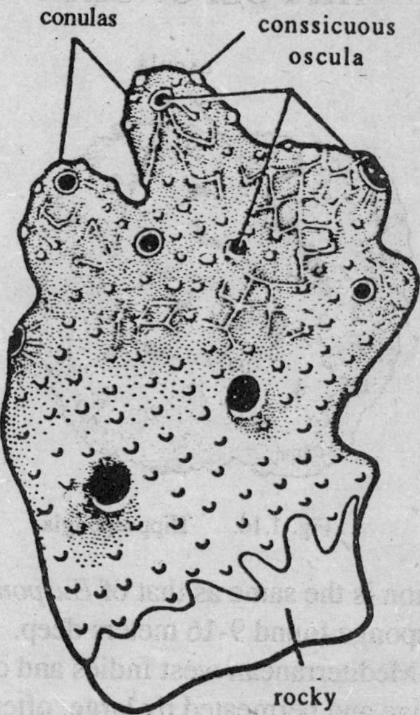

Fig. 1.17. Hircinia

EUPLECTELLA

Phylum	-	**Porifera**
Class	-	**Hexactinellida**
Order	-	**Hexasterophora**
Genus	-	***Euplectella***
Species	-	***aspergillum***

1. It is a deep sea form found near Philippine islands and has become curved due to slow water current.
2. It is commonly called *Venus' flower basket* and is a beautiful glass sponge.
3. It is made of six-rayed siliceous spicules which are held together by siliceous cement resulting in a beautiful basket-like structure.
4. The curved upper end of the body is closed by an *oscular sieve*.
5. From the base arise root spicules which are long siliceous threads which fasten the sponge to its unstable slimy substratum.
6. The spongocoel contains a pair of crustacea (*Spongicola*) which had commensal life throughout their life. For this reason *Euplectella*

is a beautiful presentation for newly married couple in some countries.

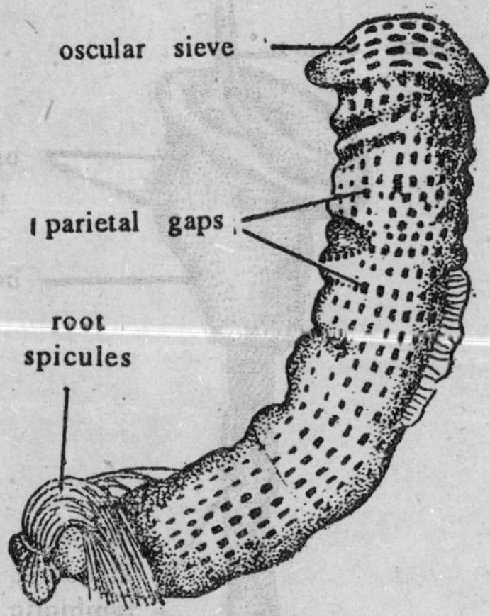

Fig. 1.18. Euplectella

HYALONEMA

Phylum	-	Porifera
Class	-	Hexactinellida
Order	-	Amphidiscophora
Genus	-	*Hyalonema*

1. It is a marine sponge found along new England coast.
2. It is commonly called as 'glassrope sponge'.
3. It has a spherical or oval about 30 cm long body fastened in the mud by a root-tuft of long twisted spicules.
4. The spicules continue through the body as an axis or *columella* and often project above it as the *gastral cone.*
5. The gastral cone bears opening of the excurrent canals.
6. The surface may be depressed to enclose a cavity, the spongocoel, or may be flat in which case there is no spongocoel and the excurrent canals open out directly.
7. It possess large amphidisc spicules like fresh-water sponges. Entire body contains small, branching, six-rayed spicules.
8. The flagellated chambers are arranged radially and in parallel planes in the sponge wall.
9. Some anemones live attached to the root-tuft and certain crustaceans live inside the body as commensals.

Fig. 1.19. Hyalonema

PHERONEMA

The classification is the same as that of *Hyalonema*.

1. It is found on the bottom of sea. It occurs in U.K., U.S.A. and Europe.
2. It has a cup or bowl-shaped body.
3. Body thick-walled with a large spongocoel, which opens above by an osculum.

4. Osculum is guarded by spicules or *marginal* prostals.
5. Glassy spicules project from the body surface, often in tufts called the *pleural prostals*, giving the appearence of glass wool.
6. Skeleton consists of amphidisc microsclers.
7. Canal system is of eurypylour leuconoid type.
8. Broad, ill-defined root-tufts made of silica, which are joined and serve for attachment to the substratum.

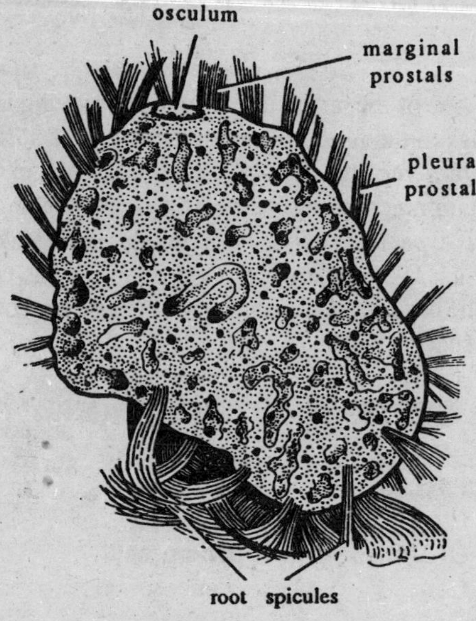

Fig. 1.20. Pheronema

LEUCOSOLENIA - AN ASCONOID SPONGE

Leucosolenia is one of the most simple sponge, growing in the sea shore water. It belongs to the family Leucosolenidae. It is an asconoid sponge which resembles very much with the olynthus stage, found in the life cycle of certain other sponges like *Scypha;* as such it is also referred to as *olynthus-type*. Near about 100 species have been recorded uptill now, out of which *L. complicata, L. botryoides*, and *L. variabilis* are quite common. The description given below is the general one and is applicable to all the species.

19-1. SYSTEMATIC POSITION

Phylum	-	**Porifera**
Class	-	**Calcarea**
Order	-	**Homocoela**
Genus	-	**Leucosolenia**

HABITAT

Leucosolenia is a small marine sponge. It occurs almost all over the world in shallow water below low tide mark on rocks, stones and warfs. Though frail, it prefers places with intense wave action to calm waters. The wave action ensures adequate food supply and well-oxygenated water.

HABITS

Leucolosenia is a colonial sponge. The colony is permanently fixed to the substratum. It feeds on microscopic animals and plants and organic particles. Digestion is intracelluar as in Protozoa. Reproduction occurs both asexually and sexually. Life-history includes a free-swimming flagellated larva for dispersal. The sponge is very sensitive to external conditions and dies if removed from its habitat. It has a good degree of power of regeneration.

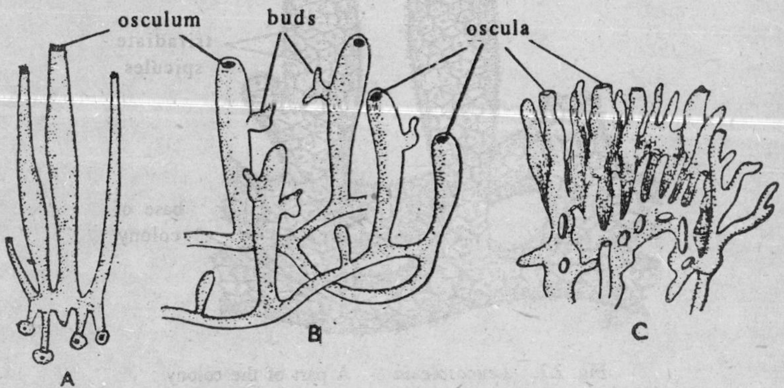

Fig. 2.1. Types of *Leucosolenia*
A—Simple, B—Branching. C—Reticulate.

STRUCTURE

TYPES OF LEUCOSOLENIA

A - Simple
B - Branched
C - Reticulate

Shape and size. The colonies of *Leucosolenia* consist of numerous cylindrical structures each of which is a complete animal. It is a radially symmetrical animal. All the cylinders of the colony are attached to each other at the base due to which the colony appears as a thick branched bush-like structure. Each cylinder is about 25 mm long and whitish or yellowish in colour.

The cylinders of *Leucosolenia* are radially symmetrical thin walled tubes which represent the *Olynthus stage*. In the centre of each cylinder is a cavity called *spongocoel* which opens to the outside by a large aperture known as *osculum*. The wall of the cylinder is perforated by numerous pores which are intracellular and called *incurrent pores or ostia*. They lead into the spongocoel.

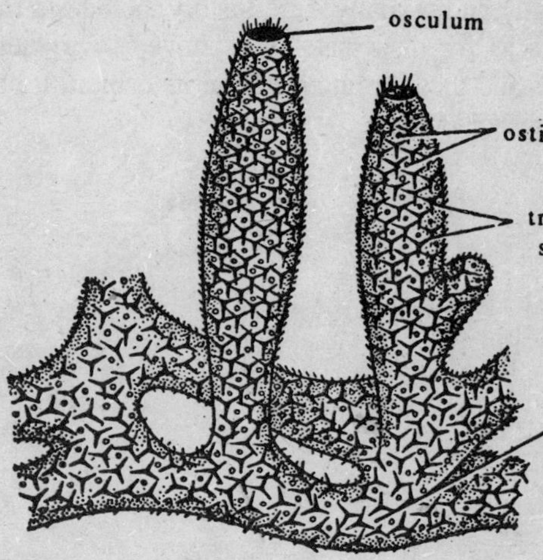

Fig. 2.2. *Leucosolenia* - A part of the colony

The *porocytes* are extremely elastic and close the ostia by contracting. The water current reaches the spongocoel through the ostia. The spongocoel is lined with choanocytes. The water of the spongocoel is expelled out through the *osculum* by the movement of the flagella of choanocytes, therefore, the structure of *Leucosolenia* is very simple.

BODY WALL

The bodywall is thin with an outer epidermis or *pinacoderm* and inner epidermis or *choanoderm* with a non-cellular mesenchyme or *mesogloea* in between. The wall is perforated by numerous pores through which water enters the spongocoel and passes out by osculum.

1. *Pinacoderm* : The outer layer consists of a single layer of thin, scale-like, flattened cells called the *pinacocytes*. The cells have thin margins and a bulging central part with a nucleus. The margins show contractility so that the sponge can increase or decrease in size to some extent.

   Specially, large and tubular *porocytes* are also present which are supposed to be modified pinacocytes. Each porocyte has a central canal-like space which communicates with the exterior through ostia or dermal pores. Internally, they open into the spongocoel. The ostia are regulated by *myocytes* which form circlets around the openings.

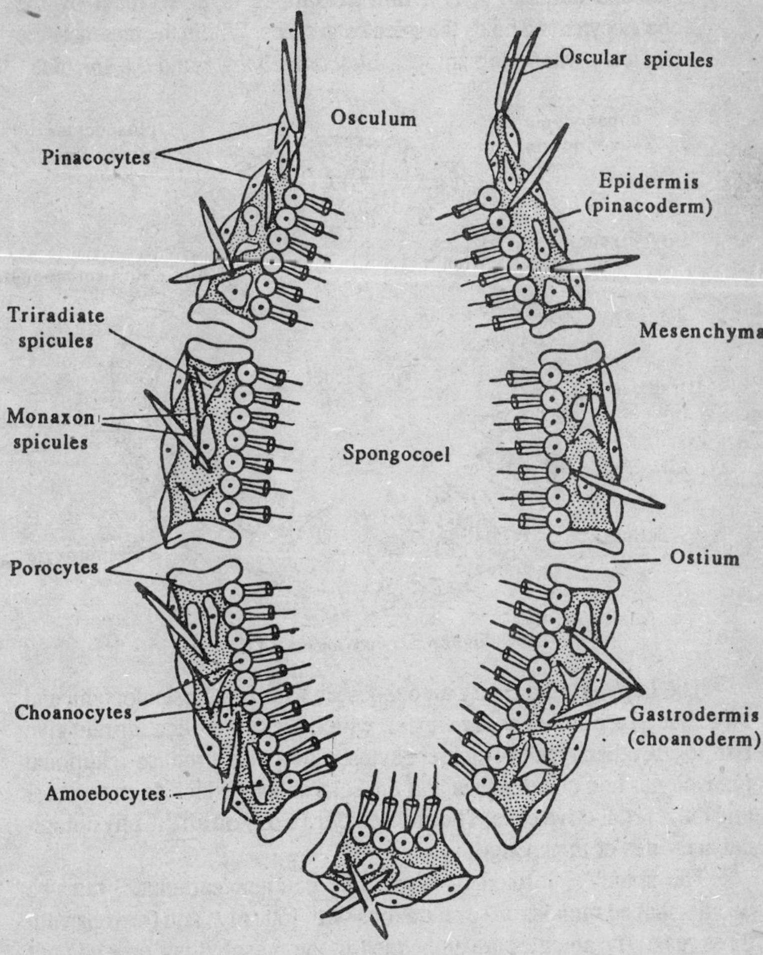

Fig. 2.3. L.S. of Leucosolenia

2. *Choanoderm* : It consists of a single layer of flagellated cells or *choanocytes* which line the spongocoel. A choanocyte cell resembles the choanoflagellate protozoa. It is an ovoid cell, its free end bearing a transparent contractile collar which surrounds a single long flagellum arising from a basal granule. The nucleus lies at the centre of the cell. These cells are used in feeding and for ensuring the flow of water within the animal's body.

3. *Mesenchyme or Mesogloea* : Mesenchyme lies between pinacoderm and choanoderm. It is a thin gelatinous layer secreted by the choanocytes and holds the spicules in space. Within the mesenchyme lie freely wandering amoeba-like, *amoebocytes*and the spicules.

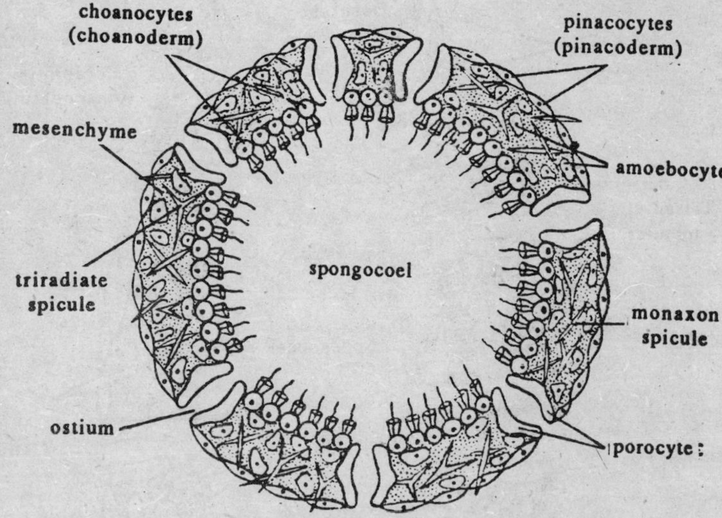

Fig. 2.4. Leucosolenia T.S.

The large sized primary amoebocytes with blunt pseudopodia and large nuclei are called *archaeocytes*, which are self replicating and give rise to all other types of amoebocytes. They also produce additional pinacocytes and choanocytes and the scleroblasts which form spicules and the reproductive cells. The archaeocytes carry on all the physiological activities of the sponge.

The spicules, formed of crystalline calcium carbonate, may be needle-shaped monoaxons or tetraxons with four rays and triaxons with three rays. The spicules are embedded in the mesohyl but protrude out of the body to give roughness to the body.

PHYSIOLOGY

The sponges being fixed animals are quiet and inactive-looking animals. Their physiology is mainly drawing a current of sea-water into their bodies.

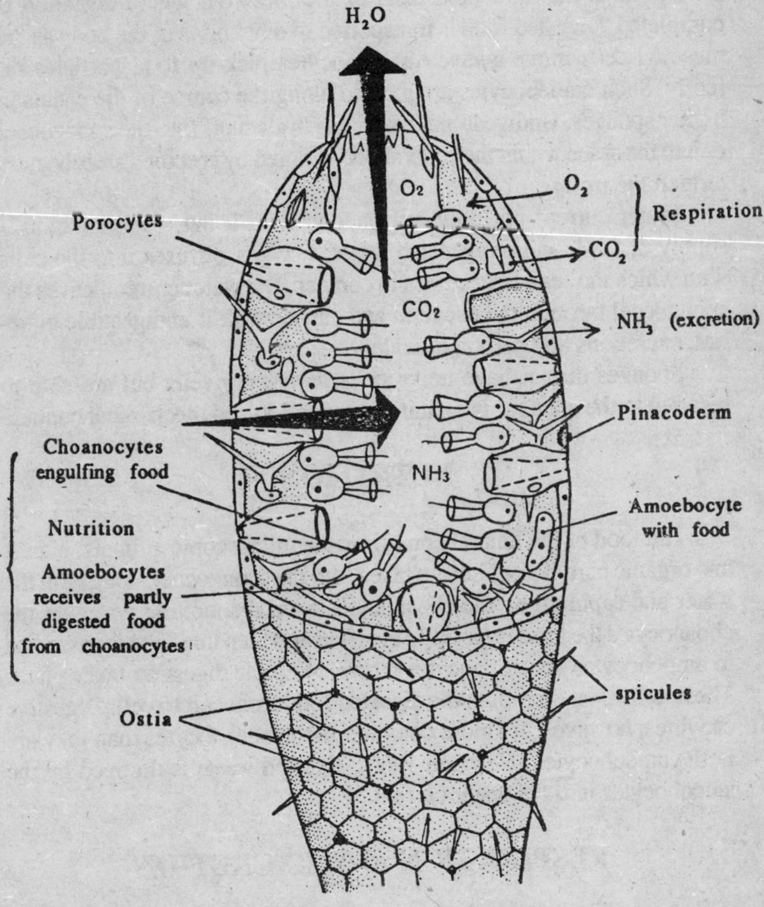

Fig. 2.5. Flow of water through the body

The movement of the flagella of collar cells create the force, which draws a current of sea-water through the numerous dermal ostia into the spongocoel. The current brings with it minute organisms that constitute the food of the sponge. Flagellum of the collar cell stirs the food particles, which slowly settle down on the surface of the collar and over its basal part. Small pseudopodia-like processes of the basal part engulf the particles and form food vacuoles. Part of whole of the digestion takes place in the basal part and the products of digestion are then passed on to the amoeboid cells of the mesohyl, where digestion is completed. Digested food is transported to other parts of the body as the mesohyl cells move away. Amoedbocytes pick up food particles directly. Such amoebocytes are located along the course of the canals in higher sponges. Undigestible material is thrown off into the spongocoel or into the space within the collar. Food is stored by specified amoebocytes termed *thesocytes*.

Water current not only brings food with it but also maintains a supply of fresh and oxygenated water. Oxygen diffuses into the cells with which the sea-water comes in contact. The water current leaves the spongocoel through the osculum and carries with it undigestible material, excretions and carbon dioxide.

Sponges do not have nerve cells or sensory cells but are able to respond to the changes in salinity illumination and mechanical contact.

NUTRITION

The food of the animal comprises the microscopic animals, plants, and organic particles which are taken into the spongocoel along with the water and captured by the choanocytes from the moving water. In the choanocytes the food is partially digested and then this food is passed on to amoebocytes i.e. trophocytes where complete digestion takes place. These cells then distribute the digested food from cell to cell. Digestive enzymes, however, are more common in the choanocytes than they are in the amoebocytes (*Hartman* 1960). Egested waste is dropped by the amoebocytes into the water current.

RESPIRATION AND EXCRETION

Gaseous exchange of oxygen and carbon dioxide occurs between the amoebocytes, choanocytes and pinacocytes and the water flowing through the canal system. Similarly nitrogenous wastes is voided.

LOCOMOTION

The sponges are fixed, as such they are incapable of movement. However, the pororcytes can open and closed according to the needs, and the amoebocytes can also move freely within the mesenchyme.

BEHAVIOUR

There is no coording nervous system in the animal. The individual cells respond independently to the external stimuli. The flagella of the choanocytes beat constantly to cause a water current. Osculum may close independently without altering the activities of other structures. Induced prolonged contraction may cause the sponge to burst, indicating little co-ordination.

REPRODUCTION

Leucosolenia reproduces both by asexual and sexual means.

(a) **Asexual reproductions.** Asexual reproduction takes place by *external budding*. Small outgrowths, called the *buds*, arise from the sides of the erect branches as well as from the upper surface of the horizonatal branches. The buds grow and, after attaining sufficient size, acquire osculum at their top, thus forming new individuals. The latter may remain attached to the parent colony and add to its size or become free by death and decay of old parts and form new colonies.

(b) **Sexual reproduction.** The sexual reproduction occurs by sperms and ova. Gamete formation and fertilization are similar in *Leucosolenia and Scypha* and will be described in the latter.

Zygote undergoes *holoblastic* cleavage and gives rise to an oval embryo called the *coeloblastula*. The latter is composed of a single layer of columnar flagellated cells with a few rounded, granular, non-flagellated cells at the posterior pole. The non-flagellated cells are thought to be the archaeocytes which will produce all future archaeocytes of the sponge. Some adjacent flagellated cells lose flagella and also migrate into the cavity. The embryo now consists of an outer layer of flagellated cells and an inner mass of amoeboid cells. It is known as the *stereogastrula or parenchymula*. It makes its way into the spongocoel and escapes through the osculum into the surrounding water, thus becoming a free-swimming larva.

The *parenchymula larva* swims about freely for a few hours, going away from the parent sponge. Finally, it fixes itself to some object by its anterior pole and flattens out. The internal amoebocyes migrate to the exterior and spread out to form a dermal layer or epidermis and mesenchyme. The flagellated cells, which get enclosed by the epidermis, change into choanocytes. A central cavity or spongocoel appears and gets lined by choanocytes. An osculum breaks through at the top and incurrent pores or ostia appear in the porocytes that differentiate in the wall of the sponge by this time. Scleroblasts shift from the epidermis into the mesenchyme and start secreting spicules. This results in a small sponge and takes only a few days.

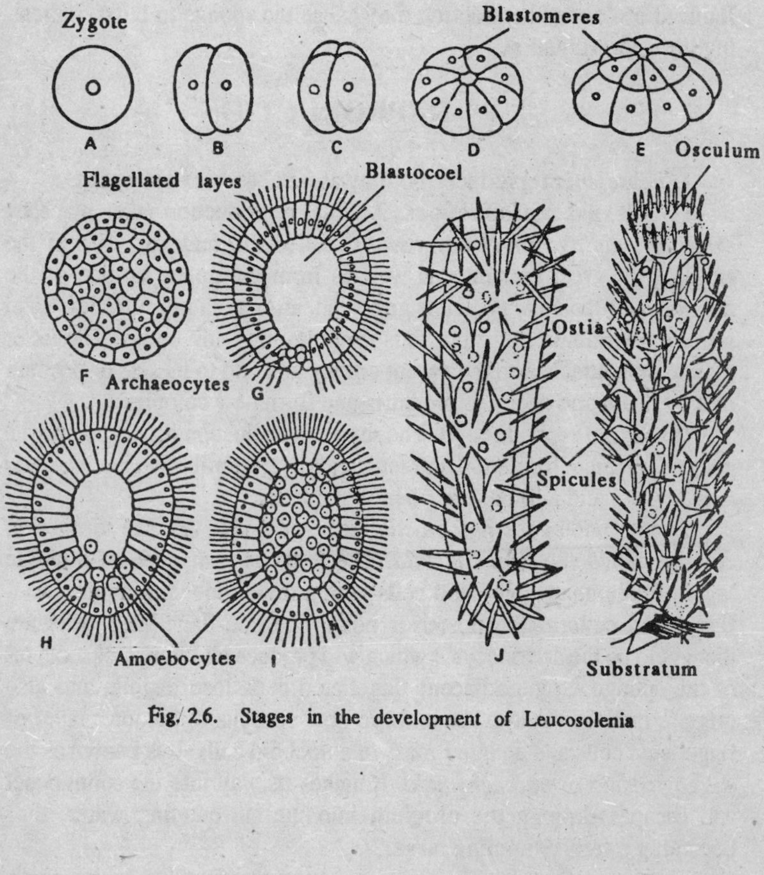

Fig. 2.6. Stages in the development of Leucosolenia

3

SCYPHA (SYCON)

SYSTEMAIC POSITION

Phylum	-	**Porifera**
Class	-	**Calcarea**
Order	-	**Heterocoela**
Genus	-	**Scypha (Sycon)**

HABIT AND HABITAT

Scypha is more complex than *Leucosolenia. Leucosolenia* shows primitive asconoid type of structure while in *Scypha*, there is folding of body wall to show syconoid type of structure where the spongocoel is reduced. *Scypha* is a small, marine sponge found attached by a sticky secretion to some submerged solid objects like rocks, shells of molluscs and corals. It is found in shallow water upto a depth of 100 meters where waves provide the animal with plenty of food and well oxygenated water. It has many species and is widely distributed but is best, known from North Atlantic shores. It may be free-living but also forms a cluster or colony by budding.

EXTERNAL MORPHOLOGY

Scypha or *sycon* possesses a vase-shaped and radially symmetrical body which measures from 1 to 3 cm in length. The colour is not specific but varies from grey to light brown. Near the attached end are found a few small tubular projections, or the *buds*, which are formed not so extensively as in *Leucosolenia* or other asconoid sponges. The free end of the vase-shaped individual bears an opening, the *osculum* (*exhalent* or *excurrent* pore), which is fringed with long, straight, needle-like monoaxon spicules. This *oscular fringe* checks small animals from entering into the body. The body surface is thrown into regularly arranged polygonal elevations from which project spear-like spicules, called *oxeotes* (monaxon spicules) that impart a bristly appearance to the body. The polygonal elevations are separated by deep grooves, bearing minute pores, i.e., the *ostia*, which lead into the central body cavity, the *spongocoel* or *paragastric cavity*, through a system of canals.

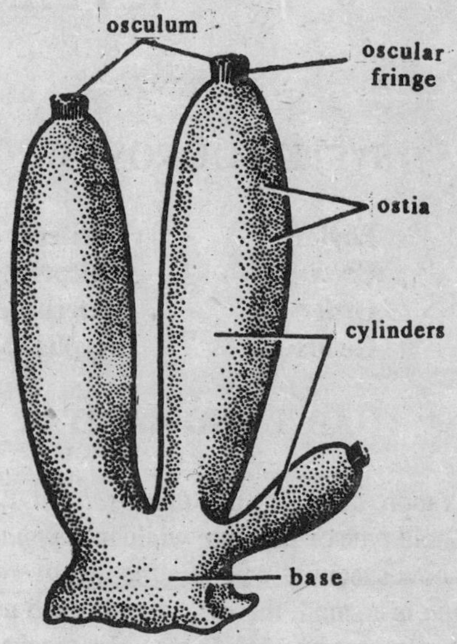

Fig. 3.1 *Scypha* (Sycon)

CANAL SYSTEM

As in other sponges, *Scypha* has the peculiarity of having canal system which consists of complex system of pores and canals. The body is traversed by numerous canals of several types. The body wall is thick and so folded as to form regularly arranged alternating invaginations and evaginations, forming sycon type of canal system. The different components of canal system of sycon are given below.

1. *Ostia or dermal pores*. The external large openings of the incurrent canals are stretched over by a thin membrane, called the *pore-membrane*. It bears two or more openings for the ingress of outside water into the body of sponge. These pores are known as *ostia* (L., *ostium*, door) or *dermal pores*. Because of the presence of contractile

cells the *myocytes*, around them the ostia can reduce in diameter and thus regulate the amount of ingressing water.

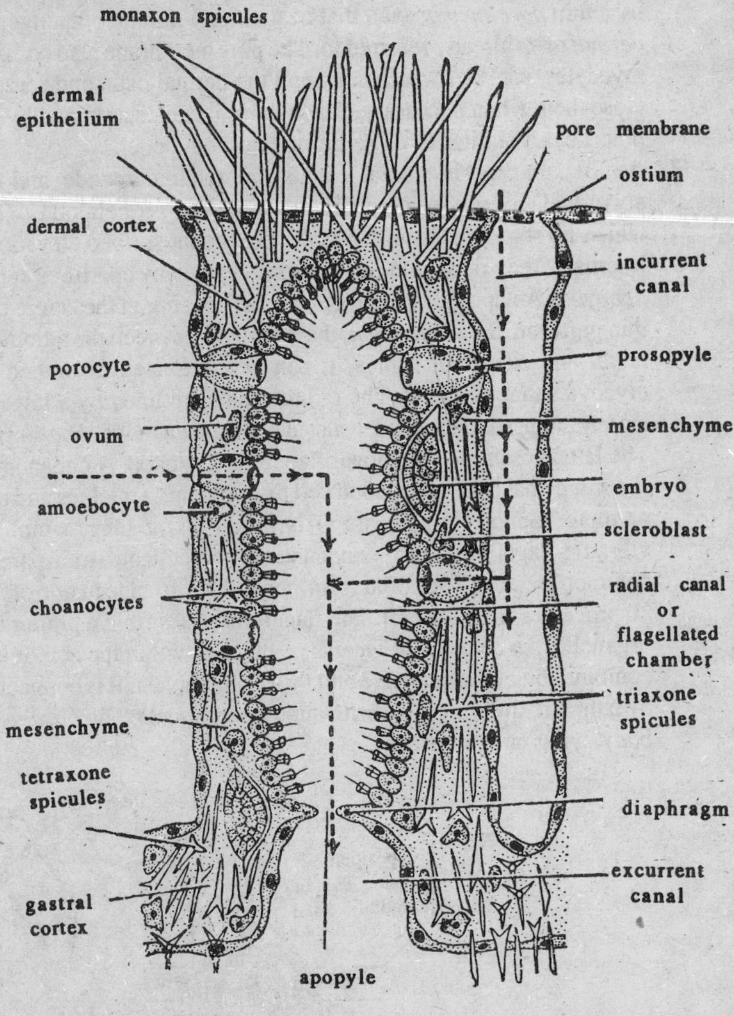

Fig. 3.2. A part of vertical section of *Scypha* showing histology and canal system.

2. *Incurrent canal.* The incurrent canals are narrow and almost squarish in cross-section. They end blindly at their inner extremities

a little distance away from the spongocoel. Externally and become somewhat dilated. The dilated part of each canal is closed externally by a thin *pore-membrane* that bears 3 or 4 minute openings, the *dermal ostia* already referred to. The pore membrane also contains myocytes, some of which surround the dermal ostia and serve to close them when necessary. The incurrent canals are lined by flat pinacocytes similar to those lining the spongocoel.

3. *Radial canals.* The radial canals are relatively wide and are polygonal in cross-section. They are closed at their outer ends which lie some distance inside the polygonal surface elevations. Internally they open into the excurrent canals by apertures called *apopyles*. An apopyle is a large circular opening in the centre of a thin partition, the *diaphragm*, that stretches between the adjoining radial and excurrent canals. It can be contracted or dialted by myocytes that border it. The radial canals are lined by a layer of loosely arranged, flagellated, collared cells known as the *choanocytes*. The latter resemble the choanoflagellate Protozoa. A choanocyte is a somewhat oval or rounded cell that contains a nucleus and one or more vacuoles and bears at its inner end a long, whiplike, vibratile flagellum, partly surrounded by a delicate, transparent, contractile collar. The collar is formed by side-to-side fusion of 20-30 processes from the cell. The flagellum arises from a prominent organelle, the *centroblepharoplast*, situated within the cell. It is a combination of the centriole and the blepharoplast. It is connected by a thread, the *rhizoplast* with another organnelle, the *parabasal body*, lying on the nucleus.

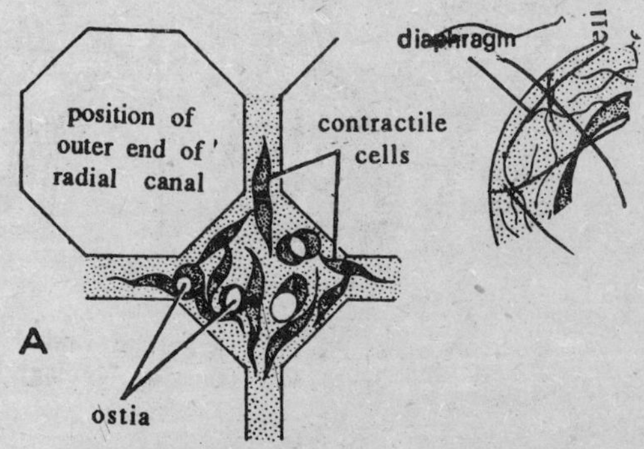

Fig. 3.3. Apopyle lined by myocytes in *Scypha*.

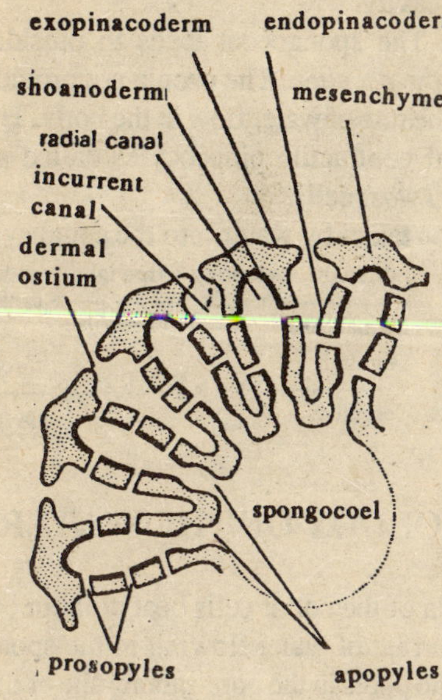

Fig. 3.4. A portion of a T.S. of Scypha

The incurrent canals communicate with the radial canal by short narrow channels called the *prosopyles*.

4. *Prosopyles*. The incurrent canals are communicated with the radial canals through intercellular spaces, called *prosopyles* (Gr., *pros*, near + *pyle*, gate). Some authors are of view that in the young *Sycon* these openings are those of the porocytes, as in *Leucosolenia*, but in adult the porocytes degenerate leaving the empty spaces or the prosopyles.
5. *Excurrent canals*. Each excurrent canal leads the radial canal into the spongocoel and forms a wide and short passage between the radial canal and spongocoel. These are lined with flat ectodermal cells like that of the spongocoel. The excurrent canal opens into the spongocoel through wide gastric ostium. Between the radial canal and excurrent canal lies a thin diaphragm perforated by a large hole known as *apopyle* which is surrounded by contractile myocytes.
6. *Spongocoel*. It is the central cavity of the body and, unlike *Leucosolenia*, is not lined by the flagellated collar cells or *choanocytes*, which have departed to line the radial canals, while the epidermal pinacocytes have immigrated into the spongocoel (Gr., *spongos*,

sponge + *koilos*, hollow) to form its lining. This cavity is comparatively a narrow cavity.

7. *Osculum*. The spongocoel leads to outside through a terminal opening, the *osculum*. The oscula are provided with *sphincters* to regulate the rate of water flow in the body. The sphincters are lined by special contractile pinacocytes called *myocytes* (Gr., *Myos*, muscle + *kytos*, cell).

The course taken by water into the canal system is as under -

Water from outside	through --------------> dermal ostia	incurrent canals	through --------------> prosophles	radial canals ↓
outside	through <------------- osculum	spongocoel	through <------------ epopyles	

PHYSIOLOGY OF THE WATER CURRENT

The flagella of the collar cells beat without coordination yet maintain a steady current of water flowing in the sponge from outside. The current enters the pores in the pore-membrane of the incurrent canals and fills the latter, it then passes into the flagellated canals through the prosopyles. From here it goes into the excurrent canals by way of apopyles. Finally, the current is thrown into the spongocoel by the large gastric ostia from where it escapes through the osculum. The osculum is also surrounded by myocytes serving to form an *oscular sphincter*, which can close the osculum.

The water current flowing through the canal system brings with it food in the form of micro-organisms and oxygen. The food, chiefly unicellular algae, bacteria and organic microfragments, is collected by the collar cells just the same way as described for *Leucosolenia*. The out going current carries with it undigestible wastes and nitrogenous excreta.

CELLULAR ORGANIZATION

Studies on structure and morphogenesis of *Scypha* have clearly revealed the presence of two types of cellular layers, the *pinacoderm* and the *choanoderm*. The former controls the interrelations between the mesenchyme and the external medium, while the later controls mainly the nutrition of the animal.

1. Pinacoderm

The pinacoderm comprises (i) the *exopinacoderm* (dermal epithelium) covering the entire body surface except the dermal ostia and osculum and (ii) the *endopinacoderm* which includes the epithelial lining of the incurrent canals and spongocoel. The pinacoderm is composed of large, flattened, polygonal cells, the *pinacocytes* (Gr., *pinako*, plank + *kytos*, cell). In profile, these cells present a central bulging containing the nucleus. The margins of adjacent cells are closely cemented together. Pinacocytes are highly contractile and they can greatly increase or reduce the surface area of the sponge body.

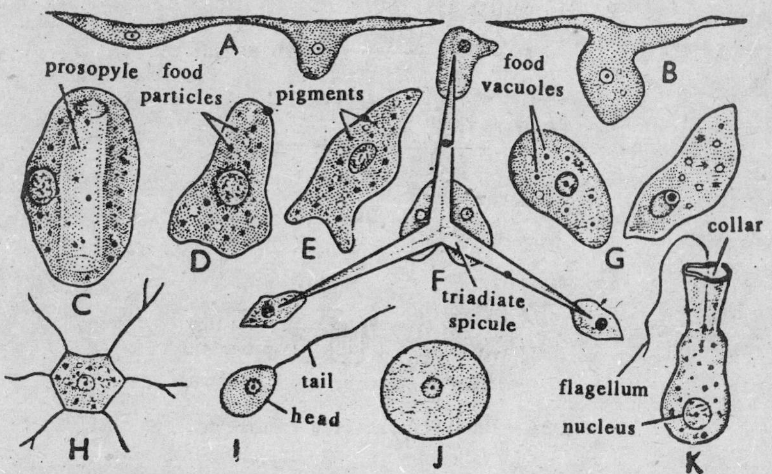

Fig. 3.5 *Scypha.* Kinds of cells. A-Pinacocytes. B-Myocyte. C-Porocyte. D- Thesocyte. E-Chromcyte. F-Sclerocytes. G-Archaeocytes. H-Collencyte. I-Sperm. J-Ovum. K-Choanocyte.

In the lining of the incurrent canals, some of the pinacocytes are modified to form tubular cells, the *porocytes*, connecting the incurrent canals with the radial canals through their intracelluar channels, the *prosopyles*. The porocytes are thin-wallled cells, open at both ends and with the nucleus present in the peripheral cytoplasm. As has already been noted, the porocytes in the adult sponge are lacking altogether. Pinacocytes surrounding the osculum, outer or dermal ostia and inner ostia or apopyles are elongated and contractile and act as muscle cells, called the *myocytes*. They form the sphincters around them to regulate these openings.

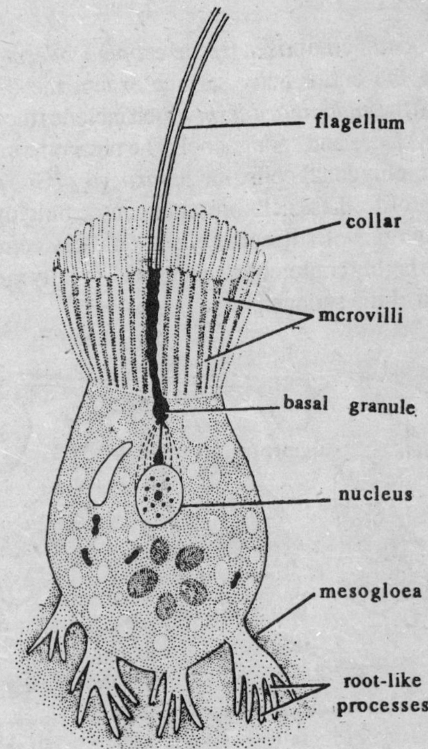

Fig. 3.6. Ultra Structure of a Choanocyte.

2. Choanoderm

This is the endoderm that forms the gastric epithelium and consists of flagellated collar cells or *choanocytes*. They are oval or rounded cells arranged upon the mesenchyme. Each cell contains a single nucleus, one or two contractile vacuoles, food vacuoles, reserve food, blepharoplast, rhizoplast and a single basal granule or kinetosome from which whip-like flagellum arises. The flagellum is surrounded at its base by a thin cytoplasmisc collar. The flagellum has two central fibrils and nine double peripheral fibrils. The collar of the flagellum is contractile and is made up of a ring of erect and closely set microvilli which project from the cell. The choanocytes line the radial canals or flagellated chambers. The movement of these flagella causes a current of water.

Reprodction

Scypha reproduces asexually as well as sexually as given below :

I. Asexual Reproduction

(i) *Budding* - It is a very common method of asexual reproduction. Masses of archalocytes migrate to the tips of spicules clumps that project from the surface in the form of bud. These buds either remain attached and grows in size, thus increasing the bulk of the sponge or it may finally breaks up and form new sponge.

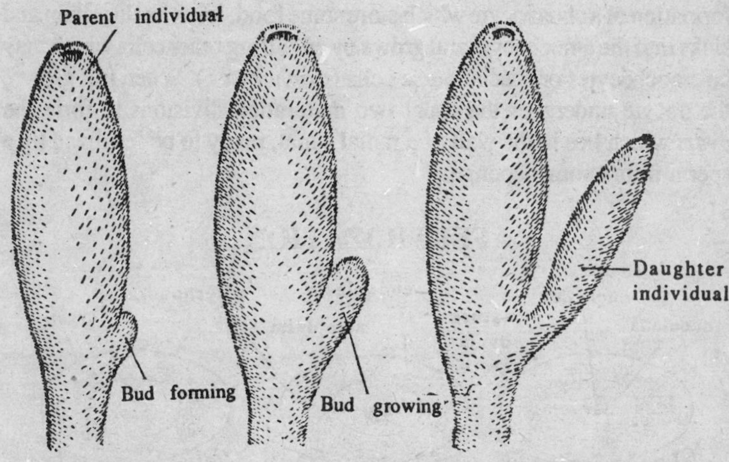

Fig. 3.9. Bud formation in *Scypha*.

(ii) *Regeneration* - Regeneration is the power to replace the parts lost bu injury. The power of regeneration has developed to such an extent that if a sponge is thoroughly teared or squeezed in the sea water so as to separate the cells from one another, it is found that the cells slowly come together and gradually form a young sponge again.

2. Sexual reproduction

Scypha is a *monocious* sponge but due to *protogyny*, only cross-fertilization occurs. Special sex organs are lacking. Male sex cells, the *sperms* and the female sex cells, the *ova,* are found in the mesenchyme. They develop from undifferentiated amoebocytes, called *archaeocytes*.

(a) *Spermatogenesis*. The sperm mother cell or *spermoatogonuim* is said to be an enlarged archaeocyte. However, it is supposed to be a modified choanocyte by *Gatenby,* who described the transformation of an entire flagellated chamber into spermatozoa. Soon the spermatogonium is surrounded by one or more flattened *cover cells* to form the *spermatocyst*. The cover cells are derived either by the division of the mother cell or from other amoebocytes. The spermatogonium undergoes two or three divisions to form the *spermatocytes* which give rise to *spermatozoa*. A mature sperm or *spermatozoon* consists of a rounded, nucleated head and vibratile tail, by the lashing movements of which it moves in water to reach other sponges.

(b) *Oogenesis*. The egg mother cell or *oocyte* is derived from a large archaeocyte with a distinct nucleus. This may sometimes arise by transformation of a choanocyte which store some food, loses its flagellum and sinks into the amoebocyte and grows by engulfing other cells which may be amoebocytes or special nurse cells (*trophocytes*). When full grown, the oocyte undergoes the usual two maturation divisions to form the *ovum* which lies in the wall of a radial canal, ready to be fertilized by a sperm from another sponge.

FERTILIZATION

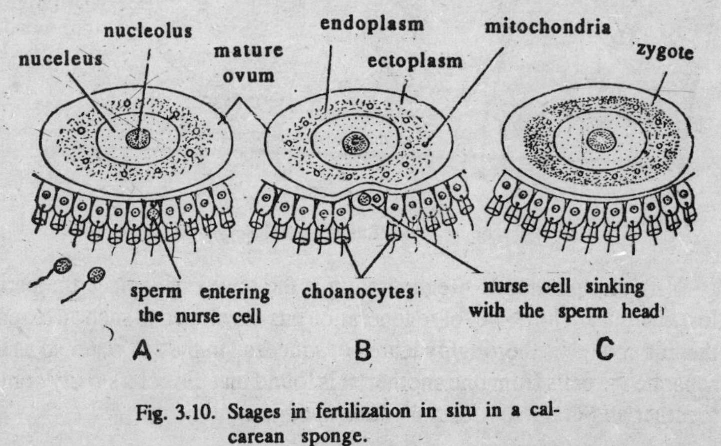

Fig. 3.10. Stages in fertilization in situ in a calcarean sponge.

Fertilization is internal. When the sperms mature, the spermatocysts rupture, discharging the sperms into the radial canals. From here, they are carried by the outgoing current of water to the exterior through the osculum. With the incoming water current, the sperms enter another sponge through this dermal ostia and reach the radial canals. Here, each sperm enters a wandering amoebocyte or a choanocyte. The invaded choanocyte loses the flagellum, withdraws the collar, becomes amoeboid and migrates into the mesogloea. The amoebocytes and choanocytes penetrated by the sperms are known as the *sperm-transit cells*, as they serve to carry the sperms to the ova through the mesogloea. A sperm-transit cell adheres to a nearby ovum, which forms a conical depression to receive it. The sperm loses its tail and its head swells up. The swollen head gets surrounded by a capsule. The capsule with the sperm head entres the ovum. This completes the act of fertilization and results in the formation of a *zygote*. The sperm transit cell now departs. The zygote gets enclosed in a *brood capsule* formed by lining up of the neighbouring amoebocytes. Early development of the zygote occurs within the brood capsule of the parent.

DEVELOPMENT

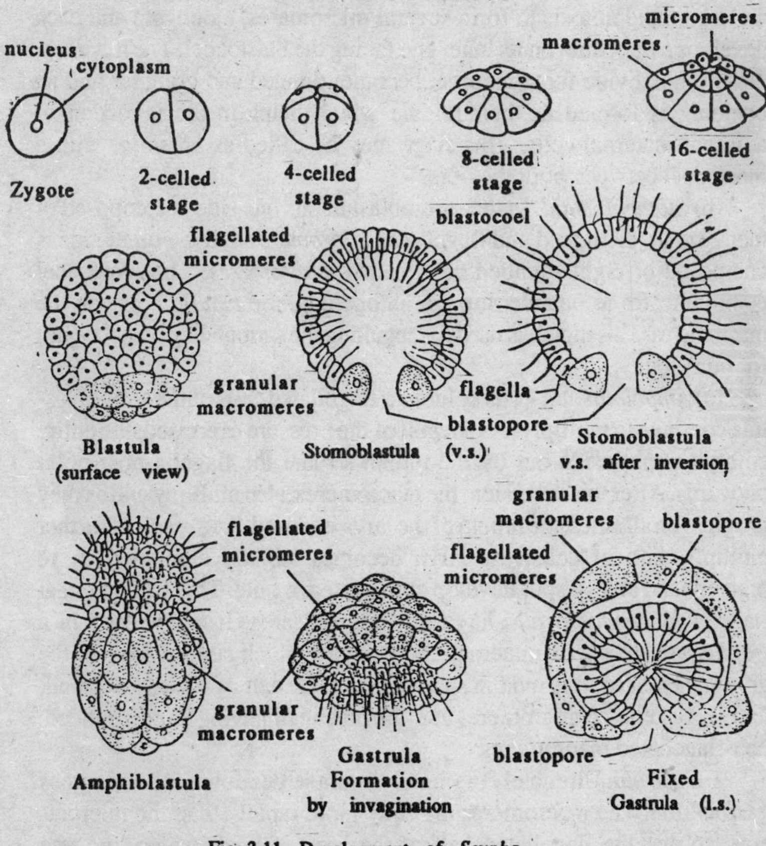

Fig. 3.11. Development of *Scypha*.

The development of *Sycon* includes the early embryonic period and larval period followed by metamorphosis.

I. Embryonic period

(a) *Cleavage.* The fertilized egg undergoes maturation and starts development within the mesenchyme of the parent. It undergoes equal and *holoblastic cleavage.* The first three cleavages are vertical forming 8 blastomeres. The horizontal division then divides the blastomeres unequally and of the 16 blastomeres produced, 8 are large cells or macromeres lying next to the parent choanoderm. They are desined to form future epidermis. Eight smaller cells or *micromeres* give rise to future choanocytes. At this stage, the embryo lies just beneath the maternal choanocyte layer as a flattened disc-shaped body. The micromeres undergo rapid mitosis to form several micromeres, elonngate and each develops a flagellum on its inner end facing the blastocoel. The macromeres do not divide for sometime, become rounded and granular and an opeining is formed in their middle which helps in the ingestion of adjacent maternal cells. This stage may be called as *blastula* with a blastocoel between both the tiers.

(b) *Stomoblastual.* In the stomoblastula, its one side is composed of many small, elongated and flagellated *micromeres* and the other side is composed of eight rounded cells called *macromeres.* The blastocoel communicates to outside through an opening, the *mouth*, between the macromeres. This mouth is used for engulfing the surrounding amoebocytes for nutrition.

(c) *Amphiblastula* - The columnar flagellated cells of the stomoblastula continue to multiply. As a result of the pressure exercised, the entire embryo turns inside out by the mouth so that the flagella now point outwards. After this inversion, the macromeres, also multiply and as they progress on all sides, the mouth of the larva is closed. As a result of further multiplication of cells, the larva becomes slightly elongated. A re pigment and an eye spot develop at the anterior pole. This is the typical flagellated *amphiblastula* larva of class Calcarea. It is enclosed in a *trophic membrane* of maternal choanocytes which supply food to the growing embryo. It is oval in shape having one half of it made of small, narrow flagellated micromeres and the other half having large, rounded, non-flagellated macromeres.

2. *Gastrula.* Ultimately the amhiblastula settles down and undergoes gastrualtion. The macromeres multiply more rapidly than the micromeres, so that the flagellated half of the larva is invaginaged into and

overgrown by the granular non-flagellated half. The larva now becomes a typical double-walled *gastrula* with a *blastopore* at the invaginated side.

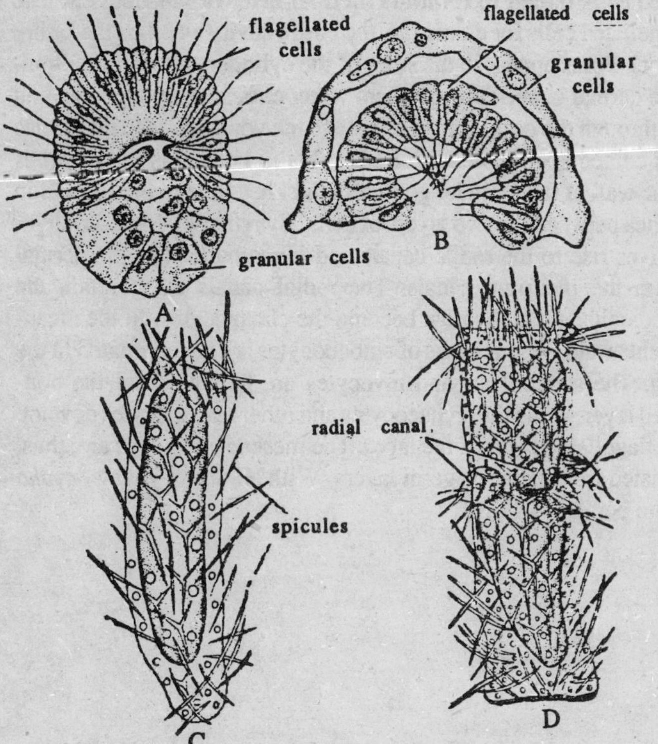

Fig. 3.12. Later stages of development of *Scypha*.

METAMORPHOSIS

After swimming for some time, the flagellated half of the larva invaginates into or is overgrown by the large non-flagellated cells. With this process, the single-layered amphiblastula larva changes into a two layered *gastrula* with an opening, the *blastopore*, at the invaginated end. Outer layer of the gastrula consists of the non-flagellated granular cells and the inner of the flagellated cells. Gastrulation, thus, in a way reverses the inversion that took place at the stomoblastual stage.

The gastrula soon fixes itself to some substratum by its blastoporal end and grows into a cylindrical structure. Its blastopore closes by the growth of both the layers of cells along the substratum. At its free end develops a perforation which becomes the osculum. The outer non-flagellated cells flatten to form the dermal layer of pinacocytes. The inner flagellated cells for a time line the entire cavity which is the future spongocoel. Pores appear in the sides of the cylinder and start functioning as the dermal ostia. Through them water entres the spongocoel and leaves it through the osculum. At this stage the young sponge resembles adjust as coined sponge *Leucosolenia*. This is known as the *olynthus stage*. The wall of the cylinder gradually thickness due to the formation of mesoglea between the two layers of cells. Evagination from the spongocoel gives rise to the radial canals and invaginations of the dermal layer form the incurrent canals. The radial canals alone retain the flagellated cells, which by now become the choanocytes. In the meantime, spicules and various types of amoebocytes are differentiated in the mesoglea. The scleroblasts and myocytes are derived from the non-flagellated layer, while the archaeocytes and other amoebocytes develop from the flagellated layer of the larva. The mesenchyme cells are, thus, differentiated from both the germ layers. With all this, a young *Scypha* comes into being.

4

ORIGIN OF PORIFERA

The evolutionary origin of sponges poses a number of intersting problems. The absence of organs and the low level of cellular differentiation and interdependence in sponges certainly seem to be primitive characteristics. But a specialized body structure built around a water canal system and lacking distint anterior and posterior ends is found in no other groups of animals. Moreover, the cellular differentiation is unlike that of other metazoans. All these features suggest that sponges are phylogenetically remote from other metazoans. There are several hypothesis to show the origin of sponges.

1. According to the ancient view, the sponges originated from sluggish gastrula-like ancestors. But this primitive view is not accepted at all.
2. According to another view the sponges originated from choanoflegellate protozoan. This theory suggested the possible origin from the colonial protozoa, *Proterospongia*. It was considered in the line of evolution from protozoa onwards to sponges. The followings are the resemblance between two :

   (i) Morphologically distinct cells.

   (ii) Presence of choanocytes or collar cells.

   (iii) Presence of amoeboid cells in jelly-like mesogloea.

   There are certain objection to this view as :

   (i) The position of choanocytes is internal in sponges whereas it is external in *Proterospongia*.

   (ii)The canal system is altogether absent in *Proterospongia*, even there is no any sign. It is characteristic of sponges.

   The above objections can be ruled out due to the fact :

   (i) The position of chanocytes in sponges is due to migration at a later stage of developement.

   (ii) The absence of alimentary canal gave rise to the evolution of elaborate canal system in sponges.

(iii) Development of spicules took place during their further evolution.

3. Due to absence of choanocytes in sponge larva a third theory have been proposed. This theory believed that the origin of sponges is from flagellate protozoan. This fomed the colony and gave rise to early sponges.

Modern workers are of the opinion that sponges are lower metazoa but they *do not stand* in the direct line of metazoan evolution. The following characters support this view :

(i) Absense of cellular layers in sponges.

(ii) Absence of division of ;labour as is found in metazoa.

(iii) Absence of organisation like tissue systems (connective or nervous) and organ systems (alimentary system, reproductive system etc.)

(iv) There is no distinct division of somatic and germinal part of the body to carry on distinct physiological functions.

(v) During development in metazoa, a gastrulla stage arises when the future germ layers are established which later on differentiate into definite tissues or systems. These are never *reversed* during later stage of development but in sponges reversal of germ layers takes place.

The above facts are enough to separate them from direct line of metazoan evolution. But at the same time there are some distinctive features in sponges which should be taken into account. These important characteristics are :

(i) In sponges although there is no tissue or organ system each cell is independent.

(ii) The large terminal opeing *osculum* can be compared with *mouth* but they are not homologous on the contrary are analogous because their origin during development is from different sources.

(iii) Although spongocoel may correspond to coelenteron of cnidaria but the presence of water canals lined by flagellated cells are not found in any other group of animals.

(iv) Poorly developed mesenchyme which is contractile in osculum region only.

(v) The presence of peculiar *amoedboid cells* which can form other cell types.

For the above facts *Hyman* considered sponges in a separate line from the direct line of metazoan evolution and the word '*Parazoa*' was given for the inclusion of sponges. Parazoa is a separate division of sub-kingdom metazoa.

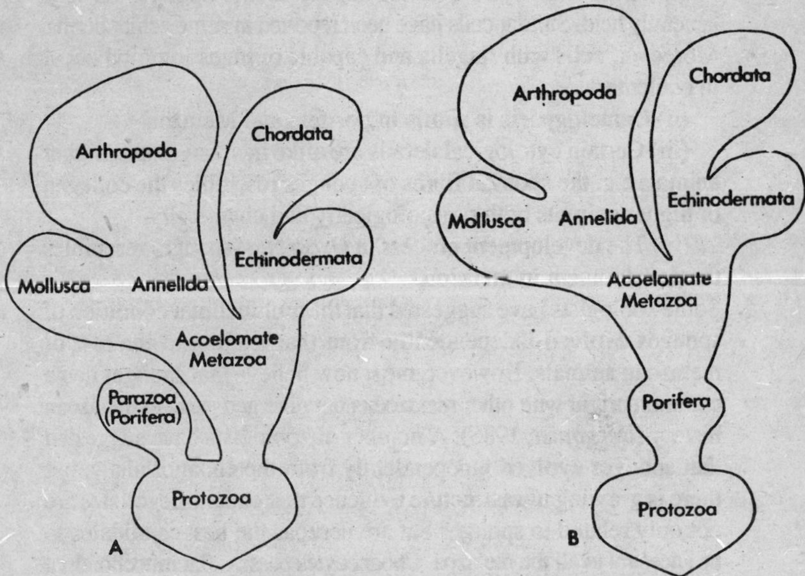

Fig. 4.1. Diagram of suggested phylogenetic position of the Porifera :- A. Connectional view of sponges as an isolated subkingdom, the Parazoa. B. Tuzet's views of sponges in the mainstream of metazoan evolution.

4. *Tuzet's Theory. Tuzet* (1963) has suggested that the sponges gave rise to other metazoa through the coelenterates. This view seems to have some weight in it.

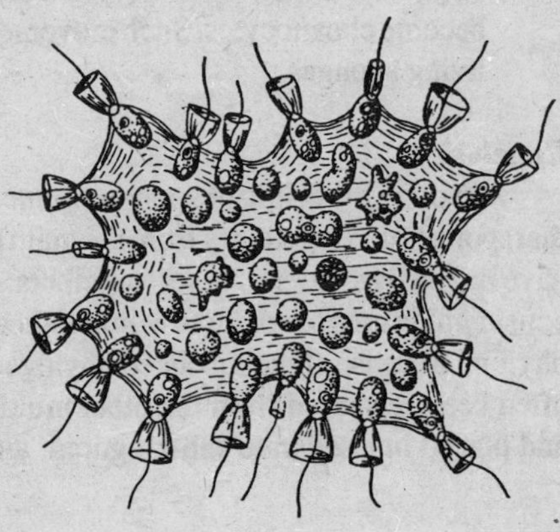

Fig. 4.2. Proterospongia

(i) The choanocytes are not unique to the sponges as it is generally held. Similar cells have been reported in some echinoderms. Moreover, cells with flagella and capable of ingesting food occur in coelenterates.

(ii) Gametogeneis is simila in porifera and Metazoa.

(iii) Certain cytological details are alike in sponges and higher animals e.g. the spongin fibres of sponges resembles the collagen of higher animals both physiologically and chemically.

(iv) The development process in *Oscarella* (sponge) resembles the development in metazoa.

5. Some zoologists have suggested that the multicellular condition of sponges evolved independently from that leading to the rest of metazoan animals. However, most now believe that sponges had a common origin with other metazoans but diverged early in metazoan history (*Bergquist,* 1985). A number of zoologists have suggested that sponges evolved independently from the choanoflallats, but there is growing ultrastructure evidence that choanoflegellates are not only related to sponges but are perhpas the best candidates to be ancestral to all the metazoa. Choanocytes possess flat mitochondrial cristae, as present in metozoans mitochondria, other protozoa have tubular cristae. Moreover, the ultrastructural details of the flegellar kinetosomes and associated structures are essentially like those of metazoan cells (*Nielsen* 1985).

Many of the specialized sponge features, such as canal system, certainly developed in conjuction with a sessile existence and the flagellated, exterior cells probably moved into the interior to become choanocytes. Such movement occurs in the embryology of living sponges.

Conclusion

Although the origin of sponges is uncertain, there can be little doubt that sponges diverged early from the main line of metazoan evolution and have given rise to no other members of the Animal Kingdom. A considerable number of facts would indicate that sponges are a dead end phylum. Because of their isolated phylogentic position, the sponges have often been separated from the other multicellular animals (Eumetazoa) and placed in a separate sub-kingdom, the *Parazoa*.

5

AFFINITIES OF PORIFERA

There has been controversy over the nature and affinities of the sponges ever since they were discovered *Aristotle* (384-322 B.C.) was the first to recognize them as animals. Since his time, there has been difference of opinion about the nature of sponges. Some naturalists regarded them as plants, while others as animals and still many others considered them to be both plants and animals.

Once sponges were considered to be plants on the basis of following characters.

1. In structure, sponges are plant-like.
2. They are irregular and sometimes branched.
3. The colours are also quite different. Some are grey, others are red, yellow, blue etc.
4. Like aquatic plants, the sponges live attched to substratum.
5. They do not show any responce to the stimuli when touched.
6. Locomotion is absent.
7. Power of regeneration.
8. Asexual reproduction by budding.

It was only 1857 that the sponges were fully recognized as animals. *Ellis* (1765) pointed out that the sponges showed sign of life by drawing in and expelling out a regualr current of water. *Robert Grant* (1825) definitely proved that sponges were animals and not plats. He found water enters into body of a sponge by means of minute pores and passes out through osculm.

Following characteristics regard sponges as animal :

1. These are either radially symmeterical or asymmetrical.
2. Presence of many canals.
3. Unique type of skeleton in sponges, consisting of spicules and spongin fibres.
4. Digestion intracellular.
5. Multicellular animals without organs and systems.

Grant (1836) studies morphology and physiology of sponges and created the group *Porifera* for them. In 19th century, other names suggested by zoologists were : Spongiae, Spongida and Spongiaria etc.

Regarding the affinities of the sponges in the animal kingdom, there are two views.

1. Affinities with Protozoa.
2. Affinities with Metazoa.

1. Affinities with Protozoa

de Blainville (1816) separated sponges into a group, Spongiaria allied to protozoa. *H. J. Clarke* (1866) first demonstrated the collar-cells and suggested that sponges are colonial protozoa. The general affinities of sponges with the colonial protoza are due to *Proterospongia*, a member of the protozoa.

The points of resemblance between the two groups are the following :

1. *Proterospongia* is a colony of collared Mastigophora placed in a mass of gelatinous matter having amoeboid zooids. These amoeboid zooids resemble to, and do the same function as, amoeboid wandering cells of sponges.
2. It is natural that such an organism may develop into a sponge, thus, *Proterospongia* may be regarded as a connecting link.
3. Cells of the colonial *Proterospongia* resemble those of middle layers of sponges. There is a little *physiological divions of labour*. Many cells of the sponges are also independent as individual protozoa.

But this stage of arrairs exists in some other members of metazoa also and, therefore, *Parker* and *Haswell* regarded these differences being of degree and not of kind.

1. The development of the sponge is also fully worked out, which has no comparision with that of colonial protozoa. The embryological investigations of *F.E. Schulze* have also belied the belief that sponges are colonial flagellated protozoa.
2. Colonial protozoa and sponges differ as sponges posses canals, skeleton and *somatic* cells.
3. *Choanocytes* are internal in *Porifera*, while they are external in *Proterospongia*. However, some zoologists are of the opinion that choanocytes become internal only at a later stage of life-history.

Considering the above points of difference, zoologists have discarded the view that sponges are colonial protozoa and that they originated from *choanoflagellate mastigophora*.

2. Affinities with Metazoa

Linnaeus, Lamarck and *Cuvier* classified sponges as zoophytes allied to anthozoan coelentrates. Following are similarties between the metazoa and porifera :

1. Sponges as well as coelenterates are both fixed and lead a fixed mode of life.
2. Both are diploblastic. Well developed mesoderm is absent in both.
3. Coelom is absent.
4. The spongocoel of sponge opens to outside by osculum where as the central cavity or coelenteron opens to ouside by mouth.
5. Same nucleic and amino acids are present.
6. Both can reproduce asexually by budding.
7. Gastrula shows radial symmetry.
8. Parenchymula larva of sponge resembles the planula larva of coelenterates.

But sponges differ from coelenterates and other metazoan in several respects :

1. In sponges the cells are less specialized and interdependent than the metazoan cells. The cells are merely aggregation of cells which can assume various forms. These cells are generally found isolated in the jelly-like mesenchyma.
2. Tissue formation in sponges is restricted to the surface epithelial lining. The organs are absent as seen in higher metazoan.
3. Sponges are diploblastic where as metazoans are triploblastic. The mesenchyme and epidermis arise in sponges from the same embryonic cells so that their mesenchyme may be considered theoritically to be an ectomesoderm.
4. Osculum in sponges developmentally does not correspond with the mouth of coelenterates. The osculum serving as exhalent opening but in metazoan the mouth is inhalent opening.
5. Sponges do not posses stinging cells or nematoblasts like coelenterates.
6. Presence of inhalent ostia, inhalent canals, flagellated chambers are peculiar characters of sponges.
7. No distinct digestive system and mouth as present in many metazoan, digestion intracellular like protozoa.
8. An anterior end is absent in sponges. The nervous system is absent to co-ordinate the various activities of sponge.
9. The development of sponges is most puzzling and difficult to interpret. In sponges during gastrulation the animal half invaginates but in metazoa it is the vegetal half.

During larval metamorphosis the outer epithelial layer becomes the inner choanocyte layer whereas the inner cell layer becomes external to perform their function of epidermis and mesenchyme. This reversal of germ layers is unknown in metazoa.

Systematic Position of Sponges

Previously there were attempt to classify sponges with protozoa but this was rejected when the workers in this field recognised the germ layers in a developing sponge and the peculiarities of these germ layers were enough to separate them from metazoa was a necessary. As such where sponges are to be placed? A suitable suggestion was given by *Sollas* who proposed a new sub-kingdom *Parazoa* for sponges. These were due to the following characteristics of sponges and was considered the main features of *Parazoa* :

(i) Absence of definite tissues or tissue layers and naturally there is no organs or organ system. This is due to the loose assemblage of cells.
(ii) Absence of any kind of symmetry.
(iii) Absence of a true alimentary canal.
(iv) Inhalent pores are ostia and internal cavities are water canals.
(v) Special type of collared flagellated cells called choanocytes are present. Spicules act as skeletal structures.
(vi) Ingoing and outgoing water serves as nutrition, respiration and excretion; this also helps in reproductive activities like carrying sperms and in fertilization.
(vii) Sponges possess high power of regeneration.
(viii) Larval forms like parenchymula and amphiblastula are present in the life-cycle.

Now the situation as regards the systematic position is concerned is as follow. Sponges are on a side branch of the direct line of evolution. Although they have some superificial resemblances with metazoa but show no embroylogical or genetic resemblances. Therefore zoologists have separated them from metazoa and have placed them in a separate sub-kingdom. They have divided the entire animal kingdom into three sub-kingdom. They have divided the entire animal kingdom into three sub-kingdoms - *Protozoa, Parazoa* and *Metazoa* (or Eumetazoa). The sub-kingdom parazoa includes the sponges. Again according to *Tuzet's* view sponges evolved from protozoa and have been placed between protozoa and cnidaria in a separate phylum Porifera.

6

HISTOLOGY OF SPONGES

In most of animals the body is composed of various tissues like epithelium, connective, muscalar, nevrous and skeletal. But in sponges there is cellular grade of organization means the body is composed of different types of cells lacking proper organization. The cells found in sponges do not organize into tissues. Sponges are diploblastic i.e. composed of two layers - outer *ectoderm* or *dermal epithelium* and inner *endoderm*. In between these layers lies *mesenchyme*. The cells of these layers are simple and modified according to their function except one called *choanocytes*.

The cells of ectoderm

These are generally called *dermal epithelium*. They form a covering of the outer body surface and the lining of incurrent canals. The following cells form this layer:

(i) Pinacocytes. These are large flat, closely cemented cells with thickened central bulge containing the nucleus. They appear continuous with adjacent cells. These are generally found to cover the outer body surface when they are called *dermal epithelium*. In syconoid sponge they also line the incurrent canals and spaces found inside the body. When they line the paragastric cavity they are also called *gastral epithelium*. These cells are highly contractile and thus can reduce the surface area.

In some sponges the epidermis is formed by epitheloid membrane which is a continuous membrane having numerous scattered nuclei.

Hexactinellid sponges have no definite epidermis.

(ii) Porocytes. In asconoid sponges (*Leucosolenia*) some of the pinacocytes are modified to form tubular cells, the *porocytes*, connecting the incurrent canals with the radial canals through their intracellular channels, the *prosopyles*. The porocytes are thin-walled cells, open at both ends and with the nucleus present in the peripheral cytoplasm. As has already been noted, the porocytes in the adult sponge are lacking altogether. Pinacocytes surrounding the osculum, outer or

dermal ostia and inner ostia or apopyles are elongated and contractile and act as muscle cells, called the *myocytes*. They form the sphincters around them to regulate these openings.

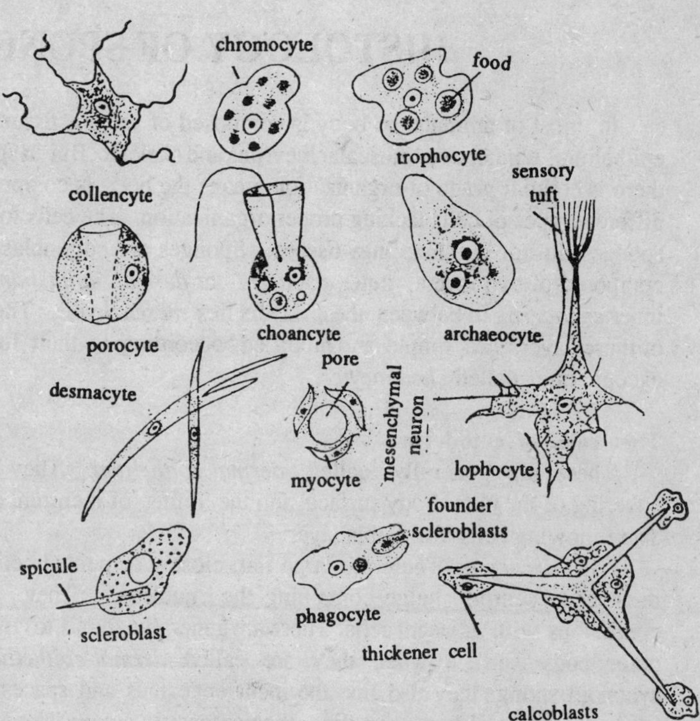

Fig. 6.1. Types of cells in a sponge.

Choanoderm

Choanoderm or endoderm forms the gastric epithelium and consists of flagellated collar cells or *choanocytes*. They are oval or rounded cells arranged upon the mesenchyme. Each cell contains a single nucleus, one or two contractile vacuoles, food vacuoles, reserve food, blepheroplast, rhizoplast and a single basal granule or kinetosome from which whip-like flagellum arises. The flagellum is surrounded at its base by a thin cytoplasmic collar. The flagellum has two central fibrils and nine double peripheral fibrils. The collar of the flagellum is contractile and is made up of a ring of erect and closely set microvilli which project from the cell. The choanocytes line the radial canals or flagellated chambers. The movement of their flagella causes a current of water

Cells present in Mesenchyme

The gelatinous, transparent matrix of mesenchyme is known as mesogloea in which are found special cells called *amoebocytes* which may be of the following type:

(i) Collenocytes. Star-shaped cells which are sometimes united to form a network.

(ii) Chromocytes. Pigmented cells with lobose pseudopodia. They probably impart colour to the sponge.

(iii) Thesocytes. Cells with lobose pseudopodia and are laden with reserve food material.

(iv) Scleroblasts. These secrete the spicules. They may be *calcoblasts* (secreting calcareous spicules), *siliceous spicules*, a *spongioblasts* (secreting spongin fibres).

(v) Trophocytes. They are loaded partly or completely with digested food. They serve to transfer food from one place to another.

(vi) Phagocytes. These cells collect food from the choanocytes through their pseudopodia and also engulf excreta and damaged tissues.

(vii) Archaeocytes. Or *undifferentiated embryonic cells* that give rise to sex cells. They have blunt pseudopodia, large nuclei, conspicuous nucleolar and cytoplasmic inclusions. They may also be associated with transport of food and waters.

(viii) Lophocytes. These are ameoboid cells with a tuft of fibrils at one or both ends and occurs in connection with the pinacocyte epithelium. Its function is unknown but it may secrete fibrillar bundles.

(ix) Desmocytes. They are long slender cells lying at the basement of the canals.

(x) Gland cells. They are derived from amoebocytes and secrete mucous.

(xi) Germ cells. Some of the amoebocytes (archaeocytes) differentiate into sex cells i.e. ova or sperms. However in some sponges, they are said to be modified choanocytes.

(xii) Myocytes. These are elongated, narrow nucleated cells arranged around the apopyles, osculum, and ostia are also present a number of narrow elongated highly contractile cells, the myocytes.

By their contractibility they can control the size of the openings.

The mesenchyme also contains the spicules which are monaxonic and triradiate in form. These serve to strengthen the body and hold it up right

7

CANAL SYSTEM IN SPONGES

All the activities of the body of the sponges depend on the current of water entering through ostia and passing out through osculum or oscula. Inside the body, water current flows through system of spaces which collectively constitute the canal system. The entire physiological activities of the animal depend on the water current as the exchanges between the body and the exterior are maintained through this water current. The food and oxygen are brought through this current while excreta and reproductive bodies are excluded through this current. The perforations of the body of the sponge by a large number of ostia is characteristic of phylum Porifera.

The arrangement, and complexity of the canal system varies considerably in different animals but has been divided into four types :

1. Ascon type
2. Sycon type
3. Rhagon type
4. Leucon type

All the living tissues of the sponges are soft and it is a dire need of the animal to maintain a constant shape with different canals or passages-ways. Therefore, deeper layers of the body are provided with supporting *spicules*. Around the osculum, spicules are long and straight, while spicules situated around ostia are short and straight; around spongocoel they are T-shaped, while trirayed in the body-wall.

Thus, all the cavities of sponges with intricate passages of canals, traversed by currents of water entering by pores and passing out by osculum are collectively termed as a *canal system of sponges*.

A typical canal system is composed by following components:

(a) Incurrent canal. It opens externally to the outside by a small pore known as *incurrent pore* or *ostium*, but internally it ends blindly.

(b) Radial canal or excurrent canal. It is closed externally but opens internally by minute pores or *apopyles* into a central cavity or *cloacal cavity* or *gastral cavity* or *spongocoel*, which cannot be compared in any way with the stomach or intestine of other animals.

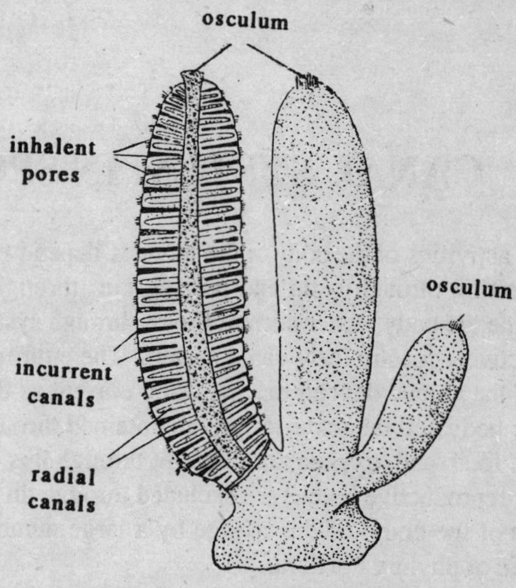

Fig. 7.1. Sycon gelatinasun showing paragastric cavity, incurrent and radial canals.

(c) Prosopyle. It is a smaller canal or passage-way connecting incurrent canal with radial canal.

The incurrent canals are lined by flat squamous cells and their functions are only to form water conduits and to form a smooth and firm surface.

The *excurrent canals* or *radial canals* are lined by *collar cells* opening at the surface and are provided with flagella or whips. The lashing movements of flagellum procure the food particles and push them into the cell-mouth. Thus, this is food-capturing arrangement of sponges.

Spongocoel or central cavity is lined by a thin gastric epithelium. It opens to the outside by an aperture, called *osculum.*

I. Ascon type.

It is the simplest type of canal system in which the body is thin-walled, bilaterally symmetrical and hollow due to the *central cavity.* This cavity is known as the *spongocoel* or *paragastric cavity* that opens to the outside by means of a circular aperture at the free distal end of the cylinder. Numerous minute pores the ostia are also present in the thin wall of the cylinder. These minutes pores are regularly disposed intracellular apertures each of which is a canal-like structure situated within a tubular porocyte. They extend from the outer surface of the body wall to the spongocoel. The water current reaches the spongocoel after passing through the ostia and goes out through the osculum. The body wall of ascon type of sponges is formed of two layers. The outer layer is called *ectoderm* and the inner layer is known as *endoderm*. The ectoderm is formed of thin and flat *pinacocytes*. The endoderm is formed by the *choanocytes* and lines the spongocoel. Between these two layers is a thin *mesoglea* formed of non-living gelatinous substance. It contains different kinds of *amoebocytes* and *triradiate spicules* formed of calcium carbonate.

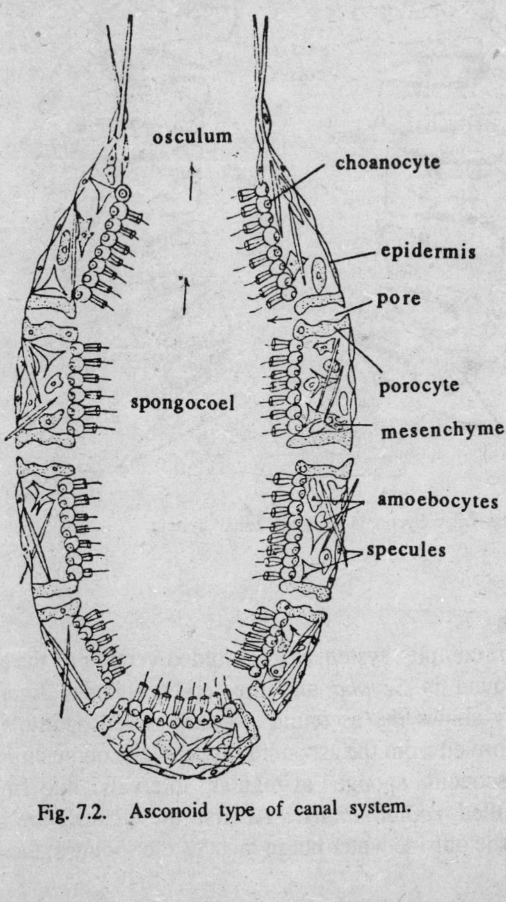

Fig. 7.2. Asconoid type of canal system.

The ascon type of canal system is also found in some calcareous sponges such as the development stages of *Clathrina*. Besides this, it is found in *Leucosolenia* and some other simple sponges.

The course of the water current is as follows:

outside water → ostium → prosopyle → paragastric cavity → osculum → outside.

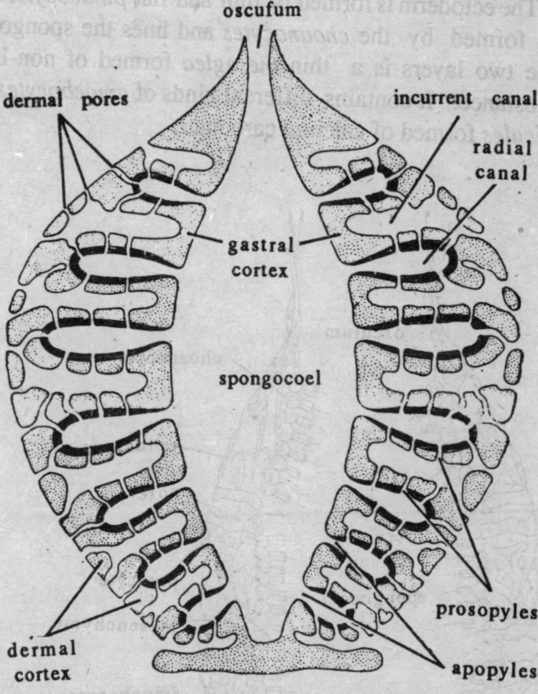

Fig. 7.3. Syconoid type of canal system.

II. Sycon type

It is a more complex system as it is a folded version of the asconoid body. It is found in *Scypha* and the embryonic development of *Scypha* clearly shows the asconoid pattern converting into syconoid pattern. It is formed from the asconoid type by the outpushings of the wall of an asconoid sponge at regular intervals into finger-like projections called *radial canals*. At first, these canals are in direct contact with the outside water but in most of the sponges, the walls of the radial canal fuse in such a manner that tubular *incurrent canals* are formed in between. These incurrent canals open to the exterior by dermal ostia or dermal pores. As these incurrent canals represent the outer-surface of asconoid type of surface, they are lined by *epidermis* while the radial canals which represent the outpushings of asconoid spongocoel are lined by *choanocytes.* The interior of the sponge in which radial canals open is a spaceous spongocoel which is lined by the flat epithelium derived from epidermis. The openings of the radial canals into spongocoel are termed *internal ostia.* The spongocoel opens to the exterior by a large single osculum. The wall between incurrent and radial canals is pierced by numerous minutes pores called *prosopyles.*

The course of water current through the canal system can be represented as follows :

dermal
Surrounding ------------------> incurrent canals prosopyles
water ostia
osculum apopyles ↓
outside <-------------- spongocoel <--------------- radial canals

In the more complicated of syconoid types of canal system, the epidermis and mesenchyme spread over the outer surface forming a thin or thick cortex having cortical spicules. The epidermis is pierced by more definite pores that lead into narrow incurrent canals.

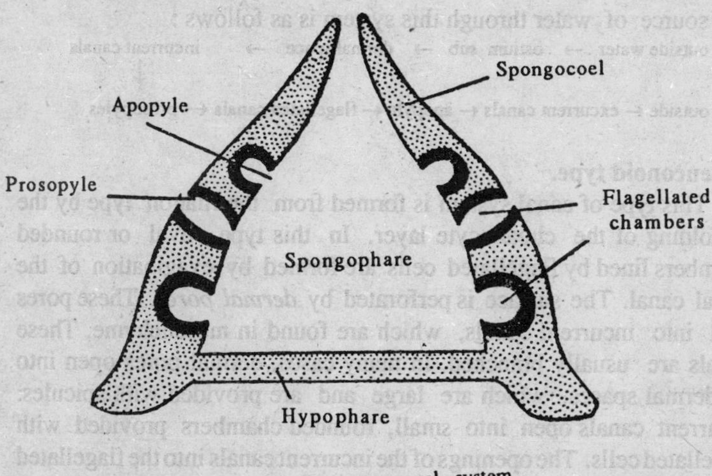

Fig. 7.4. Rhagon type canal system.

3. Rhagon type.

This type of canal system is found in the larva of Demospongiae called *rhagon* which has a broad base and is conical in shape. Due to excessive growth of mesoglea *sub-dermal spaces* are formed in its body wall.

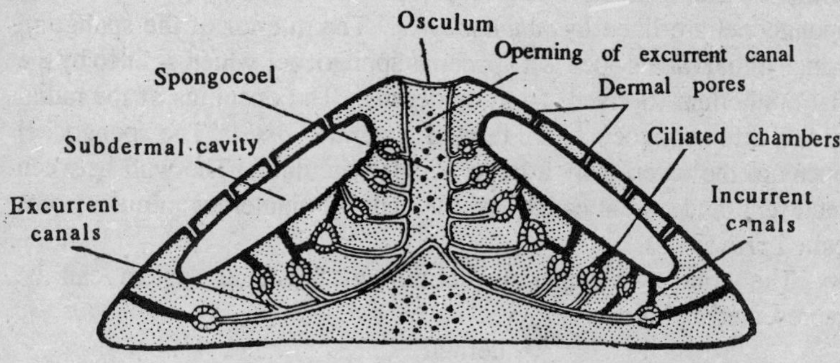

Fig. 7.5. Rhagon type in V.S. of spongilla.

The ostia open in these spaces which lead into incurrent canals. The incurrent canals open by prosopyles into flagellated canals which are lined with choanocytes. The flagellated canals open by apopyles into excurrent canals which lead into *paragastric cavity.* The paragastic cavity opens to the outside by the osculum which is present at the apex. The incurrent and excurrent canals may be complex and branched in it. The source of water through this system is as follows :

outside water → ostium sub → dermal space → incurrent canals
↓
outside ← excurrent canals ← apopyle ← flagellated canals ← prosopyles

4. Leuconoid type.

This type of canal system is formed from the rhagon type by the outfolding of the choanocyte layer. In this type, oval or rounded chambers lined by flagellated cells are formed by evagination of the radial canal. The surface is perforated by *dermal pores.* These pores lead into incurrent canals, which are found in mesenchyme. These canals are usually branched. In many cases, dermal pores open into subdermal spaces, which are large and are provided with spicules. Incurrent canals open into small, rounded chambers provided with flagellated cells. The openings of the incurrent canals into the flagellated chambers are called *prosopyles.* The flagellated chambers open into the excurrent canals by small apertures, known as *apopyles.* These excurrent canals are united to form large tubes, which open into *spongocoel.* This cavity is largely obliterated. Spongocoel opens to the outside by the *osculum.* Leuconoid type of canal system can be divided into 3 sub-types:-

(a) ***Euryphylous type*** - In this case the sponge has a flat broad base having an opening at the apex and looks like a pyramid. There are a number of flagellated chambers (*radial canals*) in the upper wall into which the prosopyles open. The lower basal wall is without flagellated chambers and is known as *hypophore* whereas the upper wall with flagellated chambers is called

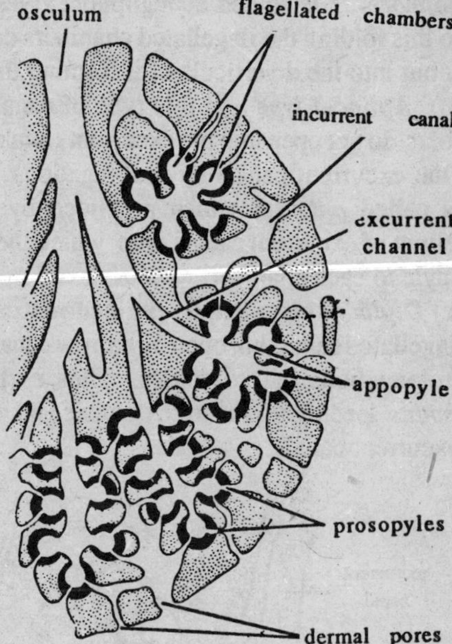

Fig. 7.6. Leuconoid type of canal system with eurypylas chambers.

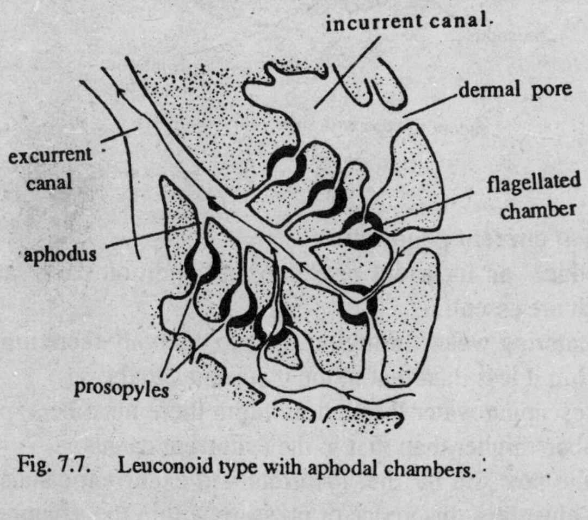

Fig. 7.7. Leuconoid type with aphodal chambers.

spongophore. The folded spongophore gives rise to incurrent canals. Due to this folding the flagellated chambers do not open into the gastral cavity but into the diverticula of it forming the excurrent canals.

(b) *Aphodal type* - In this type of canal system the flagellated chambers do not open into the excurrent canals directly but are removed from the excurrent canals by prolongations of apopyles into small canals called *aphodus* which are lined by the prolongation of the epithelium of excurrent canals into which they open. There is only one prosopyle to each chamber.

(c) *Diplodal type* - In this type incurrent canals do not open directly into flagellated canals but open by narrow canals called *prosodus* which are prolongations of prosopyles. Thus each flagellated chamber has a *prosodus* leading to incurrent canals and an *aphodus* leading from it to excurrent canals.

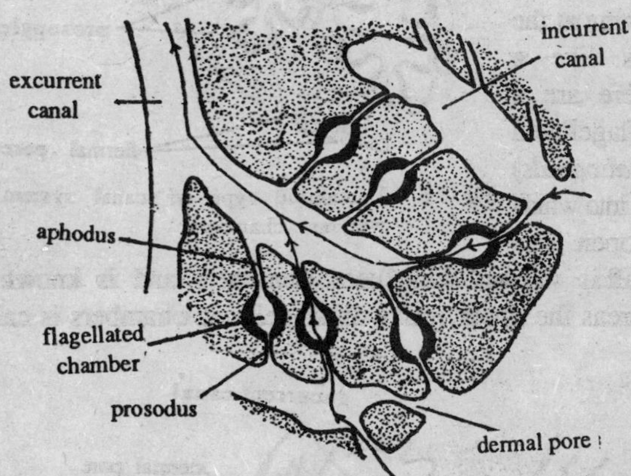

Fig. 7.8. Leuconoid type with diploidal chambers.

Mechanism of current production

To produce an incurrent or excurrent condition there are two factors which are essential :

(i) For entering water through ostia into the body there must be a pressure within it less than that in the incurrent canals.

(ii) For escaping water through osculum there must be a pressure within chambers higher than that in the excurrent canals.

But as the pressure in the incurrent and excurrent canals is the same, there must be a difference of pressure within the chamber itself and the lower pressure must be towards the periphery. Such a distribution of pressure is set up when each flagellum causes a flow of water towards the centre of the chamber.

Functions of the Canal System.

1. The canal system serves the purpose of *nutrition*. It is regarded as a highway for the food through the body cells in the radial canal with flagella, which capture the food particles. Water-currents are produced by flagella, thus, water flows into the central cavity or spongocoel. Smaller food-particles e.g., diatoms, protozoa and particles of organic debris are ingested into the cell protoplasm and digested. The digestion is *intracellular*. *Robert Grant* first of all observed the flow of water in the body-wall by adding powdered carmine to the water. Thus, canal system here does the same functions as circulatory system in higher animals.
2. In sponges, as a result of development of elaborate canal system, *massive growth* is found.
3. Streaming currents of water have dissolved air, therefore, gaseous exchange or *respiration* takes place in the cells. Oxygen is taken in by simple process of diffusion and carbon-dioxide is given out. The respiration is also intracellular.
4. The function of the canal system is also *excretory*. Currents of water, which pass outside the osculum remove the carbonic acid and other nitrogenous waste substances, which are the excretory products of the body.
5. The purpose of the canal system is also to *increase the surface area* of the animal in water. This is a characteristic point by which increase of volume is allowed by keeping the ratio of the surface to the volume.

8

SKELETON IN SPONGES

The skeleton of sponges is formed of spicules or spongin fibres or of both. The spicules are needle-like shiny structures. They have a *central axis* of organic substance around which inorganic substance are accumulated. An *organic sheath* is present outside this layer of inorganic substances. The organic substance of the spicules is in contact at places with the organic matter of the animal. The inorganic substance found in the spicules is either calcium carbonate or silica. The organic axis in silicious spicules is broad and distinct and is collected around the axis of mineral matter in the form of *concentric lamellae* which serially alternate with each other. In calcareous sponges the calcium carbonate is granular and the organic substances are absent in it.

According to size the larger spicules, constituting the main skeleton, are called *megascleres* and the smaller spicules which occur interstitially are called *microscleres*. Further, spicules may occur in several forms; they may be simple rod-like or may take the form of forks, anchors, shovels, stars, plumes, etc. These forms depend upon the number of axes and rays present.

1. Megascleres.

They are larger spicules and are named by adding *axon* in the number of their axes and *actine* or *actinal* in the number of their rays. They are of the following types :

A. Monaxon spicules. They are straight or curved rod-shaped or needle-like spicules which have only one axis. The growth of these spicules can be in one or both the directions of their axis. Thus they are of the following kinds.

(a) Monactinal monaxon. When the growth of the spicules take place at one end of the axis, it is called *style.* Its round end is called *strongylote* and pointed end is known as *oxeote.*

(i) Tylostyle. Its round end is knob-like.

(ii) Acanthostyle. Many minute spines are found on its surface.

enzymes. Spongin contains a large amount of iodine reaching 8 to 14 per cent in certain tropical species of the Spongiidae and Aplysinidae. It is interesting to know that old herb-doctors for centuries used bath-sponge as a cure for croup.

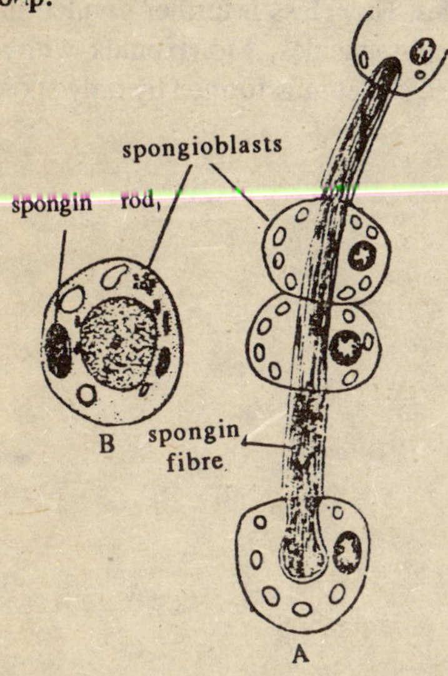

Fig. 8.4. Secretion of spongin fibres.

Spongin occurs in various forms in the class Demospongia. It may occur as a cement connecting together the siliceous spicules (many Monaxonida). More usually it is found in the form of branching or anastomosing fibres, in which siliceous spicules are embedded (Monaxonida). In Keratosa, the spicules are completely absent and the spongin alone is formed. The spongin fibres are the fine threads consisting of a soft granular axial core or *medulla* surrounded externally by concentric layers of spongin. The fibres are secreted by flask-shaped mesenchyme cells, called *spongioblasts,* which are seen coating the fibres. The spongioblasts become arranged in rows and the spongin rod secreted by each fuses with those of neighbouring cells to form a long fibre. The spongioblasts later become vacuolated and degenerate after having secreted a certain quantity of spongin.

Importance of Spicules

The structure of spicules is constant for a species so these are used in identification and form the basis of classification of the phylum depending upon their structure and chemical composition for example class Calcarea containing spicules of $CaCO_3$, class Hexactiri-ellida has six rayed siliceons spicules and class Demospongia has skeleton of spongin fibres or spongin fibres and siliceons spicules or without these spicules. This class is further divided into Tetractinellida with tetraxon siliceons spicules, Monaxonida with monaxan spicules and Keratosa whose skeleton is formed by only spongin fibres.

9

REPRODUCTION IN SPONGES

Reproduction is the multiplication of the organism resulting the production of similar offsprings being sedentary they multiply asexually as well as sexually. Asexual reproduction is more common than sexual reproduction. To tideover the unfavoruable conditions some sponges develop capsulated bodies like gemmules and reduction bodies.

In sponges following types of reproduction occur :

(A) Asexual reproduction

(i) Gemmation a budding
(ii) Fission
(iii) Gemmulation
(iv) Reduction bodies
(v) Regeneration

(B) Sexual reproduction

(A). Asexual reproduction. It is a type of reproduction in which there is no direct involvement of reproductive organs. It takes place in the following ways.

(i) Gemmation or Budding. Exogenous buds are formed at the base of the branches. During bud formation, numerous archaeocytes collect at the surface and the pinacoderms bulge out to receive them. This bud gradually grows into adult and may remain attached with the parent but more commonly, the buds constrict from the present and each forms a new sponge.

(ii) By fission. In some sponges, multiplication takes place by fission, throwing off parts of the body. The sponge is hypertrophied over a limited area, developing a line of weakness. Along this weak line, splitting occurs and a part of separated from the main parental body

which develops into a new individual. By budding this new individual takes the form of a colony.

Fig. 9.1. Showing Gemmation or Budding.

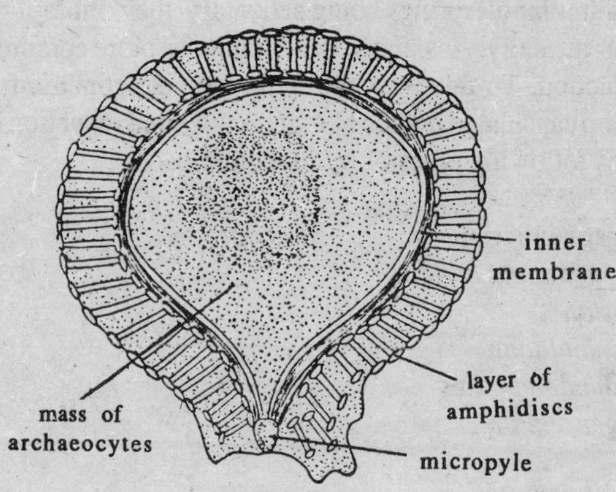

Fig. 9.2. Gennule of a sponge in section.

(iii) Internal budding or Gemmulation. In all fresh-water sponges and in some marine forms, there is a peculiar mode of asexual reproduction, which takes place by the formation of asexual reproductive bodies or *gemmules* . Gemmule formation takes place by the gathering of *amoebocytes* probably the *archaeocytes* into a ball. This ball is surrounded by a chitinous shell supported or strengthened by spicules. Archaeocytes are filled with food reserves in the form of the oval bodies consisting of *glycoprotein* or *lipoporotein.*

Gemmules of *Spongilla* possess a peripheral layer having air cells, which are helpful for floating them on the surface of the water. The spicules in these gemmules are amphidisc type, which are rod-like. Between the two membranes enclosing archaeocytes is an outlet, which

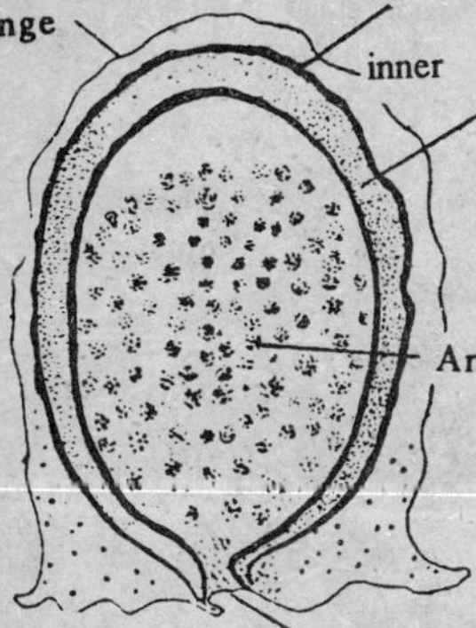

Fig. 9.3 Germinating gemmule of a sponge.

is called *micropyle*. These gemmules are produced by the sponges, when they are unable to live due to the excessive cold or drought. The fresh-water sponges form a large number or gemmules in autumn and then disintegrate; gemmules are found in the remnants of the body of sponge or they fall to the bottom. Gemmules can withstand freezing and drying. They are, therefore, generally produced by the sponge in the autumn, when adult sponges die. In India, gemmules are generally found on the approach of summer. On the return of favourable conditions in the spring, gemmules hatch. *Zeuthen* (1939) found that gemmules can hatch at any time at the temperature of 21-33°C. Late forms hatch in two weeks, while in spring they require three days to hatch. According to *Brien* (1932) hatching takes place when the contained cells of the gemmules stream into the micropyle. The large multinucleate archaeocytes divide into uninucleate ones and into small cells or *histoblasts,* which first of all originate and arrange to form an epidermis, differentiate so as to form *choanocytes, porocytes* and internal linings of spaces. Archaeocytes modified into *scleroblasts* produce spicules and after a week by differentiation and arrangement a new complete sponge is formed, surrounding the empty gemmule shell. In some cases, *forminal tube* is also formed. It is a protrusion of outer membrane in the form of a tube. Later on differentiation and arrangement of cells take place.

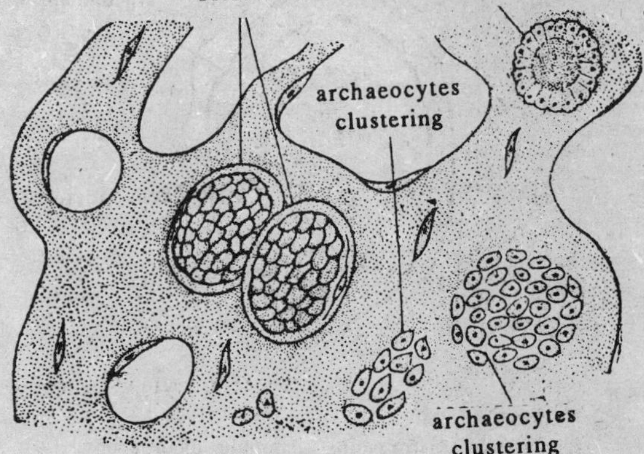

Fig. 9.4. Formation of gemmules in a marine sponge.

(iv) Formation of Reduction Bodies. These are formed during unfavourable conditions. It involves masses of many kinds of amoebocytes with epidermal coverings that persist after the disintegration of the parent sponge. These reduction bodies can develop into sponges in favourable conditions. They differ from gemmules in not having an external layer of columnar cells with spicules.

REGENERATION IN SPONGES

All animals, particularly the less specialized ones, can replace their lost or injured parts. The process is known as *regeneration.* It is greater in simple animals and in simple tissues. Thus, epithelial tissues regenerate readily, whereas highly differentiated tissues, such as muscle or nerve tissue, have very limited power of regeneration. The sponges, correlated with their low grade of organization, have a high capacity to regenerate. If an individual is cut into pieces, each piece, however, small it may be, will grow into a complete or full-size sponge, if conditions are favourable. The regenerative power of certain sponges is quite noteworthy as demonstrated by *H.V. Wilson* (1907) by a simple experiment. If a sponge is chopped into small pieces, run through a meat grinder and then squeezed through fine silk bolting cloth, its cells are separated from one another. In a suitable medium, some of these disunited cells come together by amoeboid movements and unite to form small aggregates or *spongelets*. In course of several days these reunited masses acquire canals, flagellated chambers and skeleton, thus growing up into new sponges. Cells from different species of sponges may adhere temporarily but later separate, without re-forming a sponge.

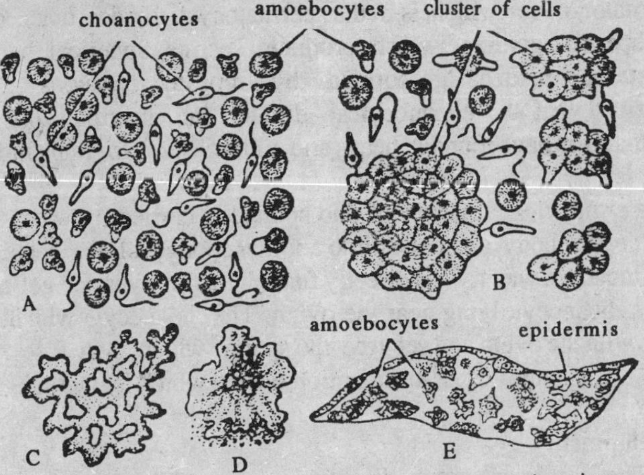

Fig. 9.5. Wilson's experiment on regeneration in sponges. A-Cells of *Microciona* separated by squeezing living sponge through bolting cloth. B-Cells aggregating in to small masses. C-A reticulate reunion mass. D-Later stage forming a young sponge or spongelet. E-Section through a stage like D.

(B) Sexual Reproduction. Sponges reproduce by sexual reproduction : by formation of ova and spermatozoa. Majority of sponges are hermaphrodite (Gr. *Hermes*, aphrodite) *i.e.*, an individual possesses both male and female reproductive organs but some are dioecious (Gr. *di*, twice; *oikos*, house) *i.e.*, having the male and female reproductive organs in separate individuals. In bisexual or hermaphorodite or monoecious forms, germ-cells of the male mature before those of the female, thus, producing sperms and ova at different periods. Eggs usually develop in the basal part of the sponge, while the sperms are formed on the apex.

Sexual reproduction in sponges, which live in shallow water, takes place at certain seasons of the year, whereas deep-sea sponges reproduce throughout the year. Sex-cells are produced from the archaeocytes but, according to some authors, sex cells originate from choanocytes. The ovum is formed by the enlarging of *amoebocytes* or *amoeboid* cells. It has a large nucleus and a distinct nucleolus. It grows and acquires food material. When full size is reached, oocyte or egg mother cell undergoes maturation divisions. In some sponges the sperm mother cell of spermatogonium is the enlarged amoebocyte covered by one or more flattened cover-cells, thus, forming spermatocyst. The spermatogonium which is in the spermatocyst, divides twice or thrice into spermatogonium, which produce sperms. Spherical bunches of minute spermatozoa are formed by repeated divisions. *Gatenby* described that the choanocytes divide into spermatozoa. These spermatozoa have rounded heads and vibratile tails for rapid movement in water.

Fertilization : Sperms of one sponge is carried by water current outside the body and enters into another sponge of the same species with incurrent water. This directly fuses with the ovum or get attached with a choanocyte lying near the ovum. The choanocyte with the sperm fuses with the ovum and set free the sperm which then fuses with it. Thus in sponge cross-fertilization occurs which is internal.

Development

The fertilized ovum is called *zygote* which develops in the parental mesenchyme for sometime and gets covered by a capsule.

(i) Cleavage or segmentation - The division of zygote is holoblastic (complete) and equal or unequal. Ultimately there appears an embryo in which 8 cells attached with parental tissue form the future *pinacoderm* and the other cells form the *choanoderm.* A cavity appears between these two and the embryo is now called *blastula* which leaves the parent body and form a free-swimming larva.

(ii) Larval forms - In all known case, during development of embryo a free-swimming ciliated larva appears but the form assumed by the larva differs in different sponges.

In case of *Clathrina* and *Leucosolenia* oval blastula changes to *coeloblastula.* This has a single layer of cells, all alike in kind - elongated, columnar and flagellated. At one pole is seen a pair of cells which are of a different character being large, rounded non-flagellated and granular which form *archaeocytes.* This changes into *stereogastrula larva* which is called *parenchymula larva.* The changes are -

(a) Some of the flagellated cells withdrawn their flagella and pass as amoeboid cells into the internal cavity.

(b) Archaeocytes also sink into the internal cavity.

(c) Soon these cells fill up the greater part of the internal cavity and the larva is now called *parenchymula larva.*

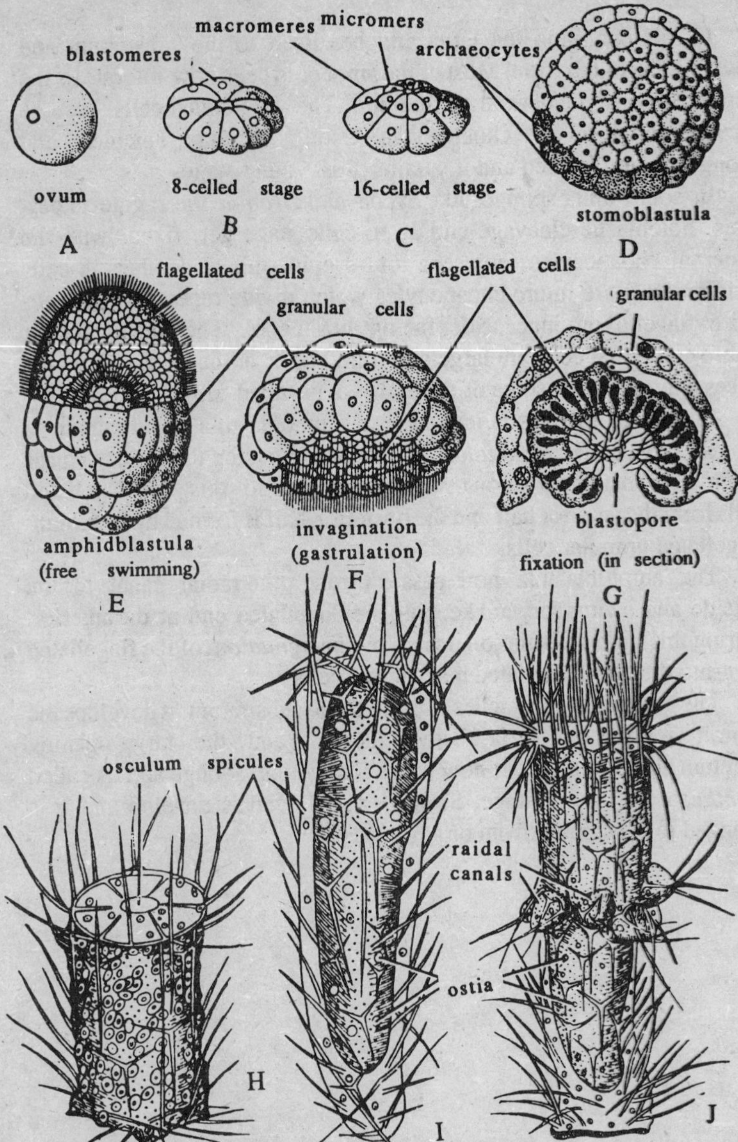

Fig. 9.6. Development of Syconoid sponge. A-Ovum, B-8-celled stage, C-16-celled stage, D-Blastula, E- Anphiblastula, F-Beginnging of invagination, G-Gastrula showing fixation (in section), H-Young sponge, I-Asconoid (olynthus) stage, J-Syconoid stage.

In this condition the larva attaches itself to the substratum and develops into adult form. Most of the amoeboid cells now migrate to the outside to form the dermal epithelium. The flagellated cells become internal and form the choanocytes. Ostia, osculum, spicules and spongocoel are formed and it attains this young stage.

In some other sponges like *Sycon* and *Grantia* the zygote undergoes holoblastic cleavage and at 16-cells stage get fixed with the maternal choanocytes represent future epidermis and other 8-cells (micromeres) are future choanocytes which divide repeatedly and acquire flagella on inner side facing blastocoel. The megameres or future epidermal cells are large and round cells at the centre of which arises a mouth. This stage of the embryo is called *stomoblastula*.

The next process is *inversion* of stomoblastula resulting in a typical larva called *amphiblastula*. During inversion the embryo turns completely inside out through the mouth. Due to this the flagellated cells form the anterior half and the posterior half is formed by large non-flagellated granular cells.

The amphiblastula now passes through the radial canal to the outside and swims forward keeping the flagellated end at the anterior. During this time gastrulation occurs by *invagination* of the flagellated part into the non-flagellated macromere cells.

The gastrula then attaches itself to some substratum. it develops the central spongocoel, ostia all over its body and the large opening osculum at the free-end. It now looks like young sponge and is called *olynthus* or asconoid stage. Slowly radial canals develop and it is changed to a syconoid from or adult stage.

10

ECONOMIC IMPORTANCE OF SPONGES

Sponges are of great economic importance, they are useful as well as harmful to mankind.

1. Useful sponges to man. The flexible and highly absorbent spongin skeleton of marine sponge, *Euspongia* and *Hippospongia* forms, after the removal of living parts, the bath sponge of commerce. The later is used for bathing, washing window pans and automobiles, mopping, ceramic work, and many more similar purposes. Today, cloth and artificial sponges of rubber or cellulose have greatly replaced the natural sponges for many purposes. Even then they are put to a wider variety of uses.

2. Sponges for Decoration and Presentation. Many people use certain glassy, transparent sponges (*Hyalonema*) for decoration purposes. Venus's-flower-basket is a beautiful sponge. It (*Euplectella*) is used for decoration and for presentation to newly wed couples symbolising a happy long married life like a pair of crustaceans which get enclosed within the body of this sponge for ever.

3. Sponges as food. Although the sponges are not eaten by other animals like fishes etc. due to their disagreeable taste and unpleasant odour, but Nudibranch molluses make a regular diet of sponges. Some parasitic crustaceans also feed on their tissues. Thus sponges have a food value to these animals.

4. Sponge fishing or sponge industry. Although synthetic sponges have become widely used, and largely replaced natural sponges yet natural sponges still have a great market, therefore, sponge industry is of great economic value. Since the days of Greece, the fibrous skeleton of bath sponge, *Spongia* is used for washing and mopping. The water holding capacity of the skeleton is due to capillary forces in the fine spaces of irregular spongin net work. The best and principal supplies of sponges occur on rocky bottoms in warm shallow waters of the Mediterranean sea and the Gulf of Mexico, Tarpon springs, Florida of United States the best commercial sponge fishing centre of U.S.A. where more than a million of dollars worth of sponges have been collected in a year. However, fungus infection reduce the total yield.

Sponges are collected by several methods, by drag hooks, by trawlers and by divers. In deeper waters, they are gathered by dredging or trawling or more commonly by divers, either with or without a diving suit known as skin divers. In shallow waters, sponges are gathered by the help of long handled hooks operated by two men while one man rows, the other looks at the bottom of the ocean through a glass-bottomed bucket. The marketable variety is taken out with a two pronged hook, fastened to a long pole.

Living sponge is black in colour, smooth and slimy mass having a leathery covering. In the inside, it resembles to a piece of raw beef liver, both in consistency and colour. After cleaning so as remove dirt, sponges are hung from the riggs of the boat and their protoplasm is allowed to decay. Later, they are well washed, beaten, bounded with a wooden mallet so as to break up any solid material such invertebrate shell and ultimately by dried in sun, cleaned sponges are cut or trimmed, sometimes bleached, graded, sorted and then auctioned. Sponges can be easily pressed for shipping due to their great elasticity. For sponge culture, the cut pieces are about 8 cu inches are fastened to slabs or concrete or other support where water current provide abundant food and oxygen.

5. Sponges and their association with other animals. Some sponges show commensalism as several crustaceans, worms, molluses and fishes inhibit and live in the internal cavities of sponges for protection against enemies. They also get a good supply of food in the circulating water. The spongocoel of *Hyalonema* and *Euplactella* is occupied by a pair of shrimp of genus *Spongicola*. It is the example of commensalism. The shrimps enter the spongocoel in small and young condition as they grow inside their exit become impossible through the sieve plate covering the osculum.

Another example is that of a small family of crabs (*Hyas, Dromia* etc.) in sponges. The crabs cover themselves with bits of sponges, algae hyroids etc. with the help of a sticky secretion. The implanted sponge grow and moulds itself. This covering act as caumoflage but also protects the crab from its predators.

6. Harmful Sponges. Sponges are not only useful to mankind but there are some sponges which are harmful to us. The boring sponge *Cliona* bore the shells of the oysters and barnacles and break them into pieces resulting their death. Some sponges produce poisonous secretions and when touched the skin is pricked by the poisonous spicules producing itching of the skin.